Automated Model Building

APPLIED LOGIC SERIES

VOLUME 31

SCOPE OF THE SERIES

Logic is applied in an increasingly wide variety of disciplines, from the traditional subjects of philosophy and mathematics to the more recent disciplines of cognitive science, computer science, artificial intelligence, and linguistics, leading to new vigor in this ancient subject. Kluwer, through its Applied Logic Series, seeks to provide a home for outstanding books and research monographs in applied logic, and in doing so demonstrates the underlying unity and applicability of logic.

The titles published in this series are listed at the end of this volume.

Automated Model Building

by

RICARDO CAFERRA
Laboratory Leibniz - IMAG, INPG,
Grenoble, France

ALEXANDER LEITSCH
Vienna University of Technology,
Austria

and

NICHOLAS PELTIER
Laboratory Leibniz - IMAG, CNRS,
Grenoble, France

KLUWER ACADEMIC PUBLISHERS
DORDRECHT / BOSTON / LONDON

A C.I.P. Catalogue record for this book is available from the Library of Congress.

ISBN 1-4020-2652-8 (HB)
ISBN 1-4020-2653-6 (e-book)

Published by Kluwer Academic Publishers,
P.O. Box 17, 3300 AA Dordrecht, The Netherlands.

Sold and distributed in North, Central and South America
by Kluwer Academic Publishers,
101 Philip Drive, Norwell, MA 02061, U.S.A.

In all other countries, sold and distributed
by Kluwer Academic Publishers,
P.O. Box 322, 3300 AH Dordrecht, The Netherlands.

Printed on acid-free paper

Printed in the Netherlands.

Contents

Preface

On the history of the book:

In the early 1990s several new methods and perspectives in automated deduction emerged. We just mention the superposition calculus, meta-term inference and schematization, deductive decision procedures, and automated model building. It was this last field which brought the authors of this book together. In 1994 they met at the Conference on Automated Deduction (CADE-12) in Nancy and agreed upon the general point of view, that semantics and, in particular, construction of models should play a central role in the field of automated deduction. In the following years the deduction groups of the laboratory LEIBNIZ at IMAG Grenoble and the University of Technology in Vienna organized several bilateral projects promoting this topic. This book emerged as a main result of this cooperation. The authors are aware of the fact, that the book does not cover all relevant methods of automated model building (also called model construction or model generation); instead the book focuses on deduction-based symbolic methods for the construction of Herbrand models developed in the last 12 years. Other methods of automated model building, in particular also finite model building, are mainly treated in the final chapter; this chapter is less formal and detailed but gives a broader view on the topic and a comparison of different approaches.

How to read this book:

In the introduction we give an overview of automated deduction in a historical context, taking into account its relationship with the human views on formal and informal proofs. We illustrate that model building (in a general sense) plays a central role in mathematics and logic, making *automated* model building a key discipline within automated reasoning. The chapters 3 (resolution-based methods) and 4 (constraint-based methods), which are largely based on former work of the authors, can be considered as the core of the book; they depend on each other and should be read in this order. Chapter 2 (preliminaries), which should be read by need only, gives the basic concepts and makes the book accessible to graduate students without a firm background in first-order logic and automated deduction. Chapter 5 (on model representation and evaluation) essentially depends on the chapters 3 and 4. Chapter 6 on finite model building is largely independent of the former chapters and provides a broader view on the topic as a whole.

Acknowledgments

We would like to thank Chris Fermüller for several substantial and fruitful discussions in the early stage of the book; important parts of the book are based on his research carried out together with one of the authors. Our special thanks go to Joseph Goguen for his encouragement and his constructive criticism; his remarks on the history, the general methodology and the philosophical background helped to improve the conceptual level of the book. He also gave us very valuable information about several other deductive methods, resulting in a clearer and more profound presentation of our approach within the field of deduction as a whole. Finally we would like to thank Dov Gabbay for efficiently promoting and supporting our work and for making its publication possible.

Chapter 1

INTRODUCTION

1. Automated Deduction

This book is on Automated Model Building. Certain keywords and domains are immediately evoked by this title. We shall consider here three of them that seem to be the most important: *model, model theory* and *automated deduction.*

The concept of *model* is a very deep one and represents a challenge to some of our strongest intellectual abilities. It is used in different fields with different intended meanings (see for example [Hermerén, 1974]). Maybe the oldest scientific meaning of "model" is that of a mathematical (not necessarily logical) object allowing to explain observational data (see for example [Davis and Hersh, 1981]). It is important to point out that the idea of non-uniqueness is implicitly accepted here: better models explain more observations more precisely. This, for example, is the concept of model used by physicists and biologists.

The other widespread meaning is that used in mathematical logic. The relationship among different uses of the notion "model" in mathematical logic and in empirical sciences did not receive very much attention in scientific research. But in [Suppes, 1966] the different concepts of model in various scientific fields (like logic, physics, social sciences) are carefully analyzed and the logical Tarskian notion is shown to be adequate to empirical sciences. In the framework of Artificial Intelligence the model building process in non-deductive inference, for example in abduction, is quite close to that of empirical sciences (and to natural language understanding). The difficulty of checking consistency of a proposed explanatory hypothesis with an existing theory and observed facts is one of the main impediments to the mechanization of the logic of discovery (see for example [Plotkin, 1971, Hájek and Havránek, 1978]).

The importance of models in the systematization of the art of reasoning was recognized very early. It is well known to logicians (see for example [Łukasiewicz, 1972]) that some explanations relating major, middle and minor premisses in Aristotle's syllogistic figures are correct only for certain concrete terms (i.e. with respect to certain *models*) and can be also be proven wrong for other concrete terms (representing *counter-models*).

The content of this book is hardly related to classical model theory, which, according to [Chang and Keisler, 1990], is "the branch of mathematical logic which deals with the relation between a formal language and its interpretations, or models", but is much closer to Automated Deduction.

Therefore we are naturally compelled to "review" automated deduction, its goals, what it is missing, ... as well as to talk about some historical and philosophical matters.

At this time Automated Deduction is about 45 years old. The field is reaching maturity and its theoretical and practical results (some of them quite striking) ensure it a firm place somewhere between Logic, Computer Science and Artificial Intelligence. Maybe the best witness of maturity is that, today, we have a much clearer view of the feasibility (or non-feasibility) of the goals implicitly defined at the beginning. Still most of the problems, identified as central already at the very beginning, lack satisfactory solutions; we just mention proof planning, the use of analogy, learning and *constructing counter-examples* (see e.g. [Bledsoe and Loveland, 1984, Wos, 1988, Ballantyne and Bledsoe, 1982, Robinson, 2000]).

In the attempt to describe the field of Automated Deduction two basic methodological questions arise quite naturally:

- What is automated deduction?
- What should (and could) automated deduction be?

A scholar way to answer the first question is to have a look at the bibliography for corresponding definitions or, at least for attempts to characterize more or less precisely the field proposed by its practitioners. Surprisingly, there are only a few informal "definitions" around automated deduction (often merged with automated theorem proving), some of them rather tautological; but all of them partially address the second point above, at least implicitly. We just mention four of them given by renowned scientists in the domain of deduction, logic and mathematics.

> *By automated theorem proving we mean the use of a computer to prove non-numerical results, i.e. determine their truth (validity). Often (but not always, e.g. decision procedures) we may also require human readable proofs rather than a simple statement: "proved". Two modes of operation are used: fully automatic proof searches and man-machine interaction ("interactive") proof searches. The label ATP will cover both.* D. Loveland [Loveland, 1984].

> *And by automated deduction I am thinking of a field broader than automatic theorem proving including also automatic processing of proofs.* D. Prawitz [Prawitz, 1980].

> *The subject of automated deduction deals with computerizing the logical aspect of mathematics, while the subject of symbolic computation deals with computerizing the computational aspects.* M. Beeson [Beeson, 1988][1].

> *The subject of automated reasoning is concerned with using computers to help the humans discover and write formal proofs* R. Constable [Constable, 1993] .

The second and fourth characterizations are the more interesting ones: they take into account features with increasing importance in reasoning systems. It should be remarked that the fourth defines automated *reasoning* (note that reasoning is a broader concept *including* deduction)[2].

The study of automated reasoning (instead of automated deduction) as well as the cooperation of systems for theorem proving and symbolic computation (compared in Beeson's characterization) should not be forgotten when answering the question about the very nature of automated deduction.

A more technical answer to the question, what automated deduction actually is, can be given by a historical analysis of the subjects that have been treated by the researchers in the field.

A few survey papers have been published ([Davis, 1983, Loveland, 1984, Wos and et al., 1985, Davis, 2001]). A compilation of early papers that have founded automated deduction in [Siekmann and Wrightson, 1983a, Siekmann and Wrightson, 1983b] is also a reference in the field. The state of the art when the domain was 25 years old is described in [Bledsoe and Loveland, 1984] (see also [Wos, 1988]).

The recent handbook [Robinson and Voronkov, 2001] bears witness of the maturity of the field.

Concerning applications, some major impacts of automated deduction on science, economy and society deserve to be mentioned: e.g. discovering of errors in chips and the role of resolution in (constraint) logic programming.

---

[1] Beeson proposes the equation Mathematics = Logic + Computation, by admitting that it is an overstatement, since there are other aspects of mathematics, as for instance the visual aspect.

[2] The term *automated reasoning* was not introduced until 1980 (see [Wos, 1988], page 114).

Recently new important potential applications have emerged. One of them is deduction in the internet, where there is a growing need for powerful inference[3] engines, a consequence of its generalized use in more and more fields and of the user's expectations of new facilities.

Directly related to the goal of the present work is the fact that, in the whole set of the surveys and handbook articles on automated deduction mentioned above, only a few pages in [Fermüller et al., 2001] are devoted to the subject of model building. In [Fermüller et al., 2001], page 1828 it is written:

> Automated model building (sometimes also called model generation) is becoming a discipline on its own and one of the more fascinating applications of automated deduction.

2. Formal and Informal Proofs

In defining the field of automated deduction, we might observe that only a few attempts have been made to build bridges between automated deduction and the deep works on the notion of "proof" by mathematicians and philosophers of mathematics. But, clearly, this notion lies at the very heart of automated deduction; its analysis seems unavoidable once the state of the art allows for treating "big" proofs. In particular *proof presentation*, *proof schemata* and *explanation of proofs* become important issues in the domain. We just mention [Robinson, 2000] and [Goguen et al., 2000] (see also [Caferra and Herment, 1995]).

At this point the question "what is a proof?" naturally arises.

The notion of proof is a very deep one, especially if "real" mathematical proofs are considered. The conditions required for accepting an object as a proof of a fact were different ones in different times. This also applies to mathematics; just consider proofs in analysis before and after Weierstrass. Etymologically, to "prove" is related to ensuring quality (note that "probus" means: of good quality, honest, faithful).

The statement below is probably supported by most mathematicians:

> The process of deducing some sentences from others is the most important part of the mathematical work. [Rasiowa and Sikorski, 1970], page 179.

Historically, the importance of proofs has been recognized first in ancient Babylon:

> ... More important than the technical algebra of these ancient Babylonians is their recognition -as shown by their work- of the necessity of *proof* in mathematics.
>
> Until recently it has been supposed that the Greeks were the first to recognize that proof is demanded for mathematical propositions. This was one of the

[3]Inference, of course, *includes* deduction.

> most important steps ever taken by human beings. Unfortunately it was taken so long ago that it led nowhere in particular so far as our own civilization is concerned -unless the Greeks followed consciously, which they may well have done. They were not particularly generous to their predecessors. [Bell, 1986], page 18.

But the concept of a rigorous mathematical proof (i.e. using logical and non logical axioms and inference rules)[4] is due to the Greeks; it is the most distinctive feature between Greek and Babylonian mathematics.

It is generally acknowledged that Parmenides was the first to propose proofs as deductive reasoning starting with irrefutable statements and using rigorous chains of deductions. He also proposed dividing proofs into parts, the premises of one part being the conclusions of previous ones (this corresponds to *structuring proofs in lemmata*), see for example [Lloyd, 1990]. The most important works in Greek science addressing the concept of proof are Aristotle's "Organon" and "Rhetorics", and of course Euclid's "Elements".

It is not surprising that, among the "intelligent" activities handed over to computers, finding proofs in a formal system was a most appealing one. Indeed, the Logic Theory Machine [Newell et al., 1983], able to produce proofs in propositional calculus, is usually considered as the first Artificial Intelligence program.

With the proof of the 4-colour theorem [Appel and Haken, 1978] computer aided theorem proving became "public domain". The computer's task in the solution of the famous problem basically consisted in testing a very high number of cases. In fact Appel and Haken tested 1482 graphs representing all possible map configurations (the computation took more than 1000 hours).

The use of computers for proving theorems inspired mathematicians, logicians and philosophers of science. Their reflections on this matter might help us in developing a better understanding of automated deduction as it is and as it should be.

In a well known paper [Tymoczko, 1979], the author writes:

> I will argue that computer-assisted proofs introduce experimental methods into pure mathematics.

Other authors formulated similar theses (see for example [Goodman, 1984, Goodman, 1990]), sometimes in a more radical way[5].

Tymoczko asks the question "What is a proof?" and gives an answer by identifying 3 main characteristics:

---

[4]The Greeks used other names: postulates, common notions.

[5]The first statement of [Goodman, 1990] is: Mathematics is a natural science.

- Proofs are convincing
- Proofs are surveyable
- Proofs are formalizable

The first point addresses one of the basic requirements of proof theory; indeed, arguments have to be convincing to be accepted as a proof[6].

The second point is of particular relevance when 'big' proofs are investigated. It is also reflected in the opinion of J.P Serre (see [Horgan, 1993]):

> Tout résultat qui est obtenu par un moyen humainement invérifiable n'est pas une démonstration.

In this context it is interesting to look at the thoughts of mathematicians concerning their own proofs (e.g. see [Lam, 1990, Lam, 1991]). Lam's analysis started with the title of an article in a newspaper: "Is a math proof a proof if no one can check it?". The alluded proof was the proof of a conjecture (by K. F. Gauss) about the projective plan of order 10 and consumed 3000 hours of a CRAY-1A[7]. Lam proposes the term "computed result" instead of "proof" for this kind of "proofs" and he contends that, in case of computed-based results, correctness is not absolute, but just almost sure[8].

The achievement of the Kepler Conjecture's proof by Thomas Hales in 1998 is particularly enlightening when trying to understand what a proof is. The full proof of this very old conjecture, belonging to the famous Hilbert's list of problems, totalised over 250 pages. This "hybrid proof" (i.e. obtained mainly by human means but with the help of computers), was solicited for publication by the very respected Annals of Mathematics, which hosted a conference in 1999 devoted to understanding the proof. A panel of 12 referees (chaired by Gabor Fejes Tóth, who made in 1953 a major breakthrough towards the final proof) reported, after 4 years, that he was 99 % certain of the correctness of the proof, but that the team had been unable to completely certify it.

The third point corresponds to the definition of proof in a formal system and the difficulties it raises are essentially technical ones. It is the underlying idea in logical frameworks, in particular in the AUTOMATH

---

[6] The need to capture human's reasoning features was present in Gentzen's natural deduction. Gentzen wanted *"to set up a formal system that came as close as possible to actual reasoning"*.

[7] At this time it was estimated that there were non detected hardware errors approximately every 1000 hours !!

[8] The author forgets that there have been "false proofs", some of them proposed by great mathematicians, long time before computers were invented.

project (see below). In 2003 a project (called Flyspeck) was launched to produce a formal proof of the Kepler Conjecture. The project is presently estimated to take as many as 20 work-years to complete. For details see http://www.math.pitt.edu/~thales.

The authors think that the relation between human proofs and computer-based proofs is made clearer by the distinction (due to Martin Löf, see [Beeson, 1985]) between *proofs* and derivations. A *proof* contains the computational information needed by a computer in order to verify a proposition. A derivation is an object that convinces us of the truth of a proposition. A similar distinction between "proofs" and "brief proofs" is done by McCarthy in his pionnier work [McCarthy, 1962]. Thus what is found in textbooks are derivations[9].

3. Proofs and Automated Deduction

Above we gave a brief description of the notion of "proof" and of the views of mathematicians, logicians and philosophers concerning computer-based proofs. But how did researchers in automated deduction look at proofs? Different approaches exist since the very beginning of research in the field. They can be characterized by the following classification (the main principles guiding the approaches are characterized by informal key phrases). Many works in automated deduction can be assigned to other categories as well (e.g. if we use the techniques as a criterion), a general problem appearing in taxonomy. Here we classify the works according to their *aim* and relegate techniques to a secondary position.

1 (**"As fast as possible"**). Designing and implementing first-order provers based on a single calculus (resolution, tableaux, model elimination,...). Research efforts are concentrated on strategies and implementation techniques. Good examples and powerful systems are very well known; we just mention the systems

- Otter (http://www-unix.mcs.anl.gov/AR/otter/)
- Setheo (http://www4.informatik.tu-muenchen.de/ letz/setheo/)
- Spass (http://spass.mpi-sb.mpg.de/)
- Vampire (http://www.cs.man.ac.uk/ riazanoa/Vampire/)
- Gandalf (http://www.math.chalmers.se/ tammet/gandalf/)

A particularly important technique, namely **rewriting**, must be mentioned here. Rewriting is considered as a *specific* domain of rea-

[9]One of the goals of the AUTOMATH project was a complete formal checking of the proofs (derivations) in Landau's "Grundlagen der Analysis" (see [De Bruijn, 1980]).

soning, closely linked to the independent research field of *Symbolic Computation*. We just mention the work of B. Buchberger [Buchberger, 1985] and the forthcoming book of J. Goguen [Goguen, 2005].

2 (**"Induction"**). Designing and implementing provers dedicated to induction. The use of induction is unavoidable in mathematical reasoning (H. Poincaré even considered it as the mathematical reasoning *par excellence*). Special induction provers have been developed (the most popular being that of Boyer and Moore [Boyer and Moore, 1979]) producing some remarkable practical results.

3 (**"As general as possible"**). Designing and implementing higher-order logical frameworks allowing the user to define any logic and any calculus and to construct and verify proofs in it (AUTOMATH ([De Bruijn, 1980, De Bruijn, 1991]), LCF ([Gordon et al., 1979, Milner, 1985]), nuPRL ([Constable, 1986]), Coq ([Coquand and Huet, 1988, The Coq Development Team, 2001]), ISABELLE ([Paulson, 1989]), KUMO (http://www.cs.ucsd.edu/groups/tatami)...). This approach is related to approaches defining consequence relations, logics and inference rules in full generality ([Avron, 1991, Barwise and Feferman, 1985, Meseguer, 1989, Fagin et al., 1992]).

4 (**"As close as possible to the human way of proving theorems"**). Designing and implementing programs in which human-style heuristics are combined with proof systems (sometimes replacing them). In general these provers are not complete (see [Bledsoe, 1977, Ballantyne and Bledsoe, 1982, Bledsoe, 1984, Bledsoe, 1986] and references therein). The logical languages used in this approach are first-order logic and subsets of second-order logic (see [Bledsoe, 1984]).

In our classification above we did not mention provers for classical propositional logic because they have only played a pioneering role (Logic Theory Machine [Newell et al., 1983]) in automated deduction; today the most essential theoretical and practical work on this logic is focused on the SAT problem (and SAT solvers) (see also chapter 6).

The work on non-classical logics can be included in 1 above. This holds for the *direct* approach (specialized and also parameterized provers have been implemented for different logics), and for *translation* approach, where the problem in non-classical logic is translated to a problem in a fragment of first-order logic (see [Catach, 1991, Ohlbach et al., 2001]).

4. Model Building in Automated Deduction

It is worth looking for more abstract classification criteria: We think that *decidability* is a good one[10]. Two kinds of problems have been mainly attacked so far: decidable– and semidecidable (but undecidable) ones.

It is revealing that the first published work reporting on a non-propositional theorem prover was that by M. Davis dealing with the mechanization of a decidable fragment of arithmetic [Davis, 1983]. Gilmore made the first step in the mechanization of a semidecidable problem [Gilmore, 1960] (that of validity in first-order logic). In this book we propose a *non–semidecidable* problem, namely that of building first-order models, as a standard topic in theorem proving[11].

It is a trivial remark that, in studying conjectures, the ability to prove or to disprove them is equally important. Some theoretical limits should be recalled when trying to automatize this process (we talk indifferently of valid formulae/countermodels (or counterexamples) or unsatisfiable formulae/models).

As the set of satisfiable first-order formulae is not recursively enumerable there can be no "universal" model building procedure. Even in this formal sense *automated model building* is substantially more complex than just proving theorems (the set of valid formulae is recursively enumerable). But despite this barrier, as we will point out below, automated model building is by no means a hopeless and futile enterprise. On the contrary, the authors believe that a more concentrated investigation of satisfiable (non-valid) problems within deduction should lead to a deeper understanding of inference as a whole, enlarging the scope of automated deduction.

Proving theorems and constructing (counter-) models for theories (i.e. set of formulae) are activities which lie at the very heart of mathematics and even of science in general.

Maybe the best example illustrating the importance of model building is the famous problem concerning the "axiom of parallel" and the discovery of non-Euclidean geometry. The inventors of these geometries proposed interpretations that are models of the other axioms and counter-models of the axiom of parallel. This example is interesting from different points of view, mathematical, historical and philosophical ones.

---

[10]This topic is of course related to that of *computation*. To enlighten more and more the relationships between deduction and computation (as for example in [Herken, 1988]) is of the highest importance in mechanizing deduction.

[11]Obviously, algorithmic methods will only work for particular subclasses of this problem.

Long time before the discovery of non-Euclidean geometry the status of the axiom of parallel (or 5th postulate) was surrounded by some "mystery" and the understanding of its nature in the structure of geometry was intriguing. It is therefore not exaggerated to say, in a modern terminology, that this axiom was an "open problem" since many centuries. In fact, already in his "Comments" to the book I of Euclid the Greek mathematician and astronomer Proclus promoted the idea of the formulation of the axiom which is most popular nowadays.

A continuous investigation of this axiom can also be found in the Arabic tradition (from the 9th through the 13th century). The mathematician and poet O. al Khayyam (9th - 12th century) and the astronomer at-Tusi were among the most important Arabic scientist who contributed to this matter. Though they did not contribute, strictly speaking, to non-Euclidean geometry they should not be forgotten from a historical point of view: al Khayyam and at-Tusi used in their research the quadrilateral applied later by G. Saccheri and J.H. Lambert (18th century), who first considered the possibility of negating the 5th postulate. The work of at-Tusi was translated in Europe at the end of the 17th century.

More or less simultaneously with A-M. Legendre, C.F. Gauss analyzed this problem in depth, but did not publish anything about his hyperbolic geometry (according to F. Klein). Independently N.I. Lobatchevski and J. Bolyai discovered hyperbolic geometry too; but the importance of their work was not recognized at their time. Indeed, Gauss seems to be the only person appreciating the work of Lobatchevski.

The other possible geometry (anticipated by Saccheri and Lambert), the so-called elliptic geometry, was defined later by B. Riemann. Concerning the history of the problem of parallels, see e.g. [Taton, 1995].

The discovery of non-Euclidean geometries has been considered by the great philosopher and logician H. Putnam in his book *Mathematics, Matter and Method* as the most important event in the history of science from an epistemological point of view, because it "shows" that mathematics and empirical sciences are not truly disjoints domains, or stated differently, the mathematical statements are not true in a pure analytical way[12]. Putnam's opinion is mentioned because this view is relevant to automated deduction when approaches like [Horgan, 1993, Lam, 1990, Lam, 1991, Goodman, 1990, Tymoczko, 1979, Epstein and Levy, 1995] are taken into account.

[12] A statement (judgment) is true in a pure analytical way if only logical analysis of the concepts involved in it is necessary to show its truth. Logical analysis does not suffice to determine the truth of synthetic judgments (as those coming from experience).

Putnam's position evolved in time from *metaphysical realism* towards (what he called) *internal or pragmatic realism.* It is interesting to mention here [Girard, 2003], page 133: *But nobody has ever seen the tail of a model, models are ideal (usually infinite) objects.* This kind of statement can be conciliated with a more realistic view of things. In fact, the ideal objects Girard talks about can perfectly correspond to the way our minds apprehend (what is called) the external reality.

In 20th century mathematics, several further important contributions of this type were made, e.g. the development of nonstandard analysis and Gödel's consistency proof of the continuum hypothesis.

Not only the construction of single models but also proving model-theoretic properties became an important technique in 20th century mathematical logic: if a first-order class has the finite model property, then it is decidable (i.e., the satisfiability problem is decidable); a decision procedure can be obtained by simultaneously applying a complete first-order refutation procedure and an enumeration of finite domain interpretations (the interested reader may consult [Bullock and Schneider, 1973]). Using essentially this principle, Gödel proved the decidability of the $\forall\forall\exists$-class. Gödel's result also yields an algorithmic (though inefficient) method of constructing models of the $\forall\forall\exists$-class.

In [Richman, 1990] the number of *intended models* is used to convey a very abstract and deep classification of mathematical theories:

> We may distinguish two types of theories, characterizing them by their intent. In the first type, exemplified by Euclidean geometry, arithmetic and set theory there is a single intended interpretation, a standard model. ... In the second type of axiomatic theory, exemplified by group theory and point-set topology, the purpose and power of the theory lie in the large number of intended models. [Richman, 1990], page 125.

The usefulness of (counter-)models can be also exemplified in simpler topics, as for example, in showing the non-truth-functionality of modal connectives such as possibility and necessity. In order to prove that some reasoning patterns valid in classical logic are not applicable in non-classical logics, examples (i.e. particular situations or models) are exposed in which, if subsentences (of a given sentence) are replaced by other sentences with the same truth value, the overall truth value of the original sentence changes.

Epistemic logic furnishes another example ("... Initially, a player may consider possible all deals consistent with the cards in her hand...", [Fagin et al., 1995], page 16).

The list of examples from different domains could be easily enlarged.

Roughly speaking counter-examples do not have the same reputation as proofs, because they do not deal with the "general case", but their mathematical and pedagogical value cannot be overestimated[13].

In particular, counter-examples serve the purpose to convince everybody of the *necessity* of each hypothesis in the statement of a theorem. One major goal of counter-examples is their irreplaceable role in the correction of wrong intuitions (see for example [Hauchecorne, 1988]) or in testing conjectures (see A. Wiles' opinion in [Horgan, 1993]). This capability is undoubtedly needed for a *deep* understanding of proofs (and therefore of mathematics), and surely is missing in present day theorem provers.

The concept of model is at the heart of the Tarskian notion of *logical consequence* (see for example [Etchemendy, 1988] and references therein). It is (implicitly) also of central importance to probability, for example in the notion of *possible event*, i.e. an event that can be *realized* in an experience, for example to have 3 points throwing a dice[14].

Already in the early days of automated deduction, *model-based inference* was recognized as a powerful tool in proof–search; we just mention the GTM geometry prover of Gelernter et al. [Gelernter, 1983] and J. Slagle's semantic resolution [Slagle, 1967]. Loveland [Loveland, 1984] remarks, that already in 1956, "Minsky made the observation that the diagram that traditionally accompanies plane geometry problems is a simple model of the theorem that could greatly prune the proof search."

Thus besides serving as counterexamples (as in non-Euclidean geo metry) models play an important role in inference itself: they allow introducing *semantics* into a basically syntactic process. Despite the prominent role of model-based techniques in the early days of automated deduction, the following phase of research in this field was characterized by optimization of refinements and the corresponding completeness proofs. Theorem provers were mainly understood as inference engines producing proofs for provable sentences; the problem of dealing with nonderivable sentences (represented by satisfiable sets of clauses in resolution theorem proving), received considerably less attention. Some striking results have been obtained on model building using techniques exploiting exclusively *deductive* capabilities of theorem provers [Winker, 1982].

---

[13]Concerning the pedagogical importance of studying topics such as *consistency, nonconsequence, independence*, the interested reader can consult [Barwise and Etchemendy, 1994].

[14]An impossible event, on the contrary, cannot be realized in an experience, for example, to have 7 points throwing a dice.

That inference systems can do more than just produce proofs was demonstrated by S.Y. Maslov [Maslov, 1968] and later by W.J. Joyner [Joyner, 1976][15]. Maslov constructed a computational calculus capable not only of proving theorems but also of deciding (the validity problem of) first-order sentences belonging to specific first-order classes. He even proved the decidability of the K-class by this method, thus obtaining a strong and new mathematical result. On formulae of the K-class, the calculus, commonly called the inverse method, produces only finitely many derivations; if none of them is a proof then the sentence under investigation is not provable at all. Indeed, an inference system can be used to prove that *sentences are **not** derivable!!* Although, in case of nonderivability, Maslov's method yields the existence of countermodels, *no such model is actually constructed.*

Joyner realized the same idea within the resolution calculus; on some decidable clause classes (among them clause forms of well-known prenex classes) specific ordering refinements of resolution terminate and thus provide decision procedures for these classes. If the refinement terminates without producing the empty clause, then the original set of clauses is satisfiable; but, as in the case of the inverse method, it does not produce a model.

Not only the inverse method and resolution, but also some Gentzen-type calculi were designed to provide decision procedures, e.g. Kleene's G3 for intuitionistic propositional calculus [Kleene, 1952].

So the next step in the "development" of inference systems can be defined as that of model construction. It is the main purpose of this book to present and analyze such inference systems and to demonstrate their value to theorem proving and to science in general.

In the same sense that proofs are more than just provability, models are more than the fact of satisfiability. Indeed, both proofs and models provide *evidence*, i.e., they show *why* statements hold or do not. This underlines the conceptual value of model construction in general. The problem remains, to which extent the construction of models can be calculized and automatized at all. As already pointed out, it is impossible, even theoretically, to realize "universal" model builders for first-order logic.

However this does not imply that algorithmic model construction is pointless *a priori*. On the contrary our aim is to show that automated model building is a reasonable and even realistic task. Sometimes this

[15]An obviously much simpler use of inference systems is proposed in [Caicedo, 1978] to *deduce* non-theorems in *classical propositional calculus*. The author gives a Hilbert system for *non-theorems*. The idea is related to that of "non consequence" (see Chapter 4).

task can be carried out by ordinary resolution theorem provers, in other cases suitable extensions of the calculi are necessary.

Basically we distinguish two types of methods for model construction:

- *enumeration (or verification) based methods:* ground instances are sequentially generated and checked.
- *deduction-based methods:* new, in general non-ground, formulae are inferred in the model building process. We include here consequence relations (in the standard sense) as well as others (which we call weakly sound).

We call the second type of methods *symbolic* (with the meaning – among other ones – of representing general patterns). In contrast, the enumeration methods are called *nonsymbolic.*

Enumeration can be considered, of course, as a very particular case of deduction; sometimes a combination of both approaches is the most effective one (e.g. EQMC method [Peltier, 1997b, Caferra and Peltier, 2000a], and hyper-linking [Chu and Plaisted, 1994a, Chu and Plaisted, 1994b])[16].

Enumerative model construction corresponds to exhaustive finite domain search for models. On the other hand, deductive methods are based on calculi producing syntactic model representations (mostly of Herbrand models) in some logical language. In this context we mainly present resolution, paramodulation and the RAMC-calculus as well as extended versions of tableaux; all these calculi are also "ordinary" refutational calculi but are modified in a way to *extract more logical information.* We do not claim that, so far, automated model building produced spectacular results; neither do we suggest that the field is fully developed and established. But we hope to demonstrate that building models is *as important as proving theorems* and that there exist reasonably efficient algorithmic methods, which can and should be used in the practice of automated theorem proving. We are convinced that tools for model construction will be part of any standard theorem prover in the near future. Already in principle, constructing models and, more generally, model–based inference are crucial to any intelligent deductive method; the investigation of these mechanisms presents a major challenge to computer science and logic.

Perhaps the most natural, systematic, and elegant way of building models (counter-examples) is defined by the tableau method (due to

[16]This is in some sense foreseeable: in order to obtain general patterns some particular features are put aside. When treating a specific problem, taking into account its particularities may greatly facilitate the solution.

Beth, Smullyan, Hintikka, ... see for example [Smullyan, 1968, Fitting, 1990]), but in practice only trivial models can be constructed using tableaux (due to nontermination).

In order to compare tableaux with the methods presented in this work, it is worth mentioning that finite models can be found by enumeration. This feature is particularly relevant when we are interested in effective mechanization.

The book is organized as follows:

The chapters 3 and 4, on deductive and symbolic model building methods, constitute the real heart of this book.

In our first approach, we present traditional resolution provers as decision procedures on particular first-order classes, where model building takes place as a postprocessing procedure. In particular we present new versions of the hyperresolution method of C. Fermüller and A. Leitsch and their extension to equational clause logic (by integrating ordered paramodulation). The model building procedures are based on deductive closure and unit selection and do not require any form of backtracking. The objects produced by these inference procedures are finite sets of atoms which can be interpreted as representations of Herbrand models. In many cases (which are syntactically characterized), these Herbrand models can be transformed to finite models. The corresponding transformations are based on the analysis of equivalence classes of the Herbrand universe and do not require any form of search.

Our second approach to symbolic model building is the constraint-based one. First we present the constraint language, equational logic, in more detail. Then we show how equational formulae can be transformed to (so called) equational problems, and we define algorithmic methods (related to those of Comon and Lescanne) to solve these problems. The constraint language provides an extension of clause logic, the so called *c-clause logic*, where equational constraints restrict the sets of ground instances. We demonstrate that in c-clause logic, constraints are not only useful to inference, but are equally valuable in defining *disinference* rules, i.e. rules characterizing instances that *cannot* be inferred from given premises (like disresolution and dissubsumption). We present the method RAMC by R. Caferra and N. Zabel, which incorporates these disinference principles into a resolution calculus for c-clauses. The results produced by RAMC are either refutations of unsatisfiable sets of c-clauses, or else satisfiable sets of unit c-clauses. We define two versions of RAMC, one with a strong capability for redundancy deletion, and compare these methods to hyperresolution. The original version of RAMC, having stronger expressive means for model construction (via the additional use of constraints), suffers from weaker termination (compared to hyperresolution). It is shown how both methods (which are incomparable) can be combined for the construction of more powerful model building algorithms.

The principle underlying constrained resolution and disinference rules can be naturally applied to tableaux. We describe the method RAMCET developed by Caferra, Zabel [Caferra and Zabel, 1993] and Peltier [Peltier, 1997c], which uses the constraint language for reducing redundancies. It is based on the use of a semantic strategy for pruning the search space and on inductive generalization for detecting potentially infinite branches. An interesting property of RAMCET, besides its soundness and refutational completeness, is its completeness w.r.t. the set of models that can be represented by constrained atoms.

In Chapter 5 we address a profound and central problem of symbolic model building, namely model representation. In enumerative finite model building this problem is trivial, at least from a theoretical point of view, as models can be represented by finite tables. In symbolic model building, a *formal expression* is needed to define a (generally infinite) model. In case of hyperresolution the expression is a conjunction of (closed) atoms, in case of RAMC a conjunction of constrained atoms. We prove that these representations enjoy all required computational properties, including algorithmic clause evaluation and decidability of the equivalence problem; in particular we present a generalized version of H-subsumption and of the corresponding evaluation algorithm by Fermüller and Leitsch [Fermüller and Leitsch, 1996]. However, atomic and constrained atomic representations are not strong enough to cover some quite simple cases of model specification. Here we mention alternative representation mechanisms, like regular tree grammars and term schematizations, and show how far the formalism can be pushed under preservation of desirable computational properties. In particular, we focus on the evaluation problem of clauses for different symbolic representations, which is an important feature of every model building formalism.

The last chapter is devoted to finite model building.

The main theoretical results, enlightening conceptual differences between finite and infinite models, are recalled. Different fields are mentioned in which techniques for finite model building are an important tool. We suggest the study of a particular class of problems that shed some light on the limits of enumeration methods and underline the importance of one of the key concepts studied in the present work, i.e. model representation formalisms. Several of the most representative approaches to finite model building are described. Finally, practical results attained with running systems for model building are evoked.

The areas in which the most important results have been obtained are Mathematics and Logic. Applications of model building to the semantic guiding of theorem provers are mentioned. The usefulness of model building as a help in correcting programs is illustrated by examples from programming teaching and simple program verification. Both domains show the potentials as well as the present limits of model building systems.

Chapter 2

PRELIMINARIES

Below we provide definitions for the basic notions of clause logic and resolution. Moreover we introduce some special terminology that we shall use throughout this book. Although we assume the reader to be familiar with first-order predicate logic and resolution theorem proving we discuss the basic concepts and review the fundamental definitions for the sake of clarity and completeness.

1. Clause Logic

1.1 Syntax of Clause Logic

Concerning the language of clause logic we assume that there are countably infinite sets of *variable symbols* V, *constant symbols* CS, *function symbols* FS, and *predicate* symbols PS. As usual we assume each function and predicate symbol to be associated with some fixed *arity* which we denote by $arity(F)$ for $F \in \mathrm{PS} \cup \mathrm{FS}$. We call a predicate or function symbol *unary* iff it is of arity 1, *binary* iff the arity is 2, and in general *n–place* for arity n. The set of n-place function- and constant symbols is denoted by FS_n and PS_n, respectively.

If S is some set of expressions, clauses or clause sets then $\mathrm{CS}(S)$, $\mathrm{FS}(S)$, and $\mathrm{PS}(S)$, denote the set of constant-, function- and predicate symbols, respectively, that occur in S.

We define the notions *term*, *atom*, *literal*, *expression* and *clause* formally:

DEFINITION 2.1 (TERM) A *term* is defined inductively as follows:

(i) Each variable and each constant is a term.

(ii) If $t_1, \ldots, t_n$ are terms and f is an n–place function symbol, then $f(t_1, \ldots, t_n)$ is also a term.

(iii) Nothing else is a term.

If a term t is of the form $f(t_1, \ldots, t_n)$ we call it *functional*; the set of arguments of t — $args(t)$ — is $\{t_1, \ldots, t_n\}$; f is called the *leading (function) symbol* of t.

The set of all terms is denoted by T. ♯

We shall make use of the following *naming conventions*: For variable symbols we use letters from the end of the alphabet (u, v, w, x, y, z); for constant symbols, letters a, b, c are used; function symbols are denoted by f, g or h; as metavariables for terms we use t or s; capital letters will denote atoms, literals, clauses or certain sets of expressions. Whenever needed these letters are augmented by indices.

Definition 2.2 (atom) If $t_1, \ldots, t_n$ are terms and P is an n-place predicate symbol then $A : P(t_1, \ldots, t_n)$ is an *atom*; P is called the *leading (predicate) symbol* of A; $args(A)$ is the set $\{t_1, \ldots, t_n\}$. ♯

Definition 2.3 (literal) A *literal* is either an atom or an atom preceded by a negation sign. ♯

Definition 2.4 (expression) An *expression* is either a term or a literal. ♯

Clauses represent universally closed disjunctions of literals. For a more economic and comfortable notation the quantifiers and the disjunctions are omitted. The order of literals is irrelevant by commutativity of $\vee$, parentheses are superfluous by associativity of $\vee$ and multiple occurrences can be dropped by the idempotence of $\vee$. These observations lead to the clause concept below:

Definition 2.5 (clause) A *clause* is a finite set of literals. The *empty clause* is denoted by $\Box$.

Although clauses are sets of literals, we mostly write them in the more legible form as disjunctions. Thus, unless there is danger of confusion, the clause $\{P(x, y), \neg Q(f(y))\}$ will be written as

$$P(x, y) \vee \neg Q(f(y)).$$

A clause C is called *positive* if all literals in C are unnegated, and *negative* if all literals in C are negated.

If $\mathcal{C}$ is a set of clauses then $POS(\mathcal{C})$ denotes the set of all positive clauses in $\mathcal{C}$. ♯

Remark: There are several different clause concepts. In most parts of the book we use the most traditional one as defined above. However in the Chapter 4 we will work with clauses as *multisets* of literals. Indeed for some formalisms the automatic reduction of multiplicity of literals to 1 is inconvenient, and we consider clauses as invariant only under associativity and commutativity.

For other applications, sets of literals are too weak and we need a concept defining larger equivalence classes. Such a concept is that of *condensed clause* which gives very compact representations. These representations are needed in Chapter 3. The formal concept of condensed clause will be given in Definition 2.42. ♯

Throughout this book we shall speak about clause sets; by this, if not stated differently, we always mean *finite* sets of clauses. When we are dealing with infinite sets of clauses this will be indicated explicitely.

DEFINITION 2.6 (DUALITY) If a literal L is unsigned, i.e. if it is identical with an atom A, then the *dual* of L — L^d — equals $\neg A$. Otherwise, if L is of the form $\neg A$ then $L^d = A$. For a set of literals $C = \{L_1, \ldots, L_n\}$ we define $C^d = \{L_1^d, \ldots, L_n^d\}$. ♯

DEFINITION 2.7 (SIGNATURE) A *signature* (usually denoted by Σ) is a subset of $\mathrm{CS} \cup \mathrm{FS} \cup \mathrm{PS}$. Frequently we restrict terms, atoms, clauses and clause sets to specific signatures. So we say that E is an expression *over* Σ if $\mathrm{CS}(E) \cup \mathrm{FS}(E) \cup \mathrm{PS}(E) \subseteq \Sigma$. For the set of terms over Σ we write $T(\Sigma)$. ♯

Additionally we introduce the following notation:

DEFINITION 2.8 C_+ is the set of positive literals of a clause C, analogously C_- denotes the set of negative literals (negated atoms) in C. ♯

DEFINITION 2.9 (HORN CLAUSE) C is a *Horn clause* if it contains at most one positive literal, i.e. $|C_+| \leq 1$. ♯

The *term depth* of an expression or a clause is defined as follows:

DEFINITION 2.10 (TERM DEPTH) The *term depth* of a term t — $\tau(t)$ — is defined by:

(i) If t is a variable or a constant, then $\tau(t) = 0$.

(ii) If $t = f(t_1, \ldots, t_n)$, where f is an n–place function symbol, then $\tau(t) = 1 + \max\{\tau(t_i) | 1 \leq i \leq n\}$.

The term depth of a literal L is defined as $\tau(L) = \max\{\tau(t) | t \in args(L)\}$. The term depth of a clause C is defined as $\tau(C) = \max\{\tau(L) | L \in C\}$. For a set S of clauses we define $\tau(S) = \max\{\tau(C) | C \in S\}$. ♯

The set of all variables occurring in an expression E is denoted by $V(E)$; if C is a clause, then

$$V(C) = \bigcup_{L \in C} V(L).$$

We define E_1 and E_2 to be *variable disjoint* iff $V(E_1) \cap V(E_2) = \emptyset$.

DEFINITION 2.11 (GROUND) An expression or a clause is called *ground* if no variables occur in it. We call it *constant free* if no constants occur in it, and *function free* if it does not contain function symbols. ♯

DEFINITION 2.12 $\tau_{\min}(t, E)$ is defined as the *minimal depth of occurrence* of a term t within an expression E. If C is a clause, then $\tau_{\min}(t, C)$ denotes the minimum of $\tau_{\min}(t, P_i)$ for all atoms P_i of C. $\tau_{\max}(t, E)$ respectively $\tau_{\max}(t, C)$ are defined in the same way. ♯

EXAMPLE 2.13 If $P_1 = P(x, f(f(y)))$, $P_2 = Q(f(x))$ and $C = \{P_1, \neg P_2\}$, then $\tau(P_1) = 2$, $\tau(P_2) = 1$, $\tau(C) = 2$, $\tau_{\min}(x, C) = 0$, $\tau_{\max}(x, C) = 1$, $\tau_{\min}(y, C) = \tau_{\max}(y, C) = 2$.

DEFINITION 2.14 The *maximal variable depth* of an expression E is defined as $\tau_v(E) = \max\{\tau_{\max}(x, E) | x \in V(E)\}$.
For clauses C we define $\tau_v(C) = \max\{\tau_v(L) | L \in C\}$;
analogously for clause sets S $\tau_v(S) = \max\{\tau_v(C) | C \in S\}$. ♯

In several inference methods, terms within expressions may be replaced by other terms; the most prominent example is paramodulation which will be introduced in Chapter 3. Below we give the formal definitions of position and replacement at a position:

DEFINITION 2.15 (POSITION) A *position* α is a finite sequence of positive integers. ϵ denotes the empty sequence and $\alpha.\beta$ denotes the concatenation of the sequences α and β.

The set of *positions in an expression* E and the expression E_α (the expression at position α in E) are inductively defined as follows:

- ϵ is a position in E and $E_\epsilon = E$;

- If $E = \neg A$ for an atom A then $\epsilon.1$ is a position in E and $E_{\epsilon.1} = A$;

- If f is an n-ary functional symbol, $i \in \{1, \ldots, n\}$ and α is a position in t_i, then $i.\alpha$ is a position in $f(t_1, \ldots, t_n)$ and $t_{i.\alpha} = (t_i)_\alpha$.
- If P is an n-ary predicate symbol, $i \in \{1, \ldots, n\}$ and α is a position in t_i, then $i.\alpha$ is a position in $P(t_1, \ldots, t_n)$ and $t_{i.\alpha} = (t_i)_\alpha$.

Instead of $\epsilon.\alpha$ we write α.

Let s, t be terms. t is said to be a *subterm of* s iff there exists an α such that $s_\alpha = t$. t is a *proper* subterm if $\alpha \neq \epsilon$. ♯

EXAMPLE 2.16 Let E be the literal $\neg P(x, f(y))$. Then the set of positions in E is

$$\{\epsilon,\ 1,\ 1.1,\ 1.2,\ 1.2.1\}.$$

The expressions at these positions are

$$E_\epsilon : \neg P(x, f(y)),\ E_1 : P(x, f(y)),\ E_{1.1} : x,\ E_{1.2} : f(y),\ E_{1.2.1} : y.\ \sharp$$

If E is an expression and α a position in E, we denote by $E[t]_\alpha$ the expression obtained by replacing (at the position α) the expression E_α by t. We give the formal definition below:

DEFINITION 2.17 Let E be an expression and α be a position in E.

- If $\alpha = \epsilon$ and t is an expression of the same type as E, then $E[t]_\alpha = t$.
- If $\alpha = 1.\beta$ and $E = \neg A$ then $E[t]_\alpha = \neg A[t]_\beta$.
- If $\alpha = i.\beta$, $E = X(t_1, \ldots, t_n)$ (where X is either a function symbol or a predicate symbol) then

$$E[t]_\alpha \quad = \quad f(t_1, \ldots, t_{i-1}, t_i[t]_\beta, t_{i+1}, \ldots, t_n).\ \sharp$$

EXAMPLE 2.18 Let E be the expression from Example 2.16. Then

$$E[g(g(y))]_{1.2} \quad = \quad \neg P(x, g(g(y))).\ \sharp$$

DEFINITION 2.19 (COMPONENT) Let L and L' be literals in a clause C; we define $L \sim_v L'$ if $V(L) \cap V(L') \neq \emptyset$ and $\sim_{v*}$ as the reflexive and transitive closure of $\sim_v$. An equivalence class under $\sim_{v*}$ is called a *component* of C. A clause is called *decomposed* if it consists of only one component and *disconnected* if every component consists of one literal only. ♯

EXAMPLE 2.20 Let $C_1 = P(x) \vee R(x, y) \vee Q(y)$, $C_2 = P(x) \vee Q(y)$ and $C_3 = P(x) \vee R(x, y) \vee Q(z)$. Then C_1 is decomposed (note that $P(x) \sim_v R(x, y)$, $R(x, y) \sim_v Q(y)$ and thus $P(x) \sim_{v*} Q(y)$). C_2 is disconnected and C_3 is neither decomposed nor disconnected. ♯

1.2 Semantics of Clause Logic

As our main topic is automated model building on clause logic, the semantics of clause sets plays a central role in this book. For this reason we present the model theoretic machinery in more detail.

Clauses and sets of clauses represent first-order formulae in the following sense:

DEFINITION 2.21 A clause in first-order logic represents a universally closed formula and sets of clauses represent conjunctions of clauses. Formally we assign a formula $F(\mathcal{C})$ to a set of clauses $\mathcal{C}$ in the following way:

- Let C be the clause $\{L_1, \ldots, L_n\}$ and $V(C)$ be $\{x_1, \ldots, x_m\}$. Then we define $F(\{C\})$ by

$$F(\{C\}) \quad = \quad (\forall x_1, \ldots, x_m)(L_1 \vee \cdots \vee L_n).$$

- Let $\mathcal{C}$ be a set of clauses. If $\mathcal{C}$ is empty then $F(\mathcal{C}) = \top$ where $\top$ represents **true**. If $\mathcal{C}$ is nonempty we define

$$F(\mathcal{C}) \quad = \quad \bigwedge\{F(\{C\}) | C \in \mathcal{C}\},$$

  i.e. $F(\mathcal{C})$ is the conjunction of the $F(\{C\})$ for $C \in \mathcal{C}$. ♯

Remark: In principle we could define the semantics of sets of clauses directly via their transformations. But for these specific forms it is convenient and comfortable to adjust the concept of first-order interpretation accordingly. As clauses are just universally closed disjunctions, $F(\{C\})$ is true in an interpretation over a domain D if for all assignments of the variables $\{x_1, \ldots, x_m\}$ to values in D the disjunction $L_1 \vee \cdots \vee L_n$ evaluates to true. ♯

DEFINITION 2.22 (INTERPRETATION) Let $\mathcal{C}$ be a set of clauses. Then $\mathcal{M} : (D, \Phi)$ is called an *interpretation* of $\mathcal{C}$ if

- D is a nonempty set.
- Φ is a mapping on $\mathrm{CS}(\mathcal{C}) \cup \mathrm{FS}(\mathcal{C}) \cup \mathrm{PS}(\mathcal{C})$ fulfilling the following conditions:

  (a) $\Phi(c) \in D$ for $c \in \mathrm{CS}(\mathcal{C})$.

  (b) For $f \in \mathrm{FS}_k(\mathcal{C})$ $\Phi(f)$ is a function of type $D^k \to D$.

  (c) For $P \in \mathrm{PS}_k(\mathcal{C})$ $\Phi(P)$ is a function of type $D^k \to \{\mathbf{true}, \mathbf{false}\}$ (i.e. $\Phi(P)$ is a k-ary predicate over D).

Let I be a mapping from V to D. Then I is called a *variable assignment* over D.

Now let $\mathcal{M} : (D, \Phi)$ be an interpretation of a clause set and I be a variable assignment over D. Then $\mathcal{M}_I : (D, \Phi, I)$ is called a *V-instance* of $\mathcal{M}$. V-instances are an auxiliary tool to define the truth values of clauses and clause sets under interpretations. ♯

DEFINITION 2.23 (TRUTH EVALUATION) Let $\mathcal{C}$ be a set of clauses and $\mathcal{M} : (D, \Phi)$ be an interpretation of $\mathcal{C}$ and $\mathcal{M}_I$ be a V-instance of $\mathcal{M}$.

We first define the evaluation of $T(\mathcal{C})$, the set of terms over the signature of $\mathcal{C}$. The corresponding evaluation function is $t_{\mathcal{M}_I}$:

- $t_{\mathcal{M}_I}(x) = I(x)$ for $x \in V$.
- $t_{\mathcal{M}_I}(c) = \Phi(c)$ for $c \in \mathrm{CS}(\mathcal{C})$.
- $t_{\mathcal{M}_I}(f(t_1, \ldots, t_k)) = \Phi(f)(t_{\mathcal{M}_I}(t_1), \ldots, t_{\mathcal{M}_I}(t_k))$ for $f \in \mathrm{FS}_k(\mathcal{C})$ and $t_1, \ldots, t_k \in T(\mathcal{C})$.

The evaluation function $v_{\mathcal{M}_I}$ is defined on atoms, literals and clauses:

- $v_{\mathcal{M}_I}(P(t_1, \ldots, t_k)) = \Phi(P)(t_{\mathcal{M}_I}(t_1), \ldots, t_{\mathcal{M}_I}(t_k))$ for $P \in \mathrm{PS}_k(\mathcal{C})$ and $t_1, \ldots, t_k \in T(\mathcal{C})$.
- $v_{\mathcal{M}_I}(\neg A) = \mathit{not}(v_{\mathcal{M}_I}(A))$ for an atom formula A over $\Sigma(\mathcal{C})$ and $\mathit{not}(\mathbf{true}) = \mathbf{false}$, $\mathit{not}(\mathbf{false}) = \mathbf{true}$.
- Let $C : \{L_1, \ldots, L_n\}$ be a clause over $\Sigma(\mathcal{C})$. Then $v_{\mathcal{M}_I}(C) = \mathbf{true}$ iff there exists an $i \in \{1, \ldots, n\}$ s.t. $v_{\mathcal{M}_I}(L_i) = \mathbf{true}$.

Now we are in possession of the tools to define the evaluation of clauses and clause sets by $v_{\mathcal{M}}$ itself:

- Let C be a clause over $\Sigma(\mathcal{C})$. Then $v_{\mathcal{M}}(C) = \mathbf{true}$ iff for all V-instances $\mathcal{M}_I$ of $\mathcal{M}$ $v_{\mathcal{M}_I}(C) = \mathbf{true}$.
- Let $\mathcal{D}$ be a set of clauses over $\Sigma(\mathcal{C})$. Then $v_{\mathcal{M}}(\mathcal{D}) = \mathbf{true}$ iff for all clauses $C \in \mathcal{D}$ $v_{\mathcal{M}}(C) = \mathbf{true}$. ♯

The concept of model is the usual one:

DEFINITION 2.24 (MODEL) Let $\mathcal{C}$ be a set of clauses and $\mathcal{M} : (D, \Phi)$ be an interpretation of $\mathcal{C}$. If $v_{\mathcal{M}}(\mathcal{C}) = \mathbf{true}$ then $\mathcal{M}$ is called a *model* of $\mathcal{C}$. $\mathcal{M}$ is called a *finite model* if $\mathcal{M}$ is a model and D is a finite set. ♯

Remark: For the evaluation of a clause C under $v_{\mathcal{M}_I}$ for an assigment I, only the values of I on $V(C)$ really count. Thus for $V(C) = \{x_1, \ldots, x_k\}$

we will frequently restrict the assignments to $V(C)$; in this case an assignment I takes the form $\{(x_1, d_1), \ldots, (x_k, d_k)\}$. ♯

DEFINITION 2.25 (LOGICAL CONSEQUENCE) Let $\mathcal{C}$ be a set of clauses and D be a clause. We say that D is a *logical consequence* of $\mathcal{C}$ if every model of $\mathcal{C}$ is also a model of D (we use the notation $\mathcal{C} \models D$). The concept can be extended to sets of clauses by: $\mathcal{C} \models \mathcal{D}$ iff for all $D \in \mathcal{D} : \mathcal{C} \models D$. ♯

Most inference systems *respect* the relation $\models$, i.e. all derived clauses are logical consequences of the original set of clauses. But we will see in Chapter 4 that, in the process of model building, the relation $\models$ may be too narrow and we have to adopt more liberal semantic criteria.

DEFINITION 2.26 (SOUNDNESS) Let $\rhd$ be a binary relation on the set of all clause sets. $\rhd$ is called *sound* if $\mathcal{C} \rhd \mathcal{D}$ implies $\mathcal{C} \models \mathcal{D}$. ♯

EXAMPLE 2.27 Let

$$\mathcal{C} \;:\; \{P(x,x),\ \neg P(x, f(y)) \vee Q(y)\}$$

be a set of two clauses and let $\mathcal{M} : (D, \Phi)$ be the following interpretation of $\mathcal{C}$:

$$\begin{aligned} D &= \{0,1\} \\ \Phi(f)(0) = 1 &\quad , \quad \Phi(f)(1) = 0, \\ \Phi(P)(0,0) = \mathbf{true} &\quad , \quad \Phi(P)(0,1) = \mathbf{false}, \\ \Phi(P)(1,0) = \mathbf{false} &\quad , \quad \Phi(P)(1,1) = \mathbf{true}, \\ \Phi(Q)(0) = \mathbf{true} &\quad , \quad \Phi(Q)(1) = \mathbf{false}. \end{aligned}$$

The clause $P(x,x)$ is true in $\mathcal{M}$: There are only two relevant variable assigments $I : \{(x,0)\}$ and $J : \{(x,1)\}$. Thus $P(x,x)$ is true because $v_{\mathcal{M}_I}(P(x,x)) = \Phi(P)(0,0) = \mathbf{true}$ and $v_{\mathcal{M}_J}(P(x,x)) = \Phi(P)(1,1) = \mathbf{true}$.

On the other hand $P(x, f(y)) \vee Q(y)$ is false. Here there are 4 relevant assigments to the variables x and y. If we take the assignment $J' : \{(x,1),(y,1)\}$ then $v_{\mathcal{M}_{J'}}(P(x,f(y))) = \Phi(P)(1, \Phi(f)(1)) = \Phi(P)(1,0) = \mathbf{false}$ and also $\Phi(Q)(1) = \mathbf{false}$. As one of the clauses is false in $\mathcal{M}$ the set $\mathcal{C}$ is false in $\mathcal{M}$ as well; thus $\mathcal{M}$ is not a model of $\mathcal{C}$. ♯

The most natural semantics to sets of clauses is provided by Herbrand interpretations. Here the domain of the interpretation is defined from the syntactic material, and the function symbols and constant symbols have a fixed meaning. Herbrand interpretations and Herbrand models are also the main objects of this book. It is this type of model which naturally occurs as a result of inference processes on clause logic.

DEFINITION 2.28 (HERBRAND UNIVERSE) The *Herbrand universe* $H(\mathcal{C})$ of a clause set $\mathcal{C}$ is the set of all ground terms over the signature of $\mathcal{C}$ – in case $\mathrm{CS}(\mathcal{C}) \neq \emptyset$. If $\mathrm{CS}(\mathcal{C}) = \emptyset$ then $H(\mathcal{C})$ is the set of all ground terms over $\Sigma(\mathcal{C}) \cup \{a\}$ where a is an arbitrary constant symbol. ♯

Frequently we will write H instead of $H(\mathcal{C})$ if the set of clauses is evident from the context.

DEFINITION 2.29 (HERBRAND INTERPRETATION) Let $\mathcal{C}$ be a set of clauses and H be the Herbrand universe of $\mathcal{C}$. An interpretation $\mathcal{M} : (H, \Phi)$ is called a *Herbrand interpretation* if

- for all $c \in \mathrm{CS}(\mathcal{C})$: $\Phi(c) = c$ and
- for all $f \in \mathrm{FS}_k(\mathcal{C})$ (and arbitrary k) and (ground terms) $t_1, \ldots, t_k \in H$

$$\Phi(f)(t_1, \ldots, t_k) \quad = \quad f(t_1, \ldots, t_k).$$

A Herbrand interpretation of $\mathcal{C}$ which is a model of $\mathcal{C}$ is called a *Herbrand model* of $\mathcal{C}$. ♯

EXAMPLE 2.30 Let $\mathcal{C} : \{P(x, x), \neg P(x, f(y)) \vee Q(y)\}$ be the clause set from Example 2.27. As $\mathrm{CS}(\mathcal{C}) = \emptyset$ we define the Herbrand universe H as the set of all ground terms over $\{P, Q, f, a\}$ (i.e. the set of ground terms over $\{f, a\}$). Therefore

$$H = \{f^n(a) | n \in \mathbb{N}\}$$

for $f^0(a) = a$, $f^{n+1}(a) = f(f^n(a))$ and $n \in \mathbb{N}$.

Now for *any* Herbrand interpretation $\mathcal{M} : (H, \Phi)$ of $\mathcal{C}$ we have $\Phi(a) = a, \Phi(f)(a) = f(a), \ldots$, so only the predicate symbols P and Q are free for interpretation. If we define

$$\begin{aligned} \Phi(P)(s, t) &= \textbf{true} \text{ iff } s = t, \\ \Phi(Q)(s) &= \textbf{true} \text{ for all } s \in H, \end{aligned}$$

then $\mathcal{M}$ is a Herbrand model of $\mathcal{C}$. ♯

The importance of Herbrand models is supported by the theorem that a set of clauses has a model iff it has a Herbrand model. This result goes back to Herbrand, for a more recent and detailed treatment see e.g. [Leitsch, 1997].

Another remarkable feature is that a clause is true in a Herbrand interpretation $\mathcal{M}$ iff all ground instances are true in $\mathcal{M}$. Thus substitution and interpretation of variables become the same thing!

By the specific nature of Herbrand interpretations $\mathcal{M} : (H, \Phi)$, $\mathcal{M}$ itself is characterized just by the values of the predicates $\Phi(P)$. But, for $t_1, \ldots, t_n \in H$, $\Phi(P)(t_1, \ldots, t_n) = v_{\mathcal{M}}(P(t_1, \ldots, t_n))$. Therefore it suffices to define the values of the ground atoms over H under the interpretation $\mathcal{M}$. This leads to the concept of ground representation:

DEFINITION 2.31 (GROUND REPRESENTATION) Let $\mathcal{C}$ be a set of clauses and $\mathcal{M} : (H, \Phi)$ be a Herbrand interpretation. Let $\mathcal{A}$ be the set of ground atoms over $\Sigma(\mathcal{C} \cup H)$ (If there are constant symbols in $\mathcal{C}$ then we may take $\Sigma(\mathcal{C})$). Then the set

$$\{A \mid A \in \mathcal{A}, v_{\mathcal{M}}(A) = \mathbf{true}\}$$

is called the *ground representation* of $\mathcal{M}$. ♯

Ground representations characterize Herbrand models in the sense, that two Herbrand interpretations of a set of clauses $\mathcal{C}$ are equal iff their ground representations are equal. Ground representations are merely an abstract theoretical tool because the set of all true ground atoms is frequently infinite. In particular, *finite* representations of these sets of ground atoms will be accentuated in this book.

The set of all ground instances of a set of clauses $\mathcal{C}'$ is of central importance to the theoretical analysis of theorem proving methods. Indeed most of the completeness results are based on Herbrand's theorem:

THEOREM 1.1 (HERBRAND'S THEOREM) *A set of clauses $\mathcal{C}$ is unsatisfiable iff there exists a* finite *unsatisfiable set $\mathcal{C}'$ of ground instances from clauses in $\mathcal{C}$.*

Proofs of this theorem can be found in any textbook on automated deduction (we just mention [Chang and Lee, 1973], [Loveland, 1978] and [Leitsch, 1997]). The importance of this theorem is the reduction of predicate logic to propositional logic. Clearly any unsatisfiable set of ground instances is unsatisfiable as a set of *propositional* clauses. Moreover the minimal size of an unsatisfiable set of ground instances (Herbrand complexity) is an inference independent measure for the complexity of first-order inference (see [Baaz and Leitsch, 1992]).

EXAMPLE 2.32 Let $\mathcal{C}$ be the set of clauses

$$\{C_1 : P(a),\ C_2 : \neg P(x) \vee P(f(x)),\ C_3 : \neg P(f(f(a)))\}.$$

An unsatisfiable set of ground instances can be obtained by producing two ground instances from C_2. The minimal set of unsatisfiable ground instances of $\mathcal{C}$ is:

$$\mathcal{C}' : \{P(a), \neg P(a) \vee P(f(a)), \neg P(f(a)) \vee P(f(f(a))), \neg P(f(f(a)))\}.$$

$\mathcal{C}'$ can be represented as a propositional clause set $\mathcal{D}$ for

$$\mathcal{D} = \{X, \neg X \vee Y, \neg Y \vee Z, \neg Z\}.\ \sharp$$

2. Substitution and Unification

DEFINITION 2.33 (SUBSTITUTION) Let V be the set of variables and T be the set of terms. A *substitution* is a mapping $\sigma : V \to T$ s.t. $\sigma(x) = x$ almost everywhere. We call the set $\{x | \sigma(x) \neq x\}$ *domain* of σ and denote it by $dom(\sigma)$; $\{\sigma(x) | x \in dom(\sigma)\}$ is called *range* of σ ($rg(\sigma)$). By ε we denote the *empty substitution*, i.e. $\varepsilon(x) = x$ for all variables x. $\sharp$

We represent substitutions as sets of assigments in the following way: Let σ be a substitution with

$$dom(\sigma) = \{x_1, \ldots, x_n\} \text{ and } \sigma(x_1) = t_1, \ldots, \sigma(x_n) = t_n.$$

Then we represent σ by the set

$$\{x_1 \leftarrow t_1, \ldots, x_n \leftarrow t_n\}.$$

DEFINITION 2.34 We say that a substitution σ is *based on a clause set* $\mathcal{C}$ if $rg(\sigma) \subseteq T(\mathcal{C})$.

σ is called a *ground substitution* if $rg(\sigma)$ consists of ground terms only. $\sharp$

DEFINITION 2.35 (INSTANCE) An expression E_1 is an *instance* of another expression E_2 iff there exists a substitution σ s.t. $E_1 = E_2\sigma$. Likewise a clause C_1 is an instance of clause C_2 iff $C_1 = C_2\sigma$ for some substitution σ. $\sharp$

We may compare expressions, substitutions and clauses using the following ordering relation.

DEFINITION 2.36 (GENERALITY) Let E_1 and E_2 be expressions, then $E_1 \leq_s E_2$ — read: *E_1 is more general than E_2* — iff there exists a substitution σ s.t. $E_1\sigma = E_2$. For substitutions ρ and θ we define analogously: $\rho \leq_s \theta$ iff there exists a substitution σ s.t. $\rho\sigma = \theta$. ♯

DEFINITION 2.37 (SUBSUMPTION) Let C, D be two clauses. We say that C *subsumes* D (written as $C \leq_{sub} D$) if there exists a substitution σ s.t. $C\sigma \subseteq D$. ♯

Remark: We have already mentioned that there are different clause concepts. Usually different clause concepts require different notions of redundancy and thus of subsumption. The definition above is in accordance with the definition of clauses as sets of literals. If clauses are defined as multisets the definition is essentially the same, with the difference that the relation $\subseteq$ has to be adjusted to multisets. In Chapter 3 we will make use of the subsumption concept in Definition 2.37, in Chapter 4 of the concept based on multisets. ♯

DEFINITION 2.38 (UNIFICATION) A nonempty set of expressions M is called *unifiable* if there exists a substitution σ with $X\sigma = Y\sigma$ for all $X, Y \in M$; the corresponding substitution σ is called a *unifier* of M. σ is called a *most general unifier (m.g.u.)* of M if for every other unifier ρ of M we have: $\sigma \leq_s \rho$.

We shall also say that E_1 *is unifiable with* E_2 if $\{E_1, E_2\}$ is unifiable. ♯

Remember that any two different m.g.u.s of a set of expressions only differ in the names of the variables.

Almost all theorem proving procedures used in theory and practice are based on the *unification principle*, i.e. they *only* produce substitutions which are *most general unifiers.* Indeed the invention of the unification principle by Robinson [Robinson, 1965b] represented the first and most decisive breakthrough in automated deduction. As the performance of theorem provers strongly depends on the efficiency of unification algorithms, many sophisticated algorithms for computing m.g.u.s have been developed so far. Besides the original algorithm in Robinson's paper we just mention the algorithm of Martelli and Montanari [Martelli and Montanari, 1982]. In this algorithm, unification is considered as the problem of finding a most general solution of term equations. This perspective is not only elegant but quite close to our treatment of equational logic in Chapter 4. Nevertheless it is not our aim to present unification theory in greater depth (for this we refer to [Baader and Snyder, 2001]); therefore we only present a motivating example, some definitions, and an algorithm.

EXAMPLE 2.39 Let

$$A_1 \ : \ P(g(x), f(x, z)) \text{ and}$$

$$A_2 \quad : \quad P(g(g((u)), v).$$

The problem of unifying A_1 and A_2 can be reduced to solving the system of equations

$$\mathcal{E}_1 : \ \{g(x) \doteq g(g(u)), \ v \doteq f(x,z)\},$$

i.e. we are searching for a subsitution θ s.t.

$$g(x)\theta = g(g(u))\theta, \ v\theta = f(x,z)\theta.$$

(called a solution) s.t. for all other solutions η we have $\theta \leq_s \eta$.

Clearly the equations cannot be "read" as a substitution directly. But we observe that θ is a solution of $\mathcal{E}_1$ iff θ is a solution of $\mathcal{E}_2$ for

$$\mathcal{E}_2 \quad = \quad \{x \doteq g(u), \ v \doteq f(x,z)\}.$$

$\mathcal{E}_1$ and $\mathcal{E}_2$ are *equivalent* because substitutions are homomorphisms on terms.

$\mathcal{E}_2$ can be "read" as a substitution $\theta : \{x \leftarrow g(u), v \leftarrow f(x,z)\}$; but θ is not a unifier. Indeed $A_1\theta \neq A_2\theta$. So we apply another transformation and interpret the equation $x \doteq g(u)$ as substitution on the system $\mathcal{E}_2$. This gives us the equivalent system

$$\mathcal{E}_3 : \{x \doteq g(u), \ v \doteq f(g(u), z)\}.$$

In $\mathcal{E}_3$ every equation is of the form $x \doteq t$, where x is a variable and t is a term, and the variables on the lefthandside of the equations occur only once in the whole system. What we have obtained is a system in *solved form.* Indeed, if we read the equations as the substitution

$$\sigma \quad : \quad \{x \leftarrow g(u), \ v \leftarrow f(g(u), z)\},$$

then σ is indeed a most general solution of $\mathcal{E}_3$ and thus of $\mathcal{E}_1$. Clearly σ is also an m.g.u. of A_1 and A_2. ♯

Thus finding a most general unifier means to find a most general solution of a system of term equations. And solving such a system of equations means to transform it into an equivalent solved form. Clearly we must also define criteria for the unsolvability of such systems and to ensure that our transformations on the systems terminate.

DEFINITION 2.40 A system of term equations is a finite set of the form

$$\mathcal{E} : \{s_1 \doteq t_1, \ldots, s_n \doteq t_n\}$$

where s_i, t_i are terms for $i = 1, \ldots, n$. A substitution θ is called a *solution* of $\mathcal{E}$ if for all $i = 1, \ldots, n : s_i\theta = t_i\theta$. A solution σ is called a *most general solution* of $\mathcal{E}$ if for all solutions θ of $\mathcal{E}$ $\sigma \leq_s \theta$.

Two systems $\mathcal{E}$ and $\mathcal{E}'$ are called *equivalent* if they have the same set of solutions.

We say that $\mathcal{E}$ is in solved form if

- $\{s_1, \ldots, s_n\} \subseteq V$ and
- every s_i occurs only (and thus exactly) once in $\mathcal{E}$ (for $i = 1, \ldots, n$).

♯

By definition a solved form represents a substitution. It is easy to see that for a solved form

$$\mathcal{E} : \{x_1 \doteq t_1, \ldots, x_n \doteq t_n\}$$

the substitution $\sigma : \{x_1 \leftarrow t_1, \ldots, x_n \leftarrow t_n\}$ is a most general solution of $\mathcal{E}$. What is needed is only the guarantee that $\mathcal{E}$ is indeed equivalent to the original system.

The algorithm below computes a solved form which represents a most general solution of the original system – provided the system is solvable at all. If the system is unsolvable the algorithm stops with failure.

function *Unification*
input: a finite set of term equations $\mathcal{S} = \{t_1 \doteq s_1, \ldots, t_n \doteq s_n\}$
output: either a substitution σ (solution of $\mathcal{S}$) or $\bot$ (no solution).
begin
% Γ denotes a finite set of equations, at the beginning $\Gamma = \emptyset$.
% $\Gamma\{x \leftarrow t\}$ means 'replace all occurrences of x in Γ by t'
For each $t_i \doteq s_i$ $(1 \leq i \leq n)$ in $\mathcal{S}$ apply the following rules:

- $\{t \doteq t\} \bigcup \Gamma \longrightarrow \Gamma$.
- if t is not a variable:
  $\{t \doteq x\} \bigcup \Gamma \longrightarrow \{x \doteq t\} \bigcup \Gamma$.
- $\{x \doteq t\} \bigcup \{x \doteq s\} \bigcup \Gamma \longrightarrow \{x \doteq t\} \bigcup \{t \doteq s\} \bigcup \Gamma$.
  $\{f(t_1, \ldots, t_n) \doteq f(s_1, \ldots, s_n)\} \bigcup \Gamma \longrightarrow$
  $\{t_1 \doteq s_1\} \bigcup, \ldots, \bigcup \{t_n \doteq s_n\} \bigcup \Gamma$.
- if $x \notin V(t)$ and $x \in V(\Gamma)$:
  $\{x \doteq t\} \bigcup \Gamma \longrightarrow \{x \doteq t\} \bigcup \Gamma\{x \leftarrow t\}$.
- if $f \neq g$:
  $\{f(t_1, \ldots, t_n) \doteq g(s_1, \ldots, s_m)\} \bigcup \Gamma \longrightarrow \bot$ (**clash**).
- if $x \in V(t)$ and t is not a variable :
  $\{x \doteq t\} \bigcup \Gamma \longrightarrow \bot$ (**cycle).**
  % i.e. infinite terms are not allowed in σ.

end.

3. Resolution

DEFINITION 2.41 (FACTOR) A *factor* of a clause C is a clause $C\theta$, where θ is a m.g.u. of a set C' with $C' \neq \emptyset$ and $C' \subseteq C$. In case $|C\theta| < |C|$ we call the factor *nontrivial.* ♯

DEFINITION 2.42 (CONDENSING) A clause C is called *condensed* if there exists no nontrivial factor of C which is a subclause of C. If C' is a condensed factor of C s.t. $C' \subseteq C$ then C' is called a *condensation* of C and is denoted by $N_c(C)$; condensations are unique up to renaming [Joyner, 1976]. ♯

EXAMPLE 2.43
$\{P(x, y), P(y, x)\}$ is condensed. $\{P(x, y), P(x, a)\}$ is not condensed; its condensation is $\{P(x, a)\}$. ♯

For the *resolvent* we retain the original definition of Robinson [Robinson, 1965b], which combines factorization and (binary) resolution. But be aware that, in some of the chapters to come, will shall locally define clauses and resolvents differently (namely as multisets of literals). It will always be clear from the context which concept we are using.

DEFINITION 2.44 If C and D are variable disjoint clauses and M and N are subsets of C and D respectively, s.t. $N^d \cup M$ is unifiable by the m.g.u. θ, then $E = (C - M)\theta \cup (D - N)\theta$ is a *Robinson–resolvent* of C and D.

If M and N are singleton sets then E is called *binary resolvent* of C and D.

The atom A of $(N^d \cup M)\theta$ is called the *resolved atom.* We also say that E is *generated via* A.

The elements of N and M are called the *literals resolved upon.* ♯

DEFINITION 2.45 For a clause set S we define $\mathcal{R}es(S)$ as the set of Robinson–resolvents of S. Additionally we define:

$$R(S) = S \cup \mathcal{R}es(S)$$

$$R^0(S) = S, \quad R^{i+1}(S) = R(R^i(S)), \qquad \text{and}$$

$$R^*(S) = \bigcup_{i \geq 0} R^i(S).$$

We say that a clause C is *derivable from* a clause set S iff $C \in \mathcal{R}^*(S)$. ♯

The set of all resolvents is too big for practical and even for theoretical purposes. In particular $R^*(\mathcal{C})$ is infinite on even very simple *satisfiable* sets of clauses; but even on unsatisfiable sets of clauses the search space generated by pure resolution is much too high. In this book the finiteness of the deductive closure on satisfiable set of clauses is a necessary condition for the extraction of Herbrand models. A very natural refinement of resolution consists in preventing resolution among two nonpositive clauses. This principle of positive resolution can be refined further to a principle producing *only* positive clauses (or $\Box$). To achieve this goal we have to replace single resolution steps by a sequence of inferences. The resulting method is called *hyperresolution*; it is the oldest refinement of resolution (invented by J.A. Robinson and proven complete in [Robinson, 1965a]). Hyperresolution still plays a prominent role in automated deduction.

DEFINITION 2.46 (POSITIVE RESOLVENT)
Let C, D be condensed clauses, where D is positive. The condensation of a binary resolvent of C and a factor of D is called a *PRF-resolvent.* (PRF abbreviates "positive, restricted factoring".) $\sharp$

DEFINITION 2.47 Let C be a non-positive clause and let the clauses D_i, for $1 \leq i \leq n$, be positive. Then the sequence $\Gamma = (C; D_1, \ldots, D_n)$ is called a *clash sequence.*

Let $C_0 = C$ and C_{i+1} be a PRF-resolvent of C_i and D_{i+1} for $i < n$. If C_n is positive or $\Box$ then it is called a *clash resolvent* or *hyperresolvent* defined by Γ. $\sharp$

Remark: The condensation of the resolvents required in Definition 2.46 improves the termination of hyperresolution on satisfiable sets of clauses. Therefore this version of hyperresolution is particular useful in using resolution as decision procedures [Fermüller et al., 1993]. As our model building procedures in Chapter 3 are based on decision procedures, condensation of resolvents is the logical choice. That this specific version of hyperresolution is complete is shown in [Leitsch, 1997]. $\sharp$

Hyperresolution exemplifies the principle of macro inference. It only produces positive clauses or the empty clause $\Box$. In variance to the standard definition of hyperresolution we have included a restriction on factoring. This type of "semi-factoring" preserves completeness for several refinements, among them positive hyperresolution.

Below, we do not need to refer to hyperresolution deductions themselves but rather are interested in the set of derived clauses. For this purpose the following operator based description of hyperresolution seems most adequate.

DEFINITION 2.48 Let $\mathcal{C}$ be a set of clauses. By $\rho_H(\mathcal{C})$ we denote the set of all clash resolvents definable by clash sequences of clauses in $\mathcal{C}$. The hyperresolution operator R_H and its closure R_H^* are defined by:

$$R_H(\mathcal{C}) = \mathcal{C} \cup \rho_H(\mathcal{C}),$$

$$R_H^0(\mathcal{C}) = \mathcal{C} \text{ and } R_H^{i+1}(\mathcal{C}) = R_H(R_H^i(\mathcal{C})) \text{ for } i \geq 0.$$

$$R_H^*(\mathcal{C}) = \bigcup_{i \geq 0} R_H^i(\mathcal{C}). \ \sharp$$

For a more detailed treatment of resolution operators see Section 1.

4. Prawitz's method

Prawitz's approach to theorem proving (see [Prawitz et al., 1960, Prawitz, 1960, Prawitz, 1969]) has been much less investigated than resolution. It contains the main ideas fully exploited afterwards in the well known methods of matings (Andrews) and connections (Bibel). It has interesting features for model building:

- it is a *global* method: it *replaces* sets of clauses by *equivalent* sets of clauses. Conditions generated by instantiations (i.e. unification problems) are treated globally. Resolution instead *adds* to the current set of clauses *logical consequences* of it and treats unification locally. Prawitz's method is therefore an approach to theorem proving, which is closer to a (naive) implementation of Herbrand's theorem than resolution.

- clauses *are not* combined.

- given a set of clauses $\mathcal{C}$, the main rule *matrix reduction* (introduced in [Prawitz, 1969]), splits clauses C and produces *unit clauses* L from literals of C. L is a model of C and therefore potentially belongs to a model of $\mathcal{C}$.

The method can be specified as follows:
Let $\mathcal{C} = \{C_1, \ldots, C_m\}$ a set of clauses with

$$C_1 : L_{11} \vee \ldots \vee L_{1n_1}$$
$$\vdots$$
$$C_m : L_{m1} \vee \ldots \vee L_{mn_m}$$

A *path* of $\mathcal{C}$ is a set of m literals: $\{L_{1i_1}, L_{2i_2}, \ldots, L_{mi_m}\}$ $(1 \leq i_1 \leq n_1, \ldots, 1 \leq i_m \leq n_m)$

The following sufficient condition for unsatisfiablity can be formulated in terms of paths:

- $\mathcal{C}$ is unsatisfiable if each path π_i of $\mathcal{C}$ contains two contradictory literals $L(\overline{s_i})$ and $\neg L(\overline{t_i})$ and the set of equations $\{\overline{s_i} \doteq \overline{t_i} \mid 1 \leq i \leq n_1 \times n_2 \times \ldots \times n_m\}$ has a solution Γ [1].

  The substitution Γ is called a *refutation set* for $\mathcal{C}$.

The condition above is not necessary (otherwise the satisfiability of clause logic would be decidable). We illustrate the problem by an example:

EXAMPLE 2.49 Let $\mathcal{C} : \{C_1, C_2, C_3\}$ for

$$\begin{aligned} C_1 &= P(a), \\ C_2 &= \neg P(x) \vee P(f(x)), \\ C_3 &= \neg P(f(f(a))). \end{aligned}$$

In fact $\mathcal{C}$ is the same set as in Example 2.32. Clearly $\mathcal{C}$ is unsatisfiable, but the different paths cannot be made contradictory simultaneously, i.e. there exists no substitution of the matrix above making the set of clauses *propositionally* unsatisfiable. Indeed we have the two path

$$\pi_1 : P(a), \neg P(x), \neg P(f(f(a))) \text{ and } \pi_2 : P(a), P(f(x)), \neg P(f(f(a))).$$

To make π_1 contradictory we need the substitution $\lambda_1 : \{x \leftarrow a\}$, for π_2 we need $\{x \leftarrow f(a)\}$. As both substitutions are ground there is no way to find a common substitution making π_1 *and* π_2 contradictory. Indeed, the system $\Gamma : \{x \doteq a, x \doteq f(a)\}$ has no solution. ♯

The reason for this phenomenon lies in the multiplicity of clauses in Herbrand's theorem. In fact *two* ground instances of C_2 in Example 2.49 are needed to obtain an unsatisfiable set of ground clauses. We can continue this game and replacing C_3 by $C_{3,n} : \neg P(f^n(a))$; it is obvious that in order to obtain an unsatisfiable set of ground instances n instances of C_2 are required. This is a major source of inefficiency of all matrix based methods. So it was shown in [Eder, 1990] and [Baaz and Leitsch, 1992] that resolution proofs may be *exponentially* shorter than proofs obtained by Prawitz's and Bibel's method.

This source of inefficiency is *clause copying* . In fact the path method sketched above can only be complete if arbitrarily many copies of clauses

[1] It should be noticed that the *definition* of the method suggests an implementation with structure sharing where Γ represents a *context*.

are produced during inference; as clause logic is undecidable there exists no recursive bound on the number of copies. Thus, besides computing the solution of sets of equations, *clause copying* is required as a first-order feature. We show below that, by using clause copying, we can indeed find a refutation set.

EXAMPLE 2.50 Let us consider the set of clauses $\mathcal{C}$ defined in Example 2.49. Instead of only listing C_1, C_2, C_3 we list $C_1, C_{2,1}, C_{2,2}, C_3$ where $C_{2,1}, C_{2,2}$ are variable disjoint variants of C_2:

$$\begin{array}{rcl} C_1 & : & P(a), \\ C_{2,1} & : & \neg P(x) \vee P(f(x)), \\ C_{2,2} & : & \neg P(y) \vee P(f(y)), \\ C_3 & : & \neg P(f(f(a)). \end{array}$$

Now we have four paths which have to be investigated:

$$\begin{array}{rcl} \pi_1 & : & P(a),\ \neg P(x),\ \neg P(y),\ \neg P(f(f(a))), \\ \pi_2 & : & P(a),\ \neg P(x),\ P(f(y)), \neg P(f(f(a))), \\ \pi_3 & : & P(a),\ P(f(x)),\ \neg P(y),\ \neg P(f(f(a))), \\ \pi_4 & : & P(a),\ P(f(x)),\ P(f(y)),\ \neg P(f(f(a))). \end{array}$$

Now we select a complementary pair in each path: $(1,2)$ in π_1 and π_2, $(2,3)$ in π_3 and $(3,4)$ in π_4. The corresponding set of equations is

$$\Gamma : \{x \doteq a, y \doteq f(x), f(y) \doteq f(f(a))\}.$$

Γ has the solution $\lambda : \{x \leftarrow a, y \leftarrow f(a)\}$. Indeed if we apply λ to the matrix of clauses above we obtain an unsatisfiable set of ground clauses. ♯

Clause copying is the main barrier to model building in first-order clause logic. In fact the method is *always* nonterminating on satifiable first-order clause sets and thus is incapable of providing *symbolic* model representations. This is the main reason for our choice of resolution in this book. Only when we add *constraints* to paths (as done for branches in the tableau method RAMCET in Chapter 4, see also [Caferra and Peltier, 2000b]) we have means to prevent nontermination.

Despite its first-order inefficiency (due to clause copying) Prawitz's method (like the other matrix based approaches) is efficient *propositionally*. It is frequently superior in cases where resolution is inefficient propositionally (e.g. in non-Horn problems). Also in finite model building (see Chapter 6) features of Prawitz' method prove to be useful.

The set of equations Γ has been introduced in [Prawitz, 1960]. The idea was that, instead of generating the instances of a set of clauses by *enumeration* (with an arbitrary order), it would be better to generate the values leading to inconsistent sets of instances *by calculation.* The translation of "by calculation" into more modern terms is "by solving sets of term equations".

The matrix reduction rules can be specified as follows:

Given a set of clauses (matrix) where the variables must be replaced by the solution of a set Γ of term equations, the matrix reduction rule is based on the equivalence:

$$(L(\overline{s}) \vee C_1) \wedge (\neg L(\overline{t}) \vee C_2) \wedge \Delta) \Leftrightarrow ((L(\overline{s}) \wedge C_2 \wedge \Delta) \vee (C_1 \wedge \neg L(\overline{t}) \wedge \Delta))$$

where $L(\overline{s})$ and $\neg L(\overline{t})$ are literals, $\overline{s} = \overline{t} \cup \Gamma$ has a solution, C_1 and C_2 are clauses and Δ set of clauses.

If $L(\overline{s})$ is the only literal in its clause, the following equivalence holds:

$$(L(\overline{s}) \wedge (\neg L(\overline{t}) \vee C_2) \wedge \Delta) \Leftrightarrow (L(\overline{s}) \wedge C_2 \wedge \Delta)$$

Similarly, if $\neg L(\overline{t})$ is the only literal in its clause, we get the equivalence:

$$(L(\overline{s}) \vee C_1) \wedge \neg L(\overline{t}) \wedge \Delta) \Leftrightarrow (C_1 \wedge \neg L(\overline{t}) \wedge \Delta)$$

The corresponding *matrix reduction rules* are defined by orienting the equivalences from left to right.

The matrix reduction rules isolate literals from clauses and reduce the number of paths to be tested. Assuming (without loss of generality) that $L(\overline{s})$ belongs to a clause containing n_1 literals and $\neg L(\overline{t})$ belongs to a clause containing n_2 literals, the total number of paths to test after application of the matrix reduction rule is:

$(n_2 - 1) \times n_3 \times \ldots \times n_m + (n_1 - 1) \times n_3 \times \ldots \times n_m = (n_1 + n_2 - 2) \times n_3 \times \ldots \times n_m$

(to be compared with $n_1 \times n_2 \times n_3 \times \ldots \times n_m$ paths to be tested if matrix reduction is not applied)

For propositional calculus the matrix reduction rule can simulate the splitting rule of Davis and Putnam.

Let us consider a matrix (set of clauses) $\mathcal{C}$:

$$\mathcal{C} = \{L \vee C_1, \ldots, L \vee C_m \vee \neg L \vee D_1, \ldots, \neg L \vee D_n\} \cup \Delta$$

where $C_i, D_j (1 \leq i \leq m, 1 \leq j \leq n)$ are clauses, L and $\neg L$ literals and Δ a set of clauses.

Davis and Putnam's splitting rule applied on $(L, \neg L)$ gives the 2 matrices:

$\mathcal{C}' = \{D_1, \ldots, D_n\} \cup \Delta$ and $\mathcal{C}" = \{C_1, \ldots, C_m\} \cup \Delta$

Prawitz's reduction rule produces the same result in $(m+n+3)$ steps (unit subsumption of several clauses is computed as 1 step).

A first application of matrix reduction gives:

$\mathcal{C}' = \{L, L \vee C_2, \ldots, L \vee C_m, D_1, \neg L \vee D_2, \ldots, \neg L \vee D_n\} \cup \Delta$ and
$\mathcal{C}'' = \{C_1, L \vee C_2, \ldots, L \vee C_m, \neg L, \neg L \vee D_2, \ldots, \neg L \vee D_n\} \cup \Delta$
by unit subsumption we obtain
$\mathcal{C}' = \{L, D_1, \neg L \vee D_2, \ldots, \neg L \vee D_n\} \cup \Delta$

$\mathcal{C}'' = \{C_1, L \vee C_2, \ldots, L \vee C_m, \neg L\} \cup \Delta$

by (n-1) applications of matrix reduction:
$\mathcal{C}' = \{L, D_1, D_2, \ldots, D_n\} \cup \Delta$ and
similarly, by (m-1) applications of matrix reduction:
$\mathcal{C}'' = \{C_1, C_2, \ldots, C_m, \neg L\} \cup \Delta$

as L and $\neg L$ are pure in $\mathcal{C}'$ and $\mathcal{C}"$ respectively

$\mathcal{C}' = \{D_1, D_2, \ldots, D_n\} \cup \Delta$ and

$\mathcal{C}'' = \{C_1, C_2, \ldots, C_m\} \cup \Delta$.

Thus we obtain the same result as that obtained by Davis and Putnam's splitting rule.

5. Reduction Systems

The process of normalization, i.e. the computation of normal forms, plays an important role (also) in automated deduction. We just mention transformation into clause form and skolemization (for a detailed treatment of normal forms in automated deduction see [Baaz et al., 2001]). The central concept of any theory of normalization is *reduction*. Informally speaking the aim of reduction is to transform a set (in our case a set of formulae) into a "simpler" one. As we use reduction on sets of clauses as well as on sets of formulae we give general definitions of the basic concepts.

DEFINITION 2.51 (REDUCTION RELATION) A *reduction relation* on a set M is is a binary relation $\rhd$ on M. Let ρ be a rule with the following property: whenever ρ transforms X_1 into X_2 then $(X_1, X_2) \in \rhd$; then ρ is called a *reduction rule* (w.r.t. $\rhd$). A set of reduction rules is called a *reduction system*. ♯

If $\rhd$ is a reduction relation we write $X \rhd Y$ for $(X, Y) \in \rhd$. In this case we say that X *reduces* to Y.

DEFINITION 2.52 Let $\triangleright$ be a reduction relation on a set M. An element $X \in M$ is in $\triangleright$-normal form (or, shorthand, in normal form) if there exists no element Y in M with $X \triangleright Y$. We also say that X is *irreducible* under $\triangleright$. ♯

EXAMPLE 2.53 The most important reduction relation on clause sets which is not a deduction relation is defined by subsumption. Indeed, in resolution theorem proving, subsumed clauses can be removed without sacrificing completeness. There are different ways to remove clauses by subsumption; we consider one of the most effective, namely replacement. Any time a new generation of resolvents is produced the whole set derived so far is "reduced" under subsumption. This kind of reduction can be expressed by the following relation $>_{sr}$ on finite sets of clauses:

- $\mathcal{C} >_{sr} \mathcal{D}$ iff there exists a clause C in $\mathcal{C}$ s.t.
  1. $\mathcal{D} = \mathcal{C} \setminus \{C\}$ and
  2. $\mathcal{D} \leq_{sub} C$.

We call $>_{sr}$ *subsumption reduction.*

E.g. let

$$\mathcal{C} : \{C_1 : P(x) \vee P(f(y)),\ C_2 : P(f(u)) \vee P(f(v)),\ C_3 : R(f(x), a)\}.$$

Then $C_1 \leq_{sub} C_2$ and $C_2 \leq_{sub} C_1$. Therefore for

$$\begin{aligned} \mathcal{D}_1 &= \{P(x) \vee P(f(y)),\ R(f(x), a)\}, \\ \mathcal{D}_2 &= \{P(f(u)) \vee P(f(v)),\ R(f(x), a)\}, \end{aligned}$$

we have $\mathcal{C} >_{sr} \mathcal{D}_1$ and $\mathcal{C} >_{sr} \mathcal{D}_2$. Both $\mathcal{D}_1$ and $\mathcal{D}_2$ are irreducible under $>_{sr}$ and thus are in normal form; $\mathcal{C}$ itself is not in normal form. ♯

If an element is not in normal form we may apply the relation $\triangleright$ iteratively till we eventually obtain such an element; this is the essence of a *normal form computation.*

DEFINITION 2.54 Let $\triangleright$ be a reduction relation and $\triangleright^*$ be the reflexive and transitive closure of $\triangleright$. Let $X \triangleright^* Y$ and Y be irreducible under $\triangleright$; then we call Y a *normal form* of X. ♯

EXAMPLE 2.55 Let $>_{sr}$ be the relation and $\mathcal{C}$ be the set of clauses as defined in Example 2.53. We have seen that $\mathcal{C} >_{sr} \mathcal{D}_1$ and $\mathcal{C} >_{sr} \mathcal{D}_2$, and that both $\mathcal{D}_1, \mathcal{D}_2$ are irreducible. Therefore both $\mathcal{D}_1$ and $\mathcal{D}_2$ are normal forms of $\mathcal{C}$. ♯

Note that the definition above is quite general; it neither requires the existence nor the uniqueness (as we have seen in Example 2.55) of normal forms.

So far the definitions above only address the syntactical part of normalization, they tell nothing about soundness. Clearly we could define $X \rhd P(a)$ for all formulae X different from $P(a)$ (and $P(a) \not\rhd Y$ for all formulae Y). According to this definition, every first-order formula X has $P(a)$ as its normal form, no matter whether X is valid, satisfiable or unsatisfiable. Obviously $\rhd$ does not make sense as it does not obey even the weakest semantical constraints; we have to ask for preservation of logical equivalence or, at least, of satisfiability or validity.

DEFINITION 2.56 Let $\rhd$ be a reduction relation on sets of first-order formulae. $\rhd$ is called *strongly sound* if, for all sets X, Y, $X \rhd Y$ implies $X \models Y$. ♯

Remark: Usually, in particular in normal form computations, we need a somewhat tighter concept of soundness; indeed for most transformations on sets of formulae we expect the property: $X \rhd Y$ implies $X \sim Y$. On the other hand, many useful transformations (some of them also presented in this book) are not strongly sound. One of the most prominent examples is skolemization. ♯

EXAMPLE 2.57 Let $>_{sr}$ be the relation of subsumption reduction as defined in Example 2.53. Then $>_{sr}$ is strongly sound: indeed, $\mathcal{C} >_{sr} \mathcal{D}$ implies $\mathcal{C} \leq_{sub} \mathcal{D}$ (by $\mathcal{D} \subseteq \mathcal{C}$) and therefore $\mathcal{C} \models \mathcal{D}$. Note that we even have $\mathcal{D} \leq_{sub} \mathcal{C}$ (by definition), thus also $\mathcal{D} \models \mathcal{C}$. In fact $\mathcal{C}$ and $\mathcal{D}$ are logically equivalent. ♯

For computational purposes it is important that any reduction chain defined by a reduction relation $\rhd$ terminates, i.e. any normal form computation yields a result. Termination is a property which, in general, cannot be achieved for reduction relations modelling arbitrary computations (e.g. β-reduction in the λ-calculus). But for meaningful logical normalizations, which typically serves as a preprocessing, termination is absolutely necessary.

DEFINITION 2.58 (TERMINATION) A reduction relation $\rhd$ is called *terminating* if there exists *no infinite* sequence of formulae $(X_i)_{i \in \mathbb{N}}$ with $X_i \rhd X_{i+1}$ for all $i \in \mathbb{N}$. ♯

The relation $>_{sr}$ defined above is also terminating; this is trivial, as the relation is defined on finite sets of clauses and $\mathcal{C} >_{sr} \mathcal{D}$ implies $|\mathcal{D}| < |\mathcal{C}|$. In general (e.g. transformation into prenex normal forms) we may have $X \triangleright^* Y$ and $X \triangleright^* Z$, where Y and Z are in normal form, but $Y \neq Z$; in this case Y and Z are different normal forms of X. In particular, different normal form computations may lead different results. For computational purposes it is necessary to make a deterministic choice, i.e. to decide for a specific normal form computation.

DEFINITION 2.59 (NORMAL FORM OPERATOR)
A normal form operator ν based on a reduction relation $\triangleright$ on a set M is a computable mapping $M \to M$ s.t. for all $X, Y \in M$ $\nu(X) = Y$ implies that Y is an $\triangleright$-normal form of X. ♯

If $\triangleright$ defines unique normal forms then $\nu(X) = Y$ iff Y is *the* normal form of X. E.g. $>_{sr}$ does not define unique normal forms.

DEFINITION 2.60 (SUBSUMPTION REDUCTION OPERATOR) We define sub to be an arbitrary $>_{sr}$-normal form operator on finite sets of clauses. ♯

Note that for all logical properties of $>_{sr}$ which play a role in this book, the order in which clauses are removed from the set is completely irrelevant. In fact any normal form algorithm and any nondeterministic choice of a normal form do the job.

Chapter 3

RESOLUTION-BASED METHODS

1. Model Building by Hyperresolution

1.1 Resolution Operators

The model building methods presented in this chapter are based on deductive saturation procedures. Instead of constructing refutation trees we consider resolution as an operator on sets of clauses. If a finite number of iterations of the operator leads to a fixed point set $\mathcal{C}$ then there are two possibilities: either $\mathcal{C}$ contains $\Box$ and we have shown that the original set of clauses is unsatisfiable, or $\Box$ is not in $\mathcal{C}$; in the latter case we may conclude that $\mathcal{C}$ is satisfiable. This is the general principle of resolution decision procedures (for a thorough treatment of this topic see [Fermüller et al., 1993]). As unrefined resolution shows a quite bad termination behavior, resolution decision procedures are mostly based on refinements. Here not only the correctness but also the completeness of refinements are required in order to ensure the *correctness* of the decision procedure. Once we have found out that the original set of clauses $\mathcal{D}$ is satisfiable we may use the computed fixed point set $\mathcal{C}$ for constructing a model of $\mathcal{D}$; this is basically the approach to model building we are following in this chapter.

DEFINITION 3.1 Let x be a resolution refinement and $\varrho_x(\mathcal{C})$ be the set of all resolvents defined by clauses in $\mathcal{C}$ obeying the restriction imposed by x. Then the resolution operator R_x is defined by

$$R_x(\mathcal{C}) = \mathcal{C} \cup \varrho_x(\mathcal{C}).$$ ♯

For unrestricted resolution we simply use the notation $\emptyset$; thus $R_\emptyset(\mathcal{C})$ is the set $\mathcal{C}$ extended by all resolvents definable by clauses in $\mathcal{C}$. For the hyperresolution refinement (see Definition 2.47) we use the letter H; in particular $\varrho_H(\mathcal{C})$ denotes the set of all hyperresolvents definable by clauses in $\mathcal{C}$.

DEFINITION 3.2 Let R_x be a resolution operator. Then the deductive closure under R_x is denoted by R_x^* and defined by

$$R_x^*(\mathcal{C}) = \bigcup_{i \in \mathbb{N}} R_x^i(\mathcal{C})$$

where $R_x^0(\mathcal{C}) = \mathcal{C}$ and $R_x^{i+1}(\mathcal{C}) = R_x(R_x^i(\mathcal{C}))$ for $i \in \mathbb{N}$. ♯

DEFINITION 3.3 Let R_x be a resolution operator. If for every unsatisfiable set of clauses $\mathcal{C}$ we have $\Box \in R_x^*(\mathcal{C})$ then we call R_x *complete*. If R_x is complete and, for a class of clauses sets Γ, $R_x^*(\mathcal{C})$ is finite for all $\mathcal{C} \in \Gamma$ then we say that R_x *decides* Γ. ♯

The operators R_x are monotone, i.e. $\mathcal{C} \subseteq R_x(\mathcal{C})$ for all sets of clauses $\mathcal{C}$, a typical feature of *deduction* operators. In automated deduction, particularly in automated model building, deletion in form of replacement plays an important role; this kind of deletion method defines nonmonotone *replacement* operators.

DEFINITION 3.4 Let $\mathcal{C}$ be a set of clauses and *sub* be the subsumption reduction operator as defined in Definition 2.60. Let R_x be a resolution operator. Then the corresponding *replacement* operator $R_{x,s}$ is defined by

$$R_{x,s}(\mathcal{C}) = sub(R_x(\mathcal{C})).$$ ♯

EXAMPLE 3.5 Let $\mathcal{C} = \{P(x,a) \vee P(a,y), \neg P(a,a)\}$. Then

$$R_\emptyset(\mathcal{C}) = \mathcal{C} \cup \{P(x,a), P(a,y), \Box\}.$$

Clearly $R_\emptyset^2(\mathcal{C}) = R_\emptyset(\mathcal{C})$ and so $R_\emptyset^*(\mathcal{C}) = R_\emptyset(\mathcal{C})$.

Now $R_\emptyset(\mathcal{C})$ contains $\Box$, and $\Box$ subsumes every clause (see Definition 2.37). This implies

$$R_{\emptyset,s}(\mathcal{C}) = sub(R_\emptyset(\mathcal{C})) = \{\Box\}.$$

♯

Example 3.5 shows that replacement operators are not monotone in general. In particular $\Box \in R_x(\mathcal{C})$ implies $R_{x,s}(\mathcal{C}) = \{\Box\}$ for every replacement operator $R_{x,s}$. The example also shows that it does not make sense to define a deductive closure operation for $R_{x,s}$ like for R_x; indeed, clauses which are removed and do not "return" at some time to stay for ever should be neglected. This leads to the following definition (for a more general treatment of redundancy and deletion operators see [Bachmair and Ganzinger, 1994]):

DEFINITION 3.6 Let $R_{x,s}$ be a replacement operator. Then the closure under R_x with replacement is defined by

$$R^*_{x,s}(\mathcal{C}) = \bigcup_{i\geq 0} \bigcap_{j\geq i} R^j_{x,s}(\mathcal{C}),$$

for sets of clauses $\mathcal{C}$ and $R^0_{x,s}(\mathcal{C}) = sub(\mathcal{C})$, $R^{i+1}_{x,s}(\mathcal{C}) = R_{x,s}(R^i_{x,s}(\mathcal{C}))$. ♯

DEFINITION 3.7 Let $R_{x,s}$ be a replacement operator. Then $R_{x,s}$ is called *complete* if for every unsatisfiable set of clauses $\mathcal{C}$ $R^*_{x,s}(\mathcal{C}) = \{\Box\}$. ♯

By definition we always have $R_{x,s}(\mathcal{C}) \subseteq R_x(\mathcal{C})$ and $R^*_{x,s}(\mathcal{C}) \subseteq R^*_x(\mathcal{C})$. However it is not easy to compute $R^*_{x,s}$ in general. Here we only investigate cases where the corresponding monotone operator converges, i.e. where $R^*_x(\mathcal{C})$ is finite. Then, clearly, also $R^*_{x,s}(\mathcal{C})$ is finite and computable. For some classes of clause sets and particular refinements x one can even guarantee $R^k_{x,s}(\mathcal{C}) = R^{k+1}_{x,s}(\mathcal{C})$ for a number k depending on $\mathcal{C}$ (see [Leitsch, 1997]); then $R^*_{x,s}(\mathcal{C}) = R^k_{x,s}(\mathcal{C})$ and the replacement closure is easily computable. $R_{x,s}$ is not always complete even if R_x is; indeed there are refinements like lock resolution which do not admit subsumption (without loss of completeness). Note that clauses might be subsumed away which are crucial to the derivation of $\Box$.

For hyperresolution the use of replacement preserves completeness and so $R_{H,s}$ is complete (for a proof see [Leitsch, 1997]). Thus if R_H decides a class Γ then $R_{H,s}$ does so too; moreover $R_{H,s}$ terminates on a larger class of clauses than R_H and even yields a better decision procedure in general.

The concept of the operators R_x and $R_{x,s}$ can easily be extended to other deduction principles like paramodulation (which will be needed for model building in equational clause logic). Generally R_x may be defined as

$$R_x(\mathcal{C}) = \mathcal{C} \cup \delta_x(\mathcal{C})$$

where δ_x represents an inference operator on (equational) clause logic. In Section 2 δ_x will be defined as an operator producing all ordered paramodulants and hyperresolvents from a set of clauses. The definitions of R^*_x and of $R^*_{x,s}$ in the general case are precisely the same as for resolution.

1.2 Horn Logic and Atomic Representations

On Horn logic the operator R_H can be interpreted as a model builder in a very natural way. By definition of hyperresolution all clauses in

$R^*_H(\mathcal{C}) \setminus \mathcal{C}$ are positive. If $\mathcal{C}$ is Horn then (by the closure of the Horn property under resolution) all clauses in $R^*_H(\mathcal{C}) \setminus \mathcal{C}$ are also unit and represent the set of all derivable *facts*. We will show that such sets of facts can be interpreted as representations of Herbrand models. Herbrand models specified in this manner possess some property of minimality which is important to the semantics of logic programs.

In Herbrand models the universe and the interpretations of function– and constant symbols are fixed a priori and only depend on the signature of the set of clauses. Therefore Herbrand models are characterized by the interpretations of the predicate symbols over the Herbrand universe. More formally a Herbrand interpretation Γ of a set of clauses $\mathcal{C}$ is characterized by the truth values of Γ on the atom set $\mathrm{AT}(\mathcal{C})$; if we define $\mathcal{M} = \{A \mid A \in \mathrm{AT}(\mathcal{C}), v_\Gamma(A) = \mathbf{T}\}$ then, clearly, this set $\mathcal{M}$ uniquely represents Γ (w.r.t. the signature of $\mathcal{C}$). Thus sets of ground atoms give straightforward representations of Herbrand models. Although such a representation is very elegant and important to the semantics of logic programming it is mostly infinite; indeed if $\mathcal{C}$ contains function symbols even very simple Herbrand interpretations have infinite representations as sets of ground atoms. But infinite sets are only of computational value if they can be *represented finitely*. This leads to the following more general definition:

DEFINITION 3.8 Let $\mathcal{A}$ be a set of atoms over the signature of a set of clauses $\mathcal{C}$. Then we call $\mathcal{A}$ an *atomic representation* (ARM) of the interpretation Γ, where Γ is the Herbrand interpretation of $\mathcal{C}$ fulfilling the property: for $A \in \mathrm{AT}(\mathcal{C})$ $v_\Gamma(A) = \mathbf{T}$ iff A is a ground instance of an atom in $\mathcal{A}$. The set of all ground instances of $\mathcal{A}$ over $H(\mathcal{C})$ is called a *ground (atomic) representation* of Γ. ♯

Note that ground interpretations are just specific cases of atomic representations. To a Herbrand interpretation Γ there exists only one ground representation, but there may be infinitely many different atomic representations.

EXAMPLE 3.9 Let $\mathcal{C} = \{P(a,a),\ \neg P(x,a) \vee P(f(x),a),\ \neg P(a,f(x))\}$ and Γ be the following Herbrand model of $\mathcal{C}$:

$$v_\Gamma(P(f^n(a),a)) = \mathbf{T} \text{ for } n \in \mathbb{N},$$

$$v_\Gamma(P(s,t)) = \mathbf{F} \text{ for all } s \in H(\mathcal{C}) \text{ and } t \in H(\mathcal{C}) - \{a\}$$

Then the set $\mathcal{A} = \{P(x,a)\}$ is an atomic representation of Γ. Moreover, for every $k \in \mathbb{N}$, the set

$$\mathcal{B}_k : \{P(a,a), \ldots, P(f^k(a),a),\ P(f^{k+1}(x),a)\}$$

is an atomic representation of Γ.

The ground representation $\mathcal{M} : \{P(f^n(a), a) \mid n \in \mathbb{N}\}$ of Γ is infinite. $\mathcal{M}$ is also the result of the application of hyperresolution to $\mathcal{C}$, i.e.

$$\mathcal{M} = POS(R_H^*(\mathcal{C})).$$

$\sharp$

Example 3.9 shows that R_H generates an atomic representation of Γ (in this case a ground representation). This is not a mere coincidence but a general principle.

THEOREM 1.1 *Let $\mathcal{C}$ be a satisfiable set of Horn clauses. Then $POS(R_H^*(\mathcal{C}))$ is an atomic representation of a minimal Herbrand model of $\mathcal{C}$.*

Proof: Let us abbreviate $POS(R_H^*(\mathcal{C}))$ by $\mathcal{P}$. Then $\mathcal{P}$ is a (possibly infinite) set of atoms; note that, due to the satisfiability of $\mathcal{C}$ and the correctness of R_H, $\Box \notin \mathcal{P}$. Now let Γ be a Herbrand model of $\mathcal{C}$ (recall that every satisfiable set of clauses has a Herbrand model). Then $v_\Gamma(A) = \mathbf{T}$ for every positive unit clause $A \in \mathcal{C}$. By the Horn property of $\mathcal{C}$, $R_H(\mathcal{C}) - \mathcal{C}$ consists of positive unit clauses only; by the correctness of R_H all clauses in $R_H(\mathcal{C}) - \mathcal{C}$ are true in Γ. An easy induction argument yields that $\mathcal{P}$ is a set of unit clauses which are true in Γ.

Let $\mathcal{M}$ be the set of all ground instances over $H(\mathcal{C})$ defined by $\mathcal{P}$. Every $A \in \mathcal{P}$ is true (as a clause!) in Γ; therefore $v_\Gamma(A\theta) = \mathbf{T}$ for all $A \in \mathcal{P}$ and ground substitutions θ, i.e. $v_\Gamma(B) = \mathbf{T}$ for all $B \in \mathcal{M}$. Consequently $\mathcal{M}$ is a (possibly proper) subset of $\mathcal{G} : \{A \mid A \in \mathrm{AT}(\mathcal{C}), v_\Gamma(A) = \mathbf{T}\}$. Let us consider $\mathcal{P}$ as atomic representation of a Herbrand interpretation Δ of $\mathcal{C}$. We have seen that Δ verifies at most the set of ground atoms verified by Γ. Thus, in case Δ is a model at all, it is also a minimal model. It remains to show that Δ is also a Herbrand *model* of $\mathcal{C}$:

We assume that Δ falsifies $\mathcal{C}$ and derive a contradiction. Let $\mathcal{B} = \{\neg A \mid A \in \mathrm{AT}(\mathcal{C}) - \mathcal{M}\}$, i.e. the "negated" set of atoms which are false in Δ.

Δ falsifies $\mathcal{C}$ iff the set $\mathcal{C} \cup \mathcal{M} \cup \mathcal{B}$ is unsatisfiable. By the compactness theorem [Boolos and Jeffrey, 1974] there exists a finite subset $\mathcal{D}$ of $\mathcal{M} \cup \mathcal{B}$ such that $\mathcal{C} \cup \mathcal{D}$ is unsatisfiable too. By definition $\mathcal{D}$ is a set of ground unit clauses $\{A_1, \ldots, A_n, \neg B_1, \ldots, \neg B_m\}$. By definition of $\mathcal{M}$ and of $\mathcal{B}$ the B_j are not ground instances of atoms in $\mathcal{P}$.

Because $\mathcal{C} \cup \mathcal{P}$ is satisfiable $\mathcal{C} \cup \{A_1, \ldots, A_n\}$ is satisfiable too. Thus, by the correctness of R_H, $\Box \notin R_H^*(\mathcal{C} \cup \{A_1, \ldots, A_n\})$.

By definition of hyperresolution, negative clauses can only serve as central clauses in clashes; moreover, in Horn logic, clashes with negative center clauses can only produce the resolvent $\Box$. Therefore we obtain

$$POS(R^*_H(\mathcal{C} \cup \mathcal{D}) \subseteq POS(R^*_H(\mathcal{C} \cup \{A_1, \ldots, A_n\})) \cup \{\Box\}.$$

We show now that $POS(R^*_H(\mathcal{C} \cup \mathcal{D}))$ does not contain $\Box$. Again we provide a proof by contradiction: Let us assume that $\Box \in POS(R^*_H(\mathcal{C} \cup \mathcal{D}))$. From $\Box \notin R^*_H(\mathcal{C} \cup \{A_1, \ldots, A_n\})$ we conclude that there exists a positive unit clause $R \in R^*_H(\mathcal{C} \cup \{A_1, \ldots, A_n\})$ and a negative unit clause $\neg B \in \{B_1, \ldots, B_m\}$ such that the clash $(\neg B, R)$ resolves to $\Box$. In particular B and R are unifiable. By definition of $\mathcal{P}$ and of the set $\{A_1, \ldots, A_n\}$ R is substitution instance of an atom $A \in \mathcal{P}$; as a consequence A and B are unifiable and B is an instance of A. By definition of $\mathcal{M}$ we get $B \in \mathcal{M}$.

On the other hand $\neg B \in \{\neg B_1, \ldots, \neg B_m\} \subseteq \mathcal{B}$ where $\mathcal{B}$ contains just all negative ground literals $\neg L$ with $L \notin \mathcal{M}$. Therefore $B \notin \mathcal{M}$ and we obtain a contradiction.

Thus we have shown that $\Box \notin R^*_H(\mathcal{C} \cup \mathcal{D})$. But, by the completeness of R_H this implies that $\mathcal{C} \cup \mathcal{D}$ is satisfiable, in contradiction to the assumption above. So we see that Δ is indeed a model of $\mathcal{C}$. $\Diamond$

Atomic representations of Herbrand models are only of computational value if they are finite. But, in general, the set $R^*_H(\mathcal{C})$ is infinite for satisfiable Horn sets and no *symbolic* model representation can be obtained; indeed, only finite sets of atoms correspond to first-order *formulae* expressing models. Here we face a principal barrier: by the undecidability of Horn logic, there is no computational procedure constructing symbolic model representations for all satisfiable sets of clauses. This does not imply that model building is a pointless task in principle, rather we have to look for decidable subclasses of Horn logic, in particular for those where R_H terminates. One such "termination class" of R_H is easy to find:

By definition of R_H the set $R^*_H(\mathcal{C})$ is only infinite if the set $POS(R^*_H(\mathcal{C}))$ is. Therefore, in Horn logic, $R^*_H(\mathcal{C})$ is finite iff $POS(R^*_H(\mathcal{C}))$ is a finite set of unit clauses. If $\mathcal{C}$ is a (finite) set of clauses not containing function symbols then, up to variants, only finitely many positive unit clauses are definable over the signature of $\mathcal{C}$. As an immediate consequence $R^*_H(\mathcal{C})$ is finite for all such clause sets $\mathcal{C}$ (note that we keep clauses in condensed variable-standard form) and, due to Theorem 1.1, produces atomic model representations for satisfiable ones. The class we just described is the Horn subclass $\mathcal{BSH}$ of the well-known *Bernays-Schönfinkel* class of finite, function-free clause sets $\mathcal{BS}$.

DEFINITION 3.10 Let $\mathcal{BS}$ be the class of all closed formulae of the form

$$(\exists x_1) \ldots (\exists x_m)(\forall y_1) \ldots (\forall y_n) M,$$

where M is quantifier–free and all terms occurring in M are variables. Then $\mathcal{BS}$ is called the Bernays–Schönfinkel class. If, in addition, M is a conjunction of Horn clauses then we obtain the subclass $\mathcal{BSH}$. ♯

Because the skolemization M' of a formula M in $\mathcal{BS}$ does not contain function symbols, all Herbrand interpretations of M' have a fixed finite domain. Thus a trivial decision procedure for satisfiability consists in evaluating M' over all its Herbrand interpretations. Here, however, we are interested in the termination behavior of resolution refinements on $\mathcal{BS}$ and $\mathcal{BSH}$. For this purpose we have to define clauses classes corresponding to $\mathcal{BS}$ and $\mathcal{BSH}$.

DEFINITION 3.11 $\mathcal{BS}^*$ is the class of all finite sets of clauses $\mathcal{C}$ such that for all $C \in \mathcal{C} : \tau(C) = 0$. $\mathcal{BSH}^*$ is the subclass of $\mathcal{BS}^*$ containing sets of Horn clauses only. ♯

The condition $\tau(C) = 0$ in definition 3.11 guarantees that there are no function symbols in $\mathcal{C}$. All constant symbols appearing in a set $\mathcal{C} \in \mathcal{BS}^*$ can be thought to have been introduced by skolemization. Thus $\mathcal{BS}^*$ is exactly the clause class corresponding to $\mathcal{BS}$; similarly $\mathcal{BSH}^*$ is the clause class corresponding to $\mathcal{BSH}$. By the arguments presented above we immediately verify that $R_H^*(\mathcal{C})$ is finite for every set of clauses $\mathcal{C} \in \mathcal{BSH}^*$. Thus hyperresolution may serve as a decision– and as a model-building procedure for $\mathcal{BSH}$. We will see in the next subsection that this behavior of hyperresolution cannot be extended to $\mathcal{BS}^*$ in a straightforward manner.

The really interesting case is that of infinite Herbrand universes; here, R_H terminates on $\mathcal{C}$ if there is a bound on term depth for derivable hyperresolvents. In [Fermüller et al., 1993] several functional Horn classes are defined where R_H terminates and thus produces finite atomic model representations for satisfiable sets of clauses. Here we present $\mathcal{VED}$, one of these classes which properly contains $\mathcal{BSH}$:

DEFINITION 3.12 $\mathcal{VED}$ (**v**ariables **in** **e**qual **d**epth) is the set of all finite sets of Horn clauses $\mathcal{C}$ with the following property:

For all $C \in \mathcal{C}$ and $x \in V(C)$: $\tau_{\min}(x, C) = \tau_{\max}(x, C)$. ♯

For $\mathcal{C} \in \mathcal{BSH}$ $\tau_{\max}(x, C) = \tau_{\min}(x, C) = 0$ for all $x \in V(C)$ and $C \in \mathcal{C}$, i.e. all variables occur at depth 0. Note that, in $\mathcal{VED}$, the term depth of variables – though equal within a clause – may be different for different clauses. Thus the set of clauses

$$\{\neg P(f(x)) \vee \neg P(g(x)),\ \neg Q(x, f(y)) \vee Q(f(y), x),\ P(f(f(a))) \vee Q(f(x), a)\}$$

is in $\mathcal{VED}$.

In order to prove the finiteness of $R^*_H(\mathcal{C})$ for $\mathcal{C} \in \mathcal{VED}$ we have to ensure that the term depth is not increased by hyperresolution. Before we prove the key lemma we have to define some technical notions:

DEFINITION 3.13 Let A and B two atoms. If the predicate symbols of A and B are different then CORR(A, B), the set of corresponding pairs in A and B, is empty; otherwise

$$\mathrm{CORR}(A, B) \quad = \quad \{(A_\alpha, B_\alpha) | \alpha \text{ is a position in } A \text{ and } B\}.$$

A pair $(s, t) \in \mathrm{CORR}(A, B)$ is called admissible (for unification) if the following conditions hold

- either s or t is a variable,
- s and t are different,
- if s (t) is a variable then s (t) does not occur in t (s). ♯

Remark: If, for two atoms A and B with the same predicate symbol, there are no admissible pairs then either A and B are equal or they are not unifiable. The unification algorithm of Robinson [Robinson, 1965b] is based on an iterative unification of corresponding pairs. Indeed the algorithm is characterized by m.g.u. sequences to be defined below. ♯

DEFINITION 3.14 Let A, B be two atoms with the same predicate symbol. We define an m.g.u. sequence $(\lambda_1, \ldots, \lambda_n)$ in the following way:

- $\lambda_0 = \epsilon$.
- Let us assume that λ_i is already defined. If there exists an admissible pair $(s, t) \in \mathrm{CORR}(A\lambda_i, B\lambda_i)$ (or (t, s)) s.t. $s \in V$ (or $t \in V$) then we define

  $$\lambda_{i+1} = \lambda_i\{s \leftarrow t\},$$

  (or $\lambda_{i+1} = \lambda_i\{t \leftarrow s\}$).
- λ_n is an m.g.u. of $\{A, B\}$. ♯

By Robinson's unification theorem in [Robinson, 1965b] A, B are unifiable iff there exists a m.g.u. sequence (λ_i) ending in λ_n, where λ_n is a most general unifier of $\{A, B\}$.

LEMMA 3.15 *Let C be a nonpositive clause and D be a positive clause such that $\{C, D\} \in \mathcal{VED}$. Then, for every binary resolvent R of C and D, the following properties hold:*

a. $\tau(R) \leq \max\{\tau(C), \tau(D)\}$ *and*

b. $\{R\} \in \mathcal{VED}$.

Proof: We may assume w.l.o.g. that $V(C) \cap V(D) = \emptyset$. If C and D are not resolvable there is nothing to prove. Otherwise C is of the form $E \vee \neg B$ for an atom B s.t. $\{D, B\}$ is unifiable by a m.g.u. σ and the binary resolvent is of the form $E\sigma$.

By Robinson's unification theorem and by Definition 3.14 there exists an m.g.u. sequence $(\lambda_1, \ldots, \lambda_n)$ s.t. $\sigma = \lambda_n$. By Definition 3.14 we have $\lambda_0 = \epsilon$, $\lambda_{i+1} = \lambda_i\{x_i \leftarrow t_i\}$ for $0 \leq i < n$, where (x_i, t_i) is an admissible pair in $\mathrm{CORR}(D\lambda_i, B\lambda_i) \cup \mathrm{CORR}(B\lambda_i, D\lambda_i)$. Now R is a propositional resolvent of $D\sigma$ and $E\sigma \vee \neg B\sigma$; therefore it suffices to prove for $i = 0, \ldots, n$:

$A[i]$: $\tau(\{D\lambda_i, C\lambda_i\}) \leq \tau(\{C, D\})$ and

$B[i]$: (1) $\tau(x, D\lambda_i)$ is defined for $x \in V(D\lambda_i)$ and $\tau(x, C\lambda_i)$ is defined for $x \in V(C\lambda_i)$.
(2) $\tau(x, D\lambda_i) = \tau(x, C\lambda_i)$ for $x \in V(C\lambda_i) \cap V(D\lambda_i)$.

Clearly $A[n]$ implies $\tau(E\sigma) \leq \tau(\{C, D\})$ and $B[n] - (1)$ implies $\{E\sigma\} \in \mathcal{VED}$. We proceed by induction on i:
$i = 0$:
$\lambda_0 = \epsilon$ and thus $A[0]$ is trivial. Moreover $B[0]$ holds by $\{C, D\} \in \mathcal{VED}$ and $V(C) \cap V(D) = \emptyset$.

(IH): Assume that $A[i]$ and $B[i]$ hold.
If $i = n$ we have succeeded. Thus we may assume that $i < n$ and $\lambda_{i+1} = \lambda_i\{x_i \leftarrow t_i\}$ for an admissible pair $(x_i, t_i) \in \mathrm{CORR}(D\lambda_i, B\lambda_i) \cup \mathrm{CORR}(B\lambda_i, D\lambda_i)$.

We first show $B[i]$:
Let $x \in V(D\lambda_{i+1})$. Then, clearly $x \neq x_i$ by $x_i \notin V(t_i)$ and even $x_i \notin V(D\lambda_{i+1}) \cup V(C\lambda_{i+1})$.

If $x \in V(D\lambda_i)$ then, by (IH), $\tau(x, D\lambda_i)$ is defined and, for $x \in V(C\lambda_i)$, $\tau(x, D\lambda_i) = \tau(x, C\lambda_i)$. Moreover, by definition of λ_{i+1}, we have $x\lambda_{i+1} = x\lambda_i$ and thus $\tau(x, D\lambda_{i+1})$ is defined and $\tau(x, D\lambda_{i+1}) = \tau(x, C\lambda_{i+1})$ if $x \in V(C\lambda_{i+1})$.

If $x \notin V(D\lambda_i)$ then, by definition of λ_{i+1}, $x \in V(t_i)$. In this case $(x_i, t_i) \in \mathrm{CORR}(D\lambda_i, B\lambda_i)$ and t_i is a subterm of $B\lambda_i$. Moreover $B\lambda_i$ is a subclause of $C\lambda_i$ and thus, by (IH), $\tau(y, C\lambda_i)$ is defined for all $y \in V(t_i)$. In particular $\tau(x, D\lambda_{i+1}) = \tau(x_i, D\lambda_i) + \tau(x, t_i)$ and $\tau(x, C\lambda_{i+1}) = \tau(x, D\lambda_{i+1}) = \tau(x, C\lambda_i)$.

The case $x \in V(C\lambda_{i+1})$ can be handled in the same way. This proves $B[i+1]$.

It remains to show $A[i+1]$:
If $x_i \in V(D\lambda_i)$ then

$$\text{(I) } \tau(D\lambda_{i+1}) = \max\{\tau(D\lambda_i), \tau_{\max}(x_i, D\lambda_i) + \tau(t_i)\}.$$

Similarly, for $x_i \in V(C\lambda_i)$ we obtain

$$\text{(II) } \tau(C\lambda_{i+1}) = \max\{\tau(C\lambda_i), \tau_{\max}(x_i, C\lambda_i) + \tau(t_i)\}.$$

On the other hand, by definition of the unification procedure,

$$\text{(III) } \tau_{\min}(x_i, D\lambda_i) + \tau(t_i) \leq \tau(C\lambda_i) \text{ for } x_i \in V(D\lambda_i),$$

$$\text{(IV) } \tau_{\min}(x_i, C\lambda_i) + \tau(t_i) \leq \tau(D\lambda_i) \text{ for } x_i \in V(C\lambda_i).$$

By (IH)

$\tau_{\min}(x_i, D\lambda_i) = \tau_{\max}(x_i, D\lambda_i)$ for $x_i \in V(D\lambda_i)$ and

$\tau_{\min}(x_i, C\lambda_i) = \tau_{\max}(x_i, C\lambda_i)$ for $x_i \in V(C\lambda_i)$.

Therefore (I) and (III) yield

$$\tau(D\lambda_{i+1}) \leq \max\{\tau(D\lambda_i), \tau(C\lambda_i)\}.$$

Similarly (II) and (IV) give

$$\tau(C\lambda_{i+1}) \leq \max\{\tau(C\lambda_i), \tau(D\lambda_i)\}.$$

By (IH) (subcase $A[i]$) $max\{\tau(D\lambda_i), \tau(C\lambda_i)\} \leq \tau(\{C, D\}$.
Putting things together we eventually obtain

$$\tau(\{D\lambda_{i+1}, C\lambda_{i+1}\}) \leq \tau(\{C, D\}).$$

◇

THEOREM 1.2 *Let $\mathcal{C}$ be a set of clauses in $\mathcal{VED}$. Then $R^*_H(\mathcal{C})$ is finite and, if $\mathcal{C}$ is satisfiable, $POS(R^*_H(\mathcal{C}))$ is an atomic representation of a Herbrand model of $\mathcal{C}$.*

Proof: It is sufficient to show that

(I) $R_H(\mathcal{C}) \in \mathcal{VED}$ and

(II) $\tau(R_H(\mathcal{C})) \leq \tau(\mathcal{C})$.

Then a straightforward induction argument yields

$$R^*_H(\mathcal{C}) \in \mathcal{VED} \quad \text{and} \quad \tau(R^*_H(\mathcal{C})) \leq \tau(\mathcal{C}).$$

But, as $R^*_H(\mathcal{C})$ is a set of Horn clauses, $R^*_H(\mathcal{C}) - \mathcal{C}$ consists of positive unit clauses C with $\tau(C) \leq \tau(\mathcal{C})$. By definition of hyperresolution and of R_H all clash resolvents are condensed and in variable normal form, i.e. for all $C \in R^*_H(\mathcal{C}) : N_c(C) = C$. But, over the signature of $\mathcal{C}$, the set of all atoms A with $\tau(N_c(A)) \leq \tau(\mathcal{C})$ is finite. Therefore $R^*_H(\mathcal{C})$ is finite. By Theorem 1.1 $POS(R^*_H(\mathcal{C}))$ is an atomic representation of a Herbrand model of $\mathcal{C}$ – provided $\mathcal{C}$ is satisfiable.

It remains to prove (I) and (II): By Lemma 3.15 binary resolvents of clauses C and D, where D is unit, are in $\mathcal{VED}$ if C and D are. But hyperresolvents are defined by finite iterations of such resolutions; therefore $R_H(\mathcal{C}) \in \mathcal{VED}$ for $\mathcal{C} \in \mathcal{VED}$. If $\Pi : (C, D_1, \ldots, D_n)$ is a clash over clauses in $\mathcal{C}$ then, again by Lemma 3.15, $\tau(E) \leq \tau(\mathcal{C})$ for all intermediary resolvents of the clash, and so $\tau(R) \leq \tau(\mathcal{C})$ also for the hyperresolvent R of Π. $\diamond$

EXAMPLE 3.16 Let F be the formula

$$[P(a, f(b)) \wedge (\forall x)(\forall y)(P(x, f(y)) \rightarrow P(f(y), x))\wedge$$

$$(\forall x)(\forall y)(\forall z)((P(x, y) \wedge P(y, z)) \rightarrow P(x, z))] \rightarrow (\exists y)P(f(y), f(a)).$$

We show that F is not valid by constructing a Herbrand model of the clausal form of $\neg F$. The transformation of $\neg F$ to clausal form yields the set

$$\mathcal{C} : \{P(a, f(b)),\ \neg P(x, f(y)) \vee P(f(y), x),\ \neg P(x, y) \vee \neg P(y, z) \vee P(x, z),$$
$$\neg P(f(y), f(a))\}.$$

Clearly $\mathcal{C} \in \mathcal{VED}$ and

$$R^*_H(\mathcal{C}) = \mathcal{C} \cup \{P(f(b), a),\ P(a, a),\ P(f(b), f(b))\}.$$

Particularly the set

$$POS(R^*_H(\mathcal{C})) = \{P(a, f(b)),\ P(f(b), a),\ P(a, a),\ P(f(b), f(b))\}$$

is an atomic representation of a Herbrand model Γ of $\mathcal{C}$. Γ is defined by

$$\begin{aligned} v_\Gamma(P(s,t)) &= \mathbf{T} \text{ if } (s,t) \in \{(a, f(b)), (f(b), a), (a, a), (f(b), f(b))\} \\ &= \mathbf{F} \text{ otherwise} \end{aligned}$$

Γ is also a countermodel of F. $\sharp$

Example 3.16 shows that countermodels of formulae F having clausal forms in $\mathcal{VED}$ can be constructed directly by application of R_H. However, the restriction to variables in $\mathcal{VED}$ is quite strong. Indeed clause sets containing quite "natural" clauses like $\neg P(x,a) \vee P(f(x),a)$ (see Example 3.9) do not belong to $\mathcal{VED}$. In particular the set of clauses

$$\mathcal{D}:\ \{P(a,a),\ \neg P(x,a) \vee P(f(x),a),\ \neg P(a,f(x))\}$$

is not in $\mathcal{VED}$. Moreover $R_H^*(\mathcal{D}) = \mathcal{D} \cup \{P(f^n(a),a) \mid n \geq 1\}$ and R_H does not terminate on $\mathcal{D}$. Consequently a direct construction of a symbolic model representation by R_H is impossible for $\mathcal{D}$. However, by changing the signs in $\mathcal{D}$ we obtain the set of clauses

$$\mathcal{D}':\ \{\neg P(a,a),\ P(x,a) \vee \neg P(f(x),a),\ P(a,f(x))\}.$$

$\mathcal{D}'$ enjoys the fortunate property $R_H^*(\mathcal{D}') = \mathcal{D}'$ and the computation by R_H tells us that $\mathcal{D}'$ is satisfiable. By $\mathcal{D} \sim_{sat} \mathcal{D}'$ we conclude that $\mathcal{D}$ is satisfiable too. Moreover we are in the position to define a Herbrand model of $\mathcal{D}$ at least "indirectly":

First $POS(R_H^*(\mathcal{D}')) = \{P(a,f(x))\}$ gives the representation of the following Herbrand model Δ of $\mathcal{D}'$:

$$\begin{aligned} v_\Delta(P(a,f(t))) &= \mathbf{T} \text{ for all } t \in H(\mathcal{D}'), \\ v_\Delta(P(s,t)) &= \mathbf{F} \text{ if } s \neq a \text{ or } t = a. \end{aligned}$$

By the sign exchange defining Δ' from $\mathcal{D}$ the following Herbrand model Γ of $\mathcal{D}$ is obtained immediately:

$$\begin{aligned} v'_\Delta(P(a,f(t))) &= \mathbf{F} \text{ for all } t \in H(\mathcal{D}'), \\ v'_\Delta(P(s,t)) &= \mathbf{T} \text{ if } s \neq a \text{ or } t = a. \end{aligned}$$

Therefore the atomic representation of Γ is the finite set

$$\mathcal{B}:\ \{P(f(x),y),\ P(z,a)\}.$$

All the steps leading to the final construction of $\mathcal{B}$ can be automatized. The first step is a check for an adequate sign renaming and the second is the (symbolic) complementation of the model representation for $\mathcal{D}'$. The next section will reveal that $\mathcal{D}$ belongs to a (non-Horn) class $\mathcal{PVD}$ which will be treated in a more general context. A method for computing complements of (linear) atomic representations will be presented in subsection 1.4.

The question remains, whether all cases of symbolic model building by hyperresolution can also be handled by finite domain search. Expressed more explicitely: *Does termination of hyperresolution on a set of Horn clauses $\mathcal{C}$ imply the finite model property of $\mathcal{C}$?*

The following proposition shows that the answer is *no*:

PROPOSITION 1.1 *There exist a satisfiable set of Horn clauses $\mathcal{C}$ with the following properties:*

1. *$\mathcal{C}$ has only infinite models,*

2. *$R_H^*(\mathcal{C})$ is finite.*

Proof: Let $\mathcal{C}$ be the set of Horn clauses [Baaz, 1996]

$$\{P(x,x),\ \neg P(f(x),f(y)) \vee P(x,y), \neg P(c,f(x))\}.$$

Hyperresolution trivially terminates on $\mathcal{C}$, i.e. $R_H^*(\mathcal{C}) = \mathcal{C}$ and thus $\{P(x,x)\}$ is an atomic representation of the minimal Herbrand model of $\mathcal{C}$. On the other hand, it is easy to see that $\mathcal{C}$ does not have finite models. ◇

Proposition 1.1 tells us that symbolic model building by hyperresolution surpasses finite model building already in principle: for the set of clauses $\mathcal{C}$ defined above there exists a very simple symbolic model representation (generated by hyperresolution), but no finite interpretation table for P and f.

1.3 The Disjunctive Case

We have seen in Subsection 1.2 that R_H terminates on $\mathcal{VED}$ and produces finite atomic model representations; in particular this also holds for its function-free subclass $\mathcal{BSH}^*$. We may ask whether hyperresolution also "works" on the class $\mathcal{BS}^*$, the clausal version of the Bernays–Schönfinkel class. Although decidability itself is trivial to prove by model theoretic means, the behavior of hyperresolution on this class is problematic.

EXAMPLE 3.17

$$\begin{aligned}\mathcal{C} &= \{P(x,x,a),\ P(x,z,u) \vee \neg P(x,y,u) \vee \neg P(y,z,u),\\ &\quad P(x,y,u) \vee P(y,z,u) \vee \neg P(x,z,u),\ \neg P(x,x,b)\}.\end{aligned}$$

$\mathcal{C}$ is non–Horn and even "essentially" non–Horn; that means there exists no sign renaming transforming $\mathcal{C}$ into a set of Horn clauses. In renaming P by $\neg P$ we only exchange the roles of a and b, otherwise $\mathcal{C}$ remains as it is. R_H neither terminates on $\mathcal{C}$ nor on the renamed form $\mathcal{C}'$. R_H produces clauses of arbitrary length on $\mathcal{C}$ – even if we add subsumption (i.e. we replace R_H by $R_{H,s}$). Thus $R_{H,s}$ + sign renaming does not terminate on $\mathcal{C}$. That means hyperresolution cannot decide the Bernays–Schönfinkel class. Moreover none of the "standard" refinements terminates on $\mathcal{C}$. There is, however, general semantic clash resolution over arbitrary models $\mathcal{M}$ as defined by J. Slagle [Slagle, 1967]; in such a refinement only clauses which are false in $\mathcal{M}$ are derivable. So, in case $\mathcal{C}$ is satisfiable, we only have to choose a model of $\mathcal{C}$; on such a model all clauses are true and thus semantic clash resolution does not produce any resolvents. This trick, however, can hardly be recommended as a method in resolution decision theory. Note that models should be the outcome of our procedures, not the starting point! $\sharp$

Of course there is the brute force method to decide $\mathcal{BS}^*$ by ground saturation. We will see later that, by an appropriate use of hyperresolution, saturation can be reduced considerably.

We show now how hyperresolution can be applied as decision procedure on functional clauses classes which are non-Horn. One of these classes can be considered as a generalization of DATALOG [Cerri et al., 1990]. Formally DATALOG is a subclass of $\mathcal{BSH}$ such that all positive clauses are ground and $V(C_+) \subseteq V(C_-)$ for all other clauses.

DEFINITION 3.18 A set of clauses $\mathcal{C}$ belongs to $\mathcal{PVD}$ (**p**ositive **v**ariable **d**ominated) if for all $C \in \mathcal{C}$:

PVD-1) $V(C_+) \subseteq V(C_-)$ (C is ground for $C_- = \Box$),

PVD-2) $\tau_{\max}(x, C_+) \leq \tau_{\max}(x, C_-)$ for all $x \in V(C_+)$. $\sharp$

$\mathcal{PVD}$ corresponds to a subclass of a class named $\mathcal{PVD}$ in [Fermüller et al., 1993], where the properties above were "relativized" under settings. That means there might be some sign renaming γ such that $\gamma(\mathcal{C}) \in \mathcal{PVD}$ even if $\mathcal{C}$ itself is not in $\mathcal{PVD}$. Take for example the set of clauses

$$\begin{aligned} \mathcal{C} &= \{P(x) \vee Q(g(x,x)),\ R(f(x),y),\ P(a),\ R(x,y) \vee \neg Q(y), \\ &\quad \neg P(x) \vee \neg P(f(x)),\ \neg R(a,a) \vee \neg R(f(b),a)\}. \end{aligned}$$

Obviously $\mathcal{C}$ is not in $\mathcal{PVD}$ (there are positive clauses containing variables and $\{R(x,y) \vee \neg Q(y)\}$ violates PVD-1)).

But let γ be the sign renaming Q to $\neg Q$, R to $\neg R$ and P remaining unchanged. Then

$$\begin{aligned} \gamma(\mathcal{C}) &= \{P(x) \vee \neg Q(g(x,x)),\ \neg R(f(x),y),\ P(a),\ Q(y) \vee \neg R(x,y), \\ &\quad \neg P(x) \vee \neg P(f(x)),\ R(a,a) \vee R(f(b),a)\} \end{aligned}$$

and $\gamma(\mathcal{C}) \in \mathcal{PVD}$.

The example above suggests the following generalization of $\mathcal{PVD}$:

DEFINITION 3.19 A set of clauses $\mathcal{C}$ belongs to $\mathcal{PVD}_r$ if there exists a sign renaming γ such that $\gamma(\mathcal{C})$ belongs to $\mathcal{PVD}$. ♯

The idea behind $\mathcal{PVD}$ is that the positive parts are always "smaller" than the negative ones. As hyperresolution produces positive clauses only, we may hope that the produced clauses are small too (i.e. small enough to achieve termination). Indeed the following theorem holds (see also [Leitsch, 1993] and [Fermüller et al., 1993]):

THEOREM 1.3 *Hyperresolution decides $\mathcal{PVD}$, i.e. for every $\mathcal{C} \in \mathcal{PVD}$ the set $R^*_H(\mathcal{C})$ is finite.*

Proof: Let $\mathcal{C}$ be a set of clauses in $\mathcal{PVD}$. Obviously $R^*_H(\mathcal{C})$ is finite if the following conditions (a), (b) hold:

(a) $R^*_H(\mathcal{C}) - \mathcal{C}$ contains only (positive) ground clauses and

(b) $\tau(D) \leq r$ for all $D \in R^*_H(\mathcal{C}) - \mathcal{C}$ and $r = \max\{\tau(C_+) \mid C \in \mathcal{C}\}$

Thus it remains to prove (a) and (b).

By Definition 3.18 $V(C_+) \subseteq V(C_-)$ for all $C \in \mathcal{C}$; in particular $POS(\mathcal{C})$ consists of ground clauses only. Now consider a clash sequence $\Gamma : (C; D_1, \ldots, D_n)$ for $C \in \mathcal{C}$ and positive ground clauses $D_1, \ldots, D_n$; let E be a clash resolvent of Γ. Then $E = C_+\lambda$ where λ is the "total" substitution corresponding to the clash resolution (in fact λ is obtained by concatenation of the m.g.u.s corresponding to the single PRF-resolvents of Γ). Let L be an arbitrary (negative) literal in C_-; then, by definition of λ $\lambda(L) = \neg A$ for some ground atom A in a clause D_i. Consequently, as all D_i are ground clauses, λ is a ground substitution. By $V(C_+) \subseteq V(C_-)$ it follows that the resolvent $E : C_+\lambda$ is a ground clause too. This proves (a).

For (b) consider, like above, a clash $\Gamma : (C; D_1, \ldots, D_n)$ and its corresponding total substitution λ. Let us assume that the D_i are all ground clauses with $\tau(D_i) \leq r$. By definition of λ we get $C_-\lambda \subseteq D_1 \cup \cdots \cup D_n$ and thus $\tau(C_-\lambda) \leq r$. We have to show $\tau(C_+\lambda) \leq r$.

If $\tau(C_+\lambda) = \tau(C_+)$ then, by definition of r, we get $\tau(C_+\lambda) \leq r$ and we are done.

If $\tau(C_+\lambda) > \tau(C_+)$ then there exists a variable $x \in V(C_+)$ with

$$\tau(C_+\lambda) = \tau_{\max}(x, C_+) + \tau(x\lambda).$$

By property PVD-2) $\tau_{\max}(x, C_+) \leq \tau_{\max}(x, C_-)$ and so

$$\tau(C_+\lambda) \leq \tau_{\max}(x, C_-) + \tau(x\lambda) \leq \tau(C_-\lambda).$$

But $\tau(C_-\lambda) \leq r$ and so $\tau(C_+\lambda) \leq r$. ◇

The decision procedure for $\mathcal{PVD}$ can easily be modified to the following decision procedure for $\mathcal{PVD}_r$:

1 Search for a renaming γ such that $\gamma(\mathcal{C})$ is in $\mathcal{PVD}$ (if there is no such γ then $\mathcal{C} \notin \mathcal{PVD}_r$).

2 Apply R_H^* to $\gamma(\mathcal{C})$.

Note that there are only finitely many sign renamings on a set of clauses and that the properties PVD-1) and PVD-2) are decidable. Thus we can always decide whether a set of clauses is in $\mathcal{PVD}_r$. Once we have found the right renaming γ, we replace $\mathcal{C}$ by $\gamma(\mathcal{C})$ and apply hyperresolution.

The class $\mathcal{PVD}$ is relatively "tight" with respect to undecidability: If we add the clause

$$T^- : P(x_1, x_2) \vee P(x_2, x_3) \vee \neg P(x_1, x_3)$$

(i.e. the transitivity of $\neg P$) we can encode the word problem of any equational theory (see [Fermüller et al., 1993] chapter 3.3). From the fact that there are equational theories with undecidable word problems (e.g. the theory of combinators [Stenlund, 1971]) it follows that

$$\Gamma : \{\mathcal{C} \cup \{T^-\} | \mathcal{C} \in \mathcal{PVD}\}$$

is an undecidable class of clause sets.

The main point in the proof of theorem 1.3 consists in showing that all positive clauses in $R_H^*(\mathcal{C})$ are ground and $\tau(R_H^*(\mathcal{C})) \leq d$ for some constant d. While the property PVD-1) is essential (note that T^- does not fulfill PVD-1)), PVD-2) can be replaced by a more general condition (term depth is only a specific complexity measure for literals and clauses). Particularly we obtain a more general decision class in replacing term depth by arbitrary *atom complexity measures* α fulfilling the following axioms:

1 $\alpha(A) \leq \alpha(A\theta)$ for substitutions θ and atom formulae A.

2 For all $k \in \mathbb{N}$ the set $\{A\theta \mid \theta \in \Theta_0, \alpha(A\theta) \leq k\}$ is finite, where Θ_0 is the set of all ground substitutions over a finite signature.

3 α is extended to literals by $\alpha(A) = \alpha(\neg A)$ and to clauses by $\alpha(\{L_1, \ldots, L_n\}) = \max\{\alpha(L_i) \mid 1 \leq i \leq n\}$.

for such an α we have to postulate that there exists a constant d such that for all ground substitutions θ either $\alpha(C_+\theta) \leq d$ or $\alpha(C_+\theta) \leq \alpha(C_-\theta)$ [Leitsch, 1993].

If we relax the condition on PVD, that positive clauses must be ground, we must add a somewhat stronger restriction on the behavior of the parts C_+ and C_- relative to each other. This idea leads to the class $\mathcal{OCCIN}$ where the positive parts of clauses are "linear" (i.e. every variable occurs only once).

DEFINITION 3.20 $\mathcal{OCCIN}$ is the set of all sets of clauses $\mathcal{C}$ such that for all $C \in \mathcal{C}$:

1 $\text{OCC}(x, C_+) = 1$ for all $x \in V(C_+)$ and

2 $\tau_{max}(x, C_+) \leq \tau_{min}(x, C_-)$ for all $x \in V(C_+) \cap V(C_-)$. ♯

Like in the case of $\mathcal{PVD}$ we can define a class $\mathcal{OCCIN}_r$ via renaming and reduce the decidability of $\mathcal{OCCIN}_r$ to that of $\mathcal{OCCIN}$.

THEOREM 1.4 *Hyperresolution decides* $\mathcal{OCCIN}$, *i.e. for every* $\mathcal{C} \in \mathcal{OCCIN}$ *the set* $R_H^*(\mathcal{C})$ *is finite.*

Proof: In [Fermüller et al., 1993].

While, for deciding $\mathcal{PVD}$, condensing of clauses is not necessary, it is required for $\mathcal{OCCIN}$ (otherwise the size of clauses diverges). Of course we can always apply an even stricter (but complete) refinement like hyperresolution with forward subsumption or replacement in order to decide $\mathcal{PVD}$ and $\mathcal{OCCIN}$.

$\mathcal{BS}^*$ is not a subclass of $\mathcal{PVD}_r$. But we will define a method to transform $\mathcal{BS}^*$ into $\mathcal{BS}^* \cap \mathcal{PVD}$ under preservation of sat–equivalence. This method is more subtle and more efficient than complete ground saturation. The basic idea is the following:

Let $\mathcal{C}$ be in $\mathcal{BS}^*$. Search for a renaming γ such that $\gamma(\mathcal{C}) \in \mathcal{PVD}$. If there is such a γ then apply R_H^* to $\gamma(\mathcal{C})$, else select some arbitrary γ and transform $\gamma(\mathcal{C})$ into a set of clauses $\mathcal{D} \in \mathcal{PVD}$ by partial saturation of the variables which violate PVD-1); afterwards apply R_H^* to $\mathcal{D}$. Let us call this procedure BSALG. Then BSALG is indeed a decision algorithm for $\mathcal{BS}^*$. For the actual performance of BSALG the right selection of a renaming is crucial; clearly one should try to select a γ for which the set $\mathcal{D}$ becomes minimal. In the next example we compare brute force saturation with BSALG. For this purpose we replace R_H by the more restrictive operator $R_{H,s}$ (see Definition 3.6). This leads to a further increase of efficiency, but without loss of correctness and termination (note that $R_{H,s}$ is complete and $R_{H,s}(\mathcal{C}) \subseteq R_H(\mathcal{C})$ for all sets of clauses $\mathcal{C}$).

EXAMPLE 3.21 We take the set of clauses from example 3.17, i.e.

$$\begin{aligned}\mathcal{C} &= \{P(x,x,a),\ P(x,z,u) \vee \neg P(x,y,u) \vee \neg P(y,z,u),\\ &\quad P(x,y,u) \vee P(y,z,u) \vee \neg P(x,z,u),\ \neg P(x,x,b)\}.\end{aligned}$$

We already know that R_H does not terminate on $\mathcal{C}$. Clearly $\mathcal{C} \notin \mathcal{PVD}$ but $\mathcal{C} \in \mathcal{BS}^*$. We compute the set $\mathcal{D}$ (without renaming the predicate symbol P) and obtain

$$\begin{aligned}\mathcal{D} &= \{P(a,a,a),\ P(b,b,a),\ P(x,z,u) \vee \neg P(x,y,u) \vee \neg P(y,z,u),\\ &\quad P(x,a,u) \vee P(a,z,u) \vee \neg P(x,z,u),\\ &\quad P(x,b,u) \vee P(b,z,u) \vee \neg P(x,z,u),\ \neg P(x,x,b)\},\end{aligned}$$

$\mathcal{D} \in \mathcal{PVD}$ and $|\mathcal{D}| = |\mathcal{C}| + 2 = 6$.

$$R^*_{H,s}(\mathcal{D}) = \mathcal{D} \cup \{P(a,b,a) \vee P(b,a,a)\}.$$

Thus $R_{H,s}$ terminates on $\mathcal{D}$ producing only one additional clause ($|R^*_{H,s}(\mathcal{D})| = 7$). Note that in the second generation of hyperresolvents we obtain the clauses

$$P(b,a,a) \vee P(a,a,a) \vee P(a,b,a),\ P(b,b,a) \vee P(b,a,a) \vee P(a,b,a)$$

which are both subsumed by the clauses of the first generation and therefore are deleted.

Using the brute force saturation method we obtain a set of ground clauses $\mathcal{D}'$ which contains 36 clauses. Moreover $\mathcal{D}'$ has still to be tested for satisfiability. Thus we see that BSALG may be much faster than the pure saturation method. Later in this subsection we will define a method for extracting models out of satisfiable clause sets of the form $R^*_{H,s}(\mathcal{D})$; this method can be considered as a postprocessing of the decision procedure defined above. ♯

DEFINITION 3.22 Let $\mathcal{C}$ be a set of condensed clauses. Then $\mathcal{C}$ is called ($R_{H,s}$–) *stable* if $R_{H,s}(\mathcal{C}) = \mathcal{C}$. ♯

If a $R_{H,s}$–sequence $(R^i_{H,s}(\mathcal{C}))_{i \in \mathbb{N}}$ converges to $R^k_{H,s}(\mathcal{C})$ then, by definition 3.20, $R^k_{H,s}(\mathcal{C})$ is stable and a fixed point of the operator $R_{H,s}$. Let us assume that an $R_{H,s}$–sequence converges and yields a (stable) set $\mathcal{C}'$ such that all positive clauses in $\mathcal{C}'$ are unit. Then, by the following lemma, which is related to Theorem 1.1, these clauses form an atomic representation of a Herbrand model of $\mathcal{C}'$; for the remaining part of this section we write "ARM" for atomic representation and "stable" for $R_{H,s}$–stable.

DEFINITION 3.23 Let $\mathcal{C}$ be a set of clauses. By $\mathrm{GAT}(\mathcal{C})$ we denote the set of all ground atoms over the signature of $\mathcal{C}$. ♯

LEMMA 3.24 *Let $\mathcal{C}$ be a finite set of nonpositive condensed clauses and $\mathcal{A}$ be a finite set of positive unit clauses such that $\mathcal{C} \cup \mathcal{A}$ is satisfiable and stable. Then $\mathcal{A}$ is an ARM of a Herbrand model of $\mathcal{C} \cup \mathcal{A}$ (over the signature of $\mathcal{C} \cup \mathcal{A}$).*

Proof: Suppose, on the contrary, that $\mathcal{A}$ is not an ARM of a Herbrand model of $\mathcal{C} \cup \mathcal{A}$. Then the interpretation induced by $\mathcal{A}$ falsifies $\mathcal{C} \cup \mathcal{A}$. Let $\mathcal{M}$ be the set of all ground instances of $\mathcal{A}$ over $H(\mathcal{C} \cup \mathcal{A})$. Because $\mathcal{A}$ falsifies $\mathcal{C} \cup \mathcal{A}$, the set

$$\mathcal{C} \cup \mathcal{A} \cup \{\neg P \mid P \in \mathrm{GAT}(\mathcal{C} \cup \mathcal{A}) - \mathcal{M}\}$$

is unsatisfiable.

By the compactness theorem of first-order logic there exists a finite subset $\mathcal{F}$ of $\{\neg P \mid P \in \mathrm{GAT}(\mathcal{C} \cup \mathcal{A}) - \mathcal{M}\}$ s.t. $\mathcal{C} \cup \mathcal{A} \cup \mathcal{F}$ is unsatisfiable; moreover, the set $\mathcal{F}$ cannot be empty as $\mathcal{C} \cup \mathcal{A}$ is satisfiable. Because $R_{H,s}$ is complete we have $\Box \in R^*_{H,s}(\mathcal{C} \cup \mathcal{A} \cup \mathcal{F})$.

By the satisfiability of $\mathcal{C} \cup \mathcal{A}$, $\Box \notin \mathcal{C} \cup \mathcal{A}$ and $\Box \notin \mathcal{F}$ by definition of $\mathcal{F}$; consequently there exists a number i s.t.

$$\Box \in R^{i+1}_{H,s}(\mathcal{C} \cup \mathcal{A} \cup \mathcal{F}) - R^{i}_{H,s}(\mathcal{C} \cup \mathcal{A} \cup \mathcal{F}).$$

Let us write $\mathcal{D}_i$ for the set $R^{i}_{H,s}(\mathcal{C} \cup \mathcal{A} \cup \mathcal{F})$.

By definition of $R_{H,s}$ we obtain $\Box \in sub(\mathcal{D}_i \cup \varrho_H(\mathcal{D}_i))$; this is only possible if $\Box \in \mathcal{D}_i \cup \varrho_H(\mathcal{D}_i)$. By assumption $\Box \notin \mathcal{D}_i$ and thus $\Box \in \varrho_H(\mathcal{D}_i)$.

By definition of hyperresolution and by the fact that $\mathcal{C} \cup \mathcal{A}$ is satisfiable, $\Box$ must be the result of a clash resolvent among $\mathcal{A} \cup \mathcal{F}$. Now $\mathcal{A}$ and $\mathcal{F}$ are both sets of unit clauses where all elements of $\mathcal{F}$ are ground. Therefore there exists a ground substitution η and an atom $A \in \mathcal{A}$ s.t. $\neg A\eta \in \mathcal{F}$. By definition of $\mathcal{M}$ we also have $A\eta \in \mathcal{M}$; on the other hand, by definition of $\mathcal{F}$ and $\neg A\eta \in \mathcal{F}$ we get $A\eta \in \mathrm{GAT}(\mathcal{C} \cup \mathcal{A}) - \mathcal{M}$ and thus $A\eta \notin \mathcal{M}$. This clearly generates a contradiction and we conclude that $\mathcal{A}$ is an ARM of a Herbrand model of $\mathcal{C} \cup \mathcal{A}$. ◇

Note that Lemma 3.24 cannot be obtained directly via Theorem 1.1: first of all $\mathcal{C} \cup \mathcal{A}$ need not be a set of Horn clauses; as a consequence $R^*_H(\mathcal{C} \cup \mathcal{A})$ might contain positive clauses which are nonunit. Although, by definition of R_H and $R_{H,s}$, $\mathcal{C} \cup \mathcal{A} \subseteq R^*_H(\mathcal{C} \cup \mathcal{A})$, $POS(R^*_H(\mathcal{C} \cup \mathcal{A}))$ need not be a representation of a Herbrand model at all. On the other hand, if $\mathcal{C}$ is a set of Horn clauses then $\mathcal{A}$ and $POS(R^*_H(\mathcal{C} \cup \mathcal{A}))$ both represent the same Herbrand model.

Lemma 3.24 suggests the following strategy of finding a model: Suppose that a $R_{H,s}$–sequence converges to $\mathcal{C}$ such that $\Box \notin \mathcal{C}$ (what is equivalent to $\mathcal{C} \neq \{\Box\}$). Then search for a finite set of atoms $\mathcal{A}$ such that $(\mathcal{C} - POS(\mathcal{C})) \cup \mathcal{A}$ is finite, satisfiable and implies $\mathcal{C}$. The resulting set $\mathcal{A}$ is a an ARM of a Herbrand model of $(\mathcal{C} - POS(\mathcal{C})) \cup \mathcal{A}$ which is also a model of $\mathcal{C}$ itself. $\mathcal{PVD}$ and $\mathcal{OCCIN}$ (see definitions 3.19 and 3.20) are not only decision classes under R_H but also under $R_{H,s}$; a characteristic feature of both classes is the *disconnected* form (see Definition 2.19) of the positive clauses (and even of all derivable positive clauses). Splitting of a set of clauses w.r.t. to a disconnected clause C (in typical applications C is a ground clause) is a standard technique in automated deduction [Weidenbach, 2001]. The principle is described in the following definition:

DEFINITION 3.25 Let $\mathcal{C}$ be a set of clauses and $C \in \mathcal{C}$ s.t. $C = C_1 \cup C_2$, $C_1 \neq \emptyset, C_2 \neq \emptyset$ and $V(C_1) \cap V(C_2) = \emptyset$. Then the set of clauses $\mathcal{C}_1 : (\mathcal{C} - \{C\}) \cup \{C_1\}$ and $\mathcal{C}_2 : (\mathcal{C} - \{C\}) \cup \{C_2\}$ are called *split problems* of $\mathcal{C}$ w.r.t. C. ♯

PROPOSITION 1.2 *Splitting is correct, i.e.: let $\mathcal{C}$ be a set of clauses, $C \in \mathcal{C}$ and $\mathcal{C}_1, \mathcal{C}_2$ split problems of $\mathcal{C}$ w.r.t. C. Then $\mathcal{C}$ is unsatisfiable iff $\mathcal{C}_1$ and $\mathcal{C}_2$ are both unsatisfiable.*

Proof: Let $F(\mathcal{C})$, $F(C)$, $F(\mathcal{C}_1)$, $F(\mathcal{C}_2)$ the first-order formulae representing $\mathcal{C}, C, \mathcal{C}_1$ and $\mathcal{C}_2$ (see Definition 2.21). Moreover let $\mathcal{D} = \mathcal{C} - \{C\}$. Then, clearly, $F(\mathcal{C}) \sim F(\mathcal{D}) \wedge F(C)$, $F(\mathcal{C}_1) \sim F(\mathcal{D}) \wedge F(C_1)$ and $F(\mathcal{C}_2) \sim F(\mathcal{D}) \wedge F(C_2)$.

By the law of distributivity

$$F(\mathcal{D}) \wedge (F(C_1) \vee F(C_2)) \sim (F(\mathcal{D}) \wedge F(C_1)) \vee (F(\mathcal{D}) \wedge F(C_2)).$$

As a disjunction is unsatisfiable iff both members are, $F(\mathcal{C}_1)$ and $F(\mathcal{C}_2)$ are unsatisfiable iff $F(\mathcal{D}) \wedge (F(C_1) \vee F(C_2))$ is unsatisfiable.

It remains to show that $F(\mathcal{C}) \sim F(\mathcal{D}) \wedge (F(C_1) \vee F(C_2))$. By $V(C_1) \cap V(C_2) = \emptyset$, the universal quantifiers in the prefix of $F(C_1 \cup C_2)$ can be shifted inwards s.t. the universal formula is transformed into a disjunction which is logically equivalent to $F(C_1) \vee F(C_2)$. Thus $F(C) \sim F(C_1) \vee F(C_2)$ and so $F(\mathcal{C}) \sim F(\mathcal{D}) \wedge (F(C_1) \vee F(C_2))$. ◇

By Proposition 1.2 the problem of refuting a set of clauses $\mathcal{C}$ can be reduced to split problems $\mathcal{C}_1$, $\mathcal{C}_2$ (provided they exist). Note that the split problems imply the original one – even if splitting is performed incorrectly (i.e. C_1 and C_2 are not variable-disjoint). If $\mathcal{C}$ is satisfiable then one of $\mathcal{C}_1$ and $\mathcal{C}_2$ (but not necessarily both of them) must be satisfiable (consider, e.g., the set of clauses $\{P(a), \neg P(a) \vee Q(b)\}$). Thus if we are searching for a model of $\mathcal{C}$ we can reduce the problem to model building on $\mathcal{C}_1$ and $\mathcal{C}_2$. But, as we do not know a priori which of $\mathcal{C}_1, \mathcal{C}_2$ is the satisfiable one, we need backtracking or parallelization. Note that, as $\mathcal{C}_1$ and $\mathcal{C}_2$ imply $\mathcal{C}$, every model of $\mathcal{C}_1$ or $\mathcal{C}_2$ is a model of $\mathcal{C}$ as well, but if the first problem we try is the unsatisfiable one this is no help. We will demonstrate below that closure under $R_{H,s}$ provides us the means to avoid backtracking, i.e. a stable set is satisfiable iff *both* split problems are satisfiable. As our aim is to obtain an atomic representation, we search for clauses containing a "disconnected" atom and replace the clause by this atom. The justification of such a step is presented by the lemma below.

DEFINITION 3.26 Let $\mathcal{C}$ be a set of clauses, C be a positive clause in $\mathcal{C}$ and A be an atom in C with $V(A) \cap V(C - \{A\}) = \emptyset$. Then the set of clauses $(\mathcal{C} - \{C\}) \cup \{A\}$ is called a *unit reduct* of $\mathcal{C}$ w.r.t. (C, A). A unit reduct is called *nontrivial* if C is not a unit clause. ♮

Note that a unit reduct is just a special form of a split problem. There is no guarantee in general that an arbitrary unit reduct of a satisfiable clause set is satisfiable too. However, the situtation changes for stable sets as will be shown in the lemma below. In the proof of the lemma we need a technical notion which will turn out useful in several contexts.

DEFINITION 3.27 Let D be an arbitrary clause. We define a relation $\leq_D$ on pairs of clauses (E, F) s.t. $E \leq_D F$ iff $N_c(E) = N_c(F)$ or there exists a renaming substitution η with $V(D\eta) \cap V(E) = \emptyset$ and $N_c(E \vee D\eta) = N_c(F)$. $\leq_D$ can be extended to sets of clauses $\mathcal{E}$, $\mathcal{F}$ by: $\mathcal{E} \leq_D \mathcal{F}$ iff for all $E \in \mathcal{E}$ there exists an $F \in \mathcal{F}$ with $E \leq_D F$. ♮

Intuitively $E \leq_D F$ indicates that either E is F (under condensing) or F differs from E at most by the clause D. For sets of clauses $\mathcal{E} \leq_D \mathcal{F}$ means that $\mathcal{F}$ contains essentially all clauses of $\mathcal{E}$ with the possible "difference" D.

LEMMA 3.28 *Let $\mathcal{C}$ be satisfiable and stable and let $\mathcal{D}$ be a unit reduct of $\mathcal{C}$. Then $\mathcal{D}$ is satisfiable and $\mathcal{D}$ implies $\mathcal{C}$.*

Proof: $\mathcal{D}$ implies $\mathcal{C}$ because $\mathcal{D}$ is a split problem of $\mathcal{C}$. It remains to prove that $\mathcal{D}$ is satisfiable.

Let $\mathcal{D} = (\mathcal{C} - \{D\}) \cup \{P\}$ for some positive clause $D \in \mathcal{D}$. If $D = \{P\}$ then $\mathcal{C} = \mathcal{D}$ and the lemma is trivial. Thus we may assume that D is a nonunit clause and $D' : D - \{P\}$ is nonempty.

We use proof by contradiction and assume that $\mathcal{D}$ is unsatisfiable. Then $\Box \in R^*_H(\mathcal{D})$.

We will show below that the clauses in $R^*_H(\mathcal{D})$ differ from those in $R^*_H(\mathcal{C})$ "at most by the clause D'". Thus, according to Definition 3.27, our aim is to prove

$$\text{(I)} \quad R^*_H(\mathcal{D}) \leq_{D'} R^*_H(\mathcal{C}).$$

We prove (I) by induction on the iteration number of R_H, i.e. we have to show

$$\text{(II)} \quad R^i_H(\mathcal{D}) \leq_{D'} R^*_H(\mathcal{C})$$

for all $i \in \mathbb{N}$.

$i = 0$:

By definition of D' and $\leq_{D'}$ we have $\{P\} \leq_{D'} D$. Moreover $\mathcal{D} - \{P\} = \mathcal{C} - \{D\}$ and therefore $\mathcal{D} \leq_{D'} \mathcal{C}$. By $\mathcal{C} \subseteq R^*_H(\mathcal{C})$ and by the definition on $\leq_{D'}$ on sets of clauses we obtain $\mathcal{D} \leq_{D'} R^*_H(\mathcal{C})$ and thus $R^0_H(\mathcal{D}) \leq_{D'} R^*_H(\mathcal{C})$.

(IH) Let us assume $R^i_H(\mathcal{D}) \leq_{D'} R^*_H(\mathcal{C})$.

Now let $E \in R^{i+1}_H(\mathcal{D})$. If $E \in R^i_H(\mathcal{D})$ there is nothing to prove; thus we only focus on the case $E \in R^{i+1}_H(\mathcal{D}) - R^i_H(\mathcal{D})$:

By definition of R_H there exists a clash sequence $\gamma :\ (C; D_1, \ldots, D_n)$ with $C \in \mathcal{D}$ and $D_1, \ldots, D_n \in R^i_H(\mathcal{D})$ s.t. the clash-resolvent of γ is E. By (IH) there are clauses $H_1, \ldots, H_n \in R^*_H(\mathcal{C})$ s.t.

$$\text{(III)} \quad D_j \leq_{D'} H_j \text{ for } j = 1, \ldots n.$$

Then, by definition of $\leq_{D'}$, either $N_c(D_j) = N_c(H_j)$ or there exists a renaming substitution η_j with $V(D'\eta_j) \cap V(D_j) = \emptyset$ and $N_c(H_j) = N_c(D_j \vee D'\eta_j)$.

Now $\delta :\ (C; H_1, \ldots, H_n)$ is a clash sequence over $R^*_H(\mathcal{C})$; note that the clash-nuclei in $\mathcal{D}$, $\mathcal{C}$ and $R^*_H(\mathcal{C})$ are all the same! By (III) δ defines a clash resolvent F with

$$N_c(F) = N_c(E \vee D'\eta_{r_1} \cdots \vee D'\eta_{r_k})$$

for some $r_1, \ldots, r_k \in \{1, \ldots, n\}$.

By definition of the condensation operator N_c there is a single renaming substitution η with $N_c(F) = N_c(E \vee D'\eta)$ or $N_c(F) = N_c(E)$. In both cases we have $E \leq_{D'} F$ and so $R^{i+1}_H(\mathcal{C}') \leq_{D'} R^*_H(\mathcal{C})$. This settles the case $i+1$ of the induction proof and thus yields (I).

By $\Box \in R^*_H(\mathcal{D})$ we thus obtain a clause $H \in R^*_H(\mathcal{C})$ where H is the condensed form of D' (note that $\Box \notin \mathcal{C}$ as $\mathcal{C}$ is satisfiable!). Since $R^*_{H,s}(\mathcal{C})$ subsumes $R^*_H(\mathcal{C})$ and since $\mathcal{C}$ is $R_{H,s}$-stable there must be a clause $G \in \mathcal{C}$ s.t. $G \leq_{sub} H$. By definition of D and D' we obtain

$D' \leq_{sub} D$ and thus $H \leq_{sub} D$; moreover $H \neq D$ – otherwise D is not condensed. Moreover D does not subsume H and thus does not subsume G (note that $\leq_{sub}$ is transitive). Therefore we obtain two clauses D and G in $\mathcal{C}$ with $G \leq_{sub} D$ and $D \not\leq_{sub} G$. This, however, contradicts the assumption that $\mathcal{C}$ is stable under subsumption. We conclude that $\mathcal{D}$ is satisfiable. ◇

The validity of lemma 3.28 is essentially based on the stability of the set of clauses $\mathcal{C}$. It is very easy to see that the result becomes wrong for nonstable sets $\mathcal{C}$:

EXAMPLE 3.29 Let $\mathcal{C} = \{P(a) \vee P(b), \neg P(a)\}$.
Trivially $\mathcal{C}$ is satisfiable. But if we replace $P(a) \vee P(b)$ by $P(a)$ we obtain the set of clauses $\mathcal{C}_1 : \{P(a), \neg P(a)\}$ which is unsatisfiable. But note that $\mathcal{C}$ is not stable; rather we have $R_{H,s}(\mathcal{C}) = \{P(b), \neg P(a)\}$ and the replacement sequence converges to the set $\{P(b), \neg P(a)\}$. ♯

Example 3.29 also shows that the theorem becomes wrong if we replace the nonmonotonic operator $R_{H,s}$ by (the monotonic) R_H: The set $\mathcal{C}' : \{P(a) \vee P(b), \neg P(a), P(b)\}$ is a fixed point under R_H, but $\mathcal{C}'_1 : \{P(a), \neg P(a), P(b)\}$ is unsatisfiable as well. Thus we see that *subsumption is necessary* to guarantee the *soundness* of the reduction!

The transformation of $\mathcal{C}$ into a unit reduct $\mathcal{D}$ can be described by an operator α which (deterministically) selects a clause D and a literal P in D with $V(P) \cap V(D - \{P\}) = \emptyset$.

DEFINITION 3.30 (ATOM SELECTION FUNCTION) An operator α which maps stable sets of clauses into sets of clauses is called an *atom selection function* if

a. $\alpha(\mathcal{C}) = \mathcal{C}$ if all unit reducts of $\mathcal{C}$ are trivial or there exists no unit reducts at all, or

b. $\alpha(\mathcal{C}) = \mathcal{D}$ for some nontrivial unit reduct $\mathcal{D}$ of $\mathcal{C}$. ♯

Note that, for an atom selection function α, $\alpha(\mathcal{C})$ need not be stable, even if $\mathcal{C}$ is stable. Therefore we have to apply $R^*_{H,s}$ in order to obtain a new stable set.

Let us assume that $R^*_{H,s}(\mathcal{C})$ is defined and a finite stable set is obtained. Then atom selection and closure can be combined in a single operator.

DEFINITION 3.31 (THE OPERATOR T) Let α be an atom selection function. We define an operator T (corresponding to α) on sets of clauses $\mathcal{C}$ by $T(\mathcal{C}) = R^*_{H,s}(\alpha(\mathcal{C}))$. T is called a URC-operator (unit reduction + closure). ♯

Lemma 3.28 shows that URC-operators are correct on stable sets of clauses (i.e. $\mathcal{C}$ and $T(\mathcal{C})$ are sat-equivalent). On the other hand, Example 3.29 shows that T, if applied to arbitrary sets of clauses, is incorrect in general.

We are now in the position to define a model building procedure based on iterations of the operator T. The predicate "unit" on sets of clauses $\mathcal{D}$ returns true iff $\mathcal{D}$ consists of unit clauses only. The following algorithm is an extension of this in [Fermüller and Leitsch, 1996]; in contrast to the algorithm in [Fermüller and Leitsch, 1996] MB can be applied to arbitrary sets of clauses.

function *MB*
input: A finite set of clauses $\mathcal{C}$.
output: *satisfiable* or *unsatisfiable* or a set of atoms.
begin
  $\mathcal{C} \leftarrow R^*_{H,s}(\mathcal{C})$;
  if $R^*_{H,s}(\mathcal{C}) = \{\Box\}$ **then return** *unsatisfiable*
    else begin
      while $\alpha(\mathcal{C}) \neq \mathcal{C}$ **do** $\mathcal{C} \leftarrow T(\mathcal{C})$
      if unit$(POS(\mathcal{C}))$ **then return** $POS(\mathcal{C})$
        else return *satisfiable*
    end
end

Remark: In contrast to other model building methods based on hyperresolution and splitting of non-unit positive clauses (see e.g. [Manthey and Bry, 1988, Bry and Yahya, 1996]) MB is *backtracking-free.* Due to the computation of *stable* sets in every iteration step splitting can be replaced by selection (in the form of a *don't care nondeterminism*). Note that the computation of the first stable set $R^*_{H,s}(\mathcal{C})$ is the result of an *ordinary* refutational deduction procedure which has nothing to do with model building! ♯

THEOREM 1.5 *MB is correct, i.e.:*

1. *If MB returns* unsatisfiable *then $\mathcal{C}$ is unsatisfiable,*

2. *If MB returns* satisfiable *then $\mathcal{C}$ is satisfiable,*

3. *If MB returns a sets of clauses $\mathcal{D}$ then $\mathcal{D}$ is an atomic representation of a Herbrand model of $\mathcal{C}$.*

Proof:
1. MB returns unsatisfiable: $R_{H,s}$ is correct.

2. MB returns satisfiable or MB returns a clause set $\mathcal{D}$:
If $R^*_{H,s}(\mathcal{C})$ terminates and $R^*_{H,s}(\mathcal{C}) \neq \{\square\}$ then, by the completeness of $R^*_{H,s}$, $R^*_{H,s}(\mathcal{C})$ is satisfiable. Let $\mathcal{C}_0$ be the set $\mathcal{C}$ before execution of the while-loop. By Lemma 3.28 and by the fact that $T(\mathcal{C})$ is stable, the property "$\mathcal{C}$ is satisfiable and $\mathcal{C}$ implies $\mathcal{C}_0$" is a loop invariant. Furthermore the while-loop terminates, otherwise MB does not return an answer at all. After execution of the while-loop $\mathcal{C}$ is satisfiable and implies $\mathcal{C}_0$. Thus the answer "satisfiable" is correct anyway. MB returns the set $\mathcal{D} : POS(\mathcal{C})$ only if it consists of unit clauses only. But then all the properties for Lemma 3.24 are fulfilled and $\mathcal{D}$ is an ARM of $\mathcal{C}$. Thus $\mathcal{D}$ is an ARM of $\mathcal{C}_0$ too. $\diamond$

The following two examples illustrate how MB works:

EXAMPLE 3.32
Consider the following set of condensed clauses

$$\begin{aligned}\mathcal{C} &= \{E(a) \vee S(a),\ Q(a) \vee R(a),\ P(x) \vee Q(x) \vee \neg R(x) \vee \neg S(x),\\ &\quad \neg P(a) \vee \neg Q(a)\}.\end{aligned}$$

$\mathcal{C}$ is in $\mathcal{PVD}$ but $\mathcal{C}$ is not stable. We know that $R^*_{H,s}(\mathcal{C})$ is defined. Indeed MB computes

$$R^*_{H,s}(\mathcal{C}) = R^1_{H,s}(\mathcal{C}) = \mathcal{C} \cup \{E(a) \vee P(a) \vee Q(a)\}.$$

Let us write $\mathcal{C}_1$ for $R^1_{H,s}(\mathcal{C})$. Then MB enters the while-loop with $\mathcal{C}_1$ for

$$\alpha(\mathcal{C}_1) = (\mathcal{C}_1 - \{E(a) \vee S(a)\}) \cup \{S(a)\}.$$

Clearly $\alpha(\mathcal{C}_1) \neq \mathcal{C}_1$ and so we compute $T(\mathcal{C}_1)$:
Note that $\alpha(\mathcal{C}_1)$ itself is not stable and we have to compute its closure.

Let $\mathcal{C}_2 = T(\mathcal{C}_1) = R^*_{H,s}(\alpha(\mathcal{C}_1))$. So we get

$$\begin{aligned}\mathcal{C}_2 &= \{Q(a) \vee R(a),\ P(a) \vee Q(a),\ P(x) \vee Q(x) \vee \neg R(x) \vee \neg S(x),\\ &\quad S(a), \neg P(a) \vee \neg Q(a)\}.\end{aligned}$$

Note that in the computation of $\mathcal{C}_2$ we obtain the new clash resolvent $P(a) \vee Q(a)$ which subsumes $E(a) \vee P(a) \vee Q(a)$. On $\mathcal{C}_2$ we define

$$\alpha(\mathcal{C}_2) = (\mathcal{C}_2 - \{P(a) \vee Q(a)\}) \cup \{Q(a)\}.$$

Again $\alpha(\mathcal{C}_2) \neq \mathcal{C}_2$ and we have to iterate the while-loop once more. The result is $\mathcal{C}_3 = R^*_{H,s}(\alpha(\mathcal{C}_2)) = T(\mathcal{C}_2) =$

$$\{S(a),\ Q(a),\ P(x) \vee Q(x) \vee \neg R(x) \vee \neg S(x),\ \neg P(a) \vee \neg Q(a)\}.$$

Clearly $\alpha(\mathcal{C}_3) = \mathcal{C}_3$ and we leave the while-loop with $T(\mathcal{C}_3) = \mathcal{C}_3$ and $\mathcal{C}_3 = T^2(\mathcal{C}_1)$ (in fact $\mathcal{C}_3$ is a fixed point of T). Now $POS(\mathcal{C}_3)$ consists of unit clauses only and MB returns $\mathcal{C}_3$ as ARM of $\mathcal{C}$. ♯

EXAMPLE 3.33 Let us consider the following set of clauses $\mathcal{D}$ which has been obtained via a transformation of the Bernays-Schönfinkel class into $\mathcal{PVD}$ in Example 3.21

$$\begin{aligned}\mathcal{D} \;=\; &\{P(a,a,a),\; P(b,b,a),\; P(x,z,u) \vee \neg P(x,y,u) \vee \neg P(y,z,u),\\ &P(x,a,u) \vee P(a,z,u) \vee \neg P(x,z,u),\\ &P(x,b,u) \vee P(b,z,u) \vee \neg P(x,z,u),\; \neg P(x,x,b)\},\end{aligned}$$

We have shown that $R_{H,s}$ terminates on $\mathcal{D}$ and

$$R^*_{H,s}(\mathcal{D}) = \mathcal{D} \cup \{P(a,b,a) \vee P(b,a,a)\}.$$

This proves that $\mathcal{D}$ is satisfiable, but we do not have an explicit representation of a model. In fact $\mathcal{D}$ is the stable set computed by MB. By nonunit($POS(\mathcal{D})$) we have to apply T to $R^*_{H,s}(\mathcal{D})$ for

$$\alpha(\mathcal{D}) = \mathcal{D} - \{P(a,b,a) \vee P(b,a,a)\} \cup \{P(a,b,a)\}.$$

Although there are a lot of hyperresolvents on $\alpha(\mathcal{D})$ all of them are subsumed by set the unit clauses

$$\mathcal{A} = \{P(a,a,a),\; P(b,b,a),\; P(a,b,a)\}.$$

and so

$$T(\mathcal{D}) = R^*_{H,s}(\alpha(\mathcal{D})) = \alpha(\mathcal{D})$$

Now $\alpha(\mathcal{D})$ is stable and all positive clauses are unit; thus MB returns $\mathcal{A}$ as an atomic representation of a Herbrand model of $\mathcal{D}$. Thus we see that by mapping $\mathcal{BS}^*$ into $\mathcal{PVD}$ and applying the T-transformation we obtain an efficient model building procedure for the Bernays-Schönfinkel class. ♯

But, unfortunately, we have not yet reached our goal. Still we do not know conditions guaranteeing termination of MB. First of all we face the problem of termination of $R^*_{H,s}$, and afterwards the termination problem of the while-loop. We know that $R^*_{H,s}$ terminates on all sets of clauses $\mathcal{C}$ where $R^*_H(\mathcal{C})$ is terminating (e.g. on $\mathcal{PVD}$ and $\mathcal{OCCIN}$). Suppose we have computed a (finite) stable set $\mathcal{C}$; does the iteration of T always terminate on $\mathcal{C}$?

DEFINITION 3.34 Let $\mathcal{C}$ be a satisfiable, finite and stable set of clauses. Then we say that the sequence $(T^i(\mathcal{C}))_{i\in\mathbb{N}}$ *converges* to a set of clauses

$\mathcal{D}$ if there exists a number k such that $T^k(\mathcal{C}) = T^{k+1}(\mathcal{C})$ and $T^k(\mathcal{C}) = \mathcal{D}$. $\sharp$

Note that the convergence of $(T^i(\mathcal{C}))_{i\in\mathbb{N}}$ does not necessarily yield an ARM of $\mathcal{C}$. If, for the limit $\mathcal{D}$, $POS(\mathcal{D})$ contains nonunit clauses then the model building procedure itself fails; all we know is that $\mathcal{C}$ is satisfiable. In order to *prove* the convergence of the sequence $(T^i(\mathcal{C}))_{i\in\mathbb{N}}$ to a stable set of clauses $\mathcal{D}$, we have to introduce a Noetherian ordering $\prec$ on sets of clauses and to show that the $T^i(\mathcal{C})$ are decreasing with respect to $\prec$.

DEFINITION 3.35 Let $\mathcal{C}$ and $\mathcal{D}$ be two finite sets of condensed clauses. We define $\mathcal{C} \prec \mathcal{D}$ iff

(1) $\mathcal{C} \leq_{sub} \mathcal{D}$,

(2) For all $C \in \mathcal{C}$ there exists a $D \in \mathcal{D}$ s.t. $C \leq_{sub} D$ and $|C| \leq |D|$,

(3) $\mathcal{D} \not\leq_{sub} \mathcal{C}$. $\sharp$

PROPOSITION 1.3 *The relation $\prec$ is irreflexive, transitive and Noetherian on finite sets of condensed clauses.*

Proof: Irreflexivity is trivial by property (3) of $\prec$ and by the fact that $\mathcal{C} \leq_{sub} \mathcal{C}$ for every set of clauses $\mathcal{C}$.

For *transitivity* consider sets of clauses $\mathcal{C}_1, \mathcal{C}_2, \mathcal{C}_3$ with $\mathcal{C}_1 \prec \mathcal{C}_2$ and $\mathcal{C}_2 \prec \mathcal{C}_3$. Then by property (1) of $\prec$ and by transitivity of subsumption we obtain $\mathcal{C}_1 \leq_{sub} \mathcal{C}_3$. We show that $(\mathcal{C}_1, \mathcal{C}_3)$ fulfills property (2) of Definition 3.35: let C_1 be a clause in $\mathcal{C}_1$. By $\mathcal{C}_1 \prec \mathcal{C}_2$ there exists a clause $C_2 \in \mathcal{C}_2$ with $C_1 \leq_{sub} C_2$ and $|C_1| \leq |C_2|$; in the same way we obtain a clause $C_3 \in \mathcal{C}_3$ with $C_2 \leq_{sub} C_3$ and $|C_2| \leq |C_3|$. Both $\leq$ and $\leq_{sub}$ are transitive and so $C_1 \leq_{sub} C_3$ and $|C_1| \leq |C_3|$. It remains to show $\mathcal{C}_3 \not\leq_{sub} \mathcal{C}_1$. Let us assume, on the contrary, that $\mathcal{C}_3 \leq_{sub} \mathcal{C}_1$. By assumption $\mathcal{C}_1 \leq_{sub} \mathcal{C}_2$ and so, by transitivity of $\leq_{sub}$, $\mathcal{C}_3 \leq_{sub} \mathcal{C}_2$; this however contradicts $\mathcal{C}_2 \prec \mathcal{C}_3$.

It remains to show that $\prec$ is *Noetherian*, i.e. there is no infinite descending chain of finite sets of condensed clauses

$$\ldots..\mathcal{C}_{i+1} \prec \mathcal{C}_i \ldots \mathcal{C}_1 \prec \mathcal{C}_0.$$

Let us assume, for proof by contradiction, that there exists such an infinite chain.

By the properties (2) and (3) of $\prec$, every $\mathcal{C}_{i+1}$ must contain a clause C s.t. $C \leq_{sub} D$ for some $D \in \mathcal{C}_i$, $|C| \leq |D|$ and $\mathcal{C}_i \not\leq_{sub} \{C\}$. Since every clause subsumes itself, C is not contained in $\mathcal{C}_i$. Property (1) of $\prec$ then gives $\mathcal{C}_i \leq_{sub} \mathcal{C}_k$ for all $k \leq i$ and thus

$\mathcal{C}_k \not\leq_{sub} \{C\}$ for all $k \leq i$.

Therefore, for every number $i+1$ we obtain a clause $C_{i+1} \in \mathcal{C}_{i+1}$, s.t.

$$\bigcup_{0 \leq k < i+1} \mathcal{C}_k \not\leq_{sub} \{C_{i+1}\}.$$

Let us consider the infinite sequence $(C_i)_{i \in \mathbb{N}}$ of these clauses and the set $\mathcal{C}_0 : \{D_1, \ldots, D_m\}$. Then, by property (2) of $\prec$ on $(\mathcal{C}_i, \mathcal{C}_0)$ (we have shown transitivity above), for every $i > 0$ there exists a $j \in \{1, \ldots, m\}$ with $C_i \leq_{sub} D_j$. By the (transfinite) pigeon hole principle there exists an $r \in \{1, \ldots, m\}$ s.t. $C_i \leq_{sub} D_r$ for all $i \in I$ where I is an *infinite* subset of $\mathbb{N}$; by construction all C_i for $i \in I$ are pairwise different.

Let $d = \max\{|D_j| \mid 1 \leq j \leq m\}$; then, by property (2) of $\prec$ on $(\mathcal{C}_i, \mathcal{C}_0)$, $|C_i| \leq d$ for all $i \in I$. Now let $i \in I$: then by $C_i \leq_{sub} D_r$ we have $\tau(C_i) \leq \tau(D_r)$. Therefore the set $\mathcal{F} = \bigcup_{i \in I}\{C_i\}$ is an infinite set of condensed clauses C with $|C| \leq d$ and $\tau(C) \leq \tau(D_r)$. But, over a fixed finite signature, there are only finitely many condensed clauses with a fixed bound on their length *and* on their term depth. This contradicts the infinity of $\mathcal{F}$ and we conclude that $\prec$ is Noetherian. $\diamond$

LEMMA 3.36 *Let $\mathcal{C}$ be a $R_{H,s}$-stable set s.t. $R^*_H(\mathcal{C})$ is finite and $\alpha(\mathcal{C}) \neq \mathcal{C}$. Then $R^*_H(T(\mathcal{C})) \prec R^*_H(\mathcal{C})$.*

Proof: First of all we reduce the problem to $R^*_H(\alpha(\mathcal{C})) \prec R^*_H(\mathcal{C})$; note that $T(\mathcal{C}) = R^*_{H,s}(\alpha(\mathcal{C}))$.

Condition (1) of Definition 3.35 is easy to prove: By definition of α we have $\alpha(\mathcal{C}) \leq_{sub} \mathcal{C}$. Because $\leq_{sub}$ is preserved under R_H we obtain $R^*_H(\alpha(\mathcal{C})) \leq_{sub} R^*_H(\mathcal{C})$.

In order to show condition (3) of Definition 3.35 we have to prove that there exists a $C \in R^*_H(\alpha(\mathcal{C}))$ s.t. $R^*_H(\mathcal{C}) \not\leq_{sub} \{C\}$. Let E be the nonunit positive clause selected by α. Then there exists an atom P in E with $\alpha(\mathcal{C}) = (\mathcal{C} - \{E\}) \cup \{P\}$. Then $P \in R^*_H(\alpha(\mathcal{C}))$ and $R^*_H(\mathcal{C}) \not\leq_{sub} \{P\}$.

Condition (2) of Definition 3.35 follows from the following more general property (which we prove by induction on n):

(*) For all $n \geq 0$ and for all $C \in R^n_H(\alpha(\mathcal{C}))$ there exists a $D \in R^*_H(\mathcal{C})$ s.t. $|C| \leq |D|$ and a renaming substitution η with $\mathrm{LIT}(C\eta) \subseteq \mathrm{LIT}(D)$. Moreover, if C is positive then D is positive, too.

$n = 0$:

If $C \in \alpha(\mathcal{C})$ then either $C \in R^*_H(\mathcal{C})$ or $C = P$. The case $C \in R^*_H(\mathcal{C})$ is trivial; if $C = P$ then we choose $D = E$.

(IH) Suppose that (*) holds for n.

Now let $C \in R_H^{n+1}(\alpha(\mathcal{C}))$. If $C \in R_H^n(\alpha(\mathcal{C}))$ then the induction hypothesis can be applied directly; thus it suffices to consider the case $C \in R_H^{n+1}(\alpha(\mathcal{C})) - R_H^n(\alpha(\mathcal{C}))$.

By definition of R_H, C must be a clash resolvent of a clash γ : $(F; D_1, \ldots, D_m)$ s.t. F is a nonpositive clause in $\mathcal{C}$ and $D_i \in R_H^n(\alpha(\mathcal{C}))$ for $i = 1, \ldots, m$. By (IH) there are positive clauses $E_1, \ldots, E_m \in R_H^*(\mathcal{C})$ and renaming substitutions $\eta_1, \ldots, \eta_m$ with $\mathrm{LIT}(D_i\eta_i) \subseteq \mathrm{LIT}(E_i)$ and $|D_i| \leq |E_i|$ for $i = 1, \ldots, m$. Let δ be the clash sequence $(F; E_1, \ldots, E_m)$ over $R_H^*(\mathcal{C})$. Because every E_i contains a variant D_i' of D_i we can simulate factoring and binary resolution in D_i within E_i. Therefore there exists a clash resolvent G of δ and a renaming substitution η with $\mathrm{LIT}(C\eta) \subseteq \mathrm{LIT}(G)$ and, trivially, $|C| \leq |G|$. By $\varrho_H(R_H^*(\mathcal{C})) \subseteq R_H^*(\mathcal{C})$ we also obtain $G \in R_H^*(\mathcal{C})$. This settles the case $n+1$ and thus (*) is proven.

This eventually yields $R_H^*(\alpha(\mathcal{C})) \prec R_H^*(\mathcal{C})$. It remains to show $R_H^*(T(\mathcal{C})) \prec R_H^*(\mathcal{C})$. By definition of T we have

$$R_H^*(T(\mathcal{C})) = R_H^*(R_{H,s}^*(\alpha(\mathcal{C}))) \subseteq R_H^*(R_H^*(\alpha(\mathcal{C}))) = R_H^*(\alpha(\mathcal{C})).$$

Moreover, by definition of R_H^* and $R_{H,s}^*$ and the subsumption relation for sets of clauses we also obtain

$$T(\mathcal{C}) = R_{H,s}^*(\alpha(\mathcal{C})) \subseteq R_H^*(\alpha(\mathcal{C}))$$

and $R_H^*(T(\mathcal{C})) \subseteq R_H^*(\alpha(\mathcal{C}))$. By this last property and by $R_H^*(T(\mathcal{C})) \leq_{sub} R_H^*(\alpha(\mathcal{C}))$, $R_H^*(\alpha(\mathcal{C})) \prec R_H^*(\mathcal{C})$ we eventually obtain

$$R_H^*(T(\mathcal{C})) \prec R_H^*(\mathcal{C}).$$

◇

We are now in the position to prove our first result on termination of MB.

THEOREM 1.6 *Let $\mathcal{C}$ be a stable and satisfiable set of clauses s.t. $R_H^*(\mathcal{C})$ is finite. Then the sequence $(T^i(\mathcal{C}))_{i\in\mathbb{N}}$ converges.*

Proof: We have to prove the existence of an iteration number i with $T^i(\mathcal{C}) = T^{i+1}(\mathcal{C})$. From Lemma 3.36 we know that, for every satisfiable, $R_{H,s}$-stable set $\mathcal{D}$ with $\alpha(\mathcal{D}) \neq \mathcal{D}$ $R_H^*(T(\mathcal{D})) \prec R_H^*(\mathcal{D})$. Then Lemma 3.28 implies that $\alpha(\mathcal{D})$ is satisfiable and implies $\mathcal{D}$; clearly the same holds for $T(\mathcal{D})$. By iterating this argument we obtain a descending chain of satisfiable sets of clauses

$$\cdots \prec R_H^*(T^{k+1}(\mathcal{C})) \prec R_H^*(T^k(\mathcal{C})) \prec \cdots R_H^*(\mathcal{C}).$$

Because $\prec$ is Noetherian on finite sets of clauses this chain is finite and there exists a minimal element $R_H^*(T^i(\mathcal{C}))$. We have to show that $T^i(\mathcal{C}) = T^{i+1}(\mathcal{C})$:

Let us assume that $T^i(\mathcal{C}) \neq T^{i+1}(\mathcal{C})$; then, by definition of T, $\alpha(T^i(\mathcal{C})) \neq T^i(\mathcal{C})$. But then Lemma 3.36 yields $R_H^*(T^{i+1}(\mathcal{C})) \prec R_H^*(T^i(\mathcal{C}))$ contradicting the assumption of minimality. This implies $T^i(\mathcal{C}) = T^{i+1}(\mathcal{C})$. $\diamond$

If, due to $\alpha(\mathcal{C}) = \mathcal{C}$, MB terminates it might be the case that there are nonunit clauses in $POS(\mathcal{C})$ which cannot be used for unit reduction. In this case MB fails to produce a model. Therefore it is relevant to give some syntactic criteria which garantee termination *and* the construction of an ARM.

DEFINITION 3.37 Let $\mathcal{C}$ be a finite set of clauses. We call $\mathcal{C}$ positively disconnected if $R_H^*(\mathcal{C})$ is finite and all clauses in $POS(R_H^*(\mathcal{C}))$ are disconnected (see Definition 2.19). The class of all positively disconnected sets is denoted by $\mathcal{PDC}$. $\sharp$

Note that $\mathcal{PVD} \subseteq \mathcal{PDC}$ and $\mathcal{OCCIN} \subseteq \mathcal{PDC}$. If $\mathcal{C}$ is disconnected and D is a positive nonunit clause in $\mathcal{C}$ then there exists a nontrivial unit reduct $\mathcal{D}$ of $\mathcal{C}$, i.e. $\alpha(\mathcal{C}) \neq \mathcal{C}$. Therefore $\alpha(\mathcal{C}) = \mathcal{C}$ implies that all clauses in $POS(\mathcal{C})$ are unit! This, eventually, enforces the success of the model building procedure.

LEMMA 3.38 *Let $\mathcal{C} \in \mathcal{PDC}$ then $R_{H,s}^*(\mathcal{C}) \in \mathcal{PDC}$.*

Proof: By definition of $\mathcal{PDC}$, $R_H^*(\mathcal{C})$ is finite and all clauses in $POS(R_H^*(\mathcal{C}))$ are disconnected. By definition of $R_{H,s}^*$ we have $R_{H,s}^*(\mathcal{C}) \subseteq R_H^*(\mathcal{C})$ and thus $R_{H,s}^*(\mathcal{C})$ is finite too. Moreover

$$R_H^*(R_{H,s}^*(\mathcal{C})) \subseteq R_H^*(R_H^*(\mathcal{C})) = R_H^*(\mathcal{C})$$

and therefore $R_{H,s}^*(\mathcal{C}) \in \mathcal{PDC}$. $\diamond$

LEMMA 3.39 *Let $\mathcal{C}$ be a stable set in $\mathcal{PDC}$; Then $T(\mathcal{C})$ is in $\mathcal{PDC}$ too.*

Proof: If $\alpha(\mathcal{C}) = \mathcal{C}$ then $T(\mathcal{C}) = R_{H,s}^*(\mathcal{C})$ and the result follows from Lemma 3.38. If $\alpha(\mathcal{C}) \neq \mathcal{C}$ then $\alpha(\mathcal{C}) = (\mathcal{C} - \{C\}) \cup \{P\}$ for some nonunit clause $C \in \mathcal{C}$ and $P \in C$. Let $\mathcal{D} = \alpha(\mathcal{C})$; then, by $\mathcal{C} \in \mathcal{PDC}$, all clauses in $POS(\mathcal{D})$ are disconnected. We have to show that $R_H^*(\mathcal{D})$ is finite and all clauses in $POS(R_H^*(\mathcal{D}))$ are disconnected. From the proof of Lemma 3.28 we know that $R_H^*(\mathcal{D}) \leq_E R_H^*(\mathcal{C})$ for $E = C - \{P\}$. But E is disconnected and all positive clauses in $R_H^*(\mathcal{C})$ are disconnected; consequently $POS(R_H^*(\mathcal{D}))$ consists of disconnected clauses only.

Moreover, as $R^*_H(\mathcal{D})$ consists of condensed clauses only, $R^*_H(\mathcal{D})$ is finite. Therefore $\mathcal{D} \in \mathcal{PDC}$. By definition of T, $T(\mathcal{C}) = R^*_{H,s}(\mathcal{D})$; but then, by Lemma 3.38, $T(\mathcal{C}) \in \mathcal{PDC}$. $\diamond$

The second result on termination guarantees convergence to an atomic representation of a Herbrand model:

THEOREM 1.7 *Let $\mathcal{C}$ be a stable and satisfiable set in $\mathcal{PDC}$; then the sequence $(T^i(\mathcal{C}))_{i\in\mathbb{N}}$ converges to a set $\mathcal{D}$ s.t. $POS(\mathcal{D})$ consists of unit clauses only. Moreover the algorithm MB terminates on $\mathcal{C}$ and returns an ARM of a Herbrand model of $\mathcal{C}$.*

Proof: By definition of $\mathcal{PDC}$, $\mathcal{C} \in \mathcal{PDC}$ implies that $R^*_H(\mathcal{C})$ is finite. Therefore the sequence $(T^i(\mathcal{C}))_{i\in\mathbb{N}}$ converges by Theorem 1.6. Let $\mathcal{D}$ be the limes of the sequence. By Lemma 3.39 we get $\mathcal{D} \in \mathcal{PDC}$. We have to show that $POS(\mathcal{D})$ consists of unit clauses only. So let us assume that there exists a nonunit clause D in $POS(\mathcal{D})$. As D is disconnected there exists a unit reduct of $\mathcal{D}$ w.r.t. D, i.e. $\alpha(\mathcal{D}) \neq \mathcal{D}$. By Lemma 3.36 $R^*_H(T(\mathcal{D})) \prec R^*_H(\mathcal{D})$; in particular, $T(\mathcal{D}) \neq \mathcal{D}$. But this contradicts the assumption that $\mathcal{D}$ is the limit of $(T^i(\mathcal{C}))_{i\in\mathbb{N}}$.

By definition, MB returns $POS(\mathcal{D})$ if $\mathcal{D}$ is obtained after execution of the while-loop and $POS(\mathcal{D})$ consists of unit clauses only. By Theorem 1.5 $POS(\mathcal{D})$ is an ARM of $\mathcal{C}$. $\diamond$

The whole model building method can be extended to the classes $\mathcal{PVD}_r$ and $\mathcal{OCCIN}_r$ (see definition 3.19): Given a set of clauses $\mathcal{C}$, search for a sign renaming γ such that $\gamma(\mathcal{C}) \in \mathcal{PVD}$ and then apply MB to $\gamma(\mathcal{C})$. MB then yields an AR of a Herbrand model of $\gamma(\mathcal{C})$; by changing the signs backwards one obtains an atomic representation of a Herbrand model of $\mathcal{C}$.

EXAMPLE 3.40 We define the following satisfiable set of clauses:

$$\mathcal{C} = \{P(b),\ P(f(x)) \vee \neg P(x),\ \neg P(a) \vee \neg P(f(a))\}.$$

$\mathcal{C}$ is not in $\mathcal{PVD}$ and $(R^i_{H,s}(\mathcal{C}))_{i\in\mathbb{N}}$ is divergent. Note that for all $i \geq 1$:

$$P(f^i(b)) \in R^i_{H,s}(\mathcal{C}) - R^{i-1}_{H,s}(\mathcal{C}).$$

But $\mathcal{C} \in \mathcal{PVD}_r$ as can be seen by computing $\gamma(\mathcal{C})$ for the renaming γ exchanging P and $\neg P$. Indeed

$$\{\neg P(b),\ P(x) \vee \neg P(f(x)),\ P(a) \vee P(f(a))\}$$

is in $\mathcal{PVD}$.

By setting $\mathcal{C}_1 = \gamma(\mathcal{C})$ we obtain

$$R^0_{H,s}(\mathcal{C}_1) = \mathcal{C}_1 \text{ and } R^1_{H,s}(\mathcal{C}_1) = \{\neg P(b),\ P(x) \vee \neg P(f(x)),\ P(a)\}.$$

Clearly $\varrho_H(R^1_{H,s}(\mathcal{C}_1)) = \emptyset$ and $R^2_{H,s}(\mathcal{C}_1) = R^1_{H,s}(\mathcal{C}_1)$.

We see that $(R^i_{H,s}(\mathcal{C}))_{i\in\mathbb{N}}$ converges and

$$R^*_{H,s}(\mathcal{C}_1) = \{\neg P(b),\ P(x) \vee \neg P(f(x)),\ P(a)\}.$$

By lemma 3.24 (and clearly visible in this case) $\mathcal{A} : \{P(a)\}$ is an atomic representation of a Herbrand model of $\mathcal{C}_1$ ($\mathcal{A}$ is also a ground representation of this model).

Therefore $\mathcal{M} : \{P(b)\} \cup \{P(f(t)) | t \in H(\mathcal{C})\}$ is a ground representation of a Herbrand model of $\mathcal{C}$. By the principle of forward chaining ($\mathcal{C}$ is in Horn form) the set $POS(R^*_H(\mathcal{C}))$ must be an atomic representation of a Herbrand model of $\mathcal{C}$. Indeed $POS(R^*_H(\mathcal{C})) = \{P(f^n(b)) \mid n \in \mathbb{N}\} \subseteq \mathcal{M}$, but as the set is infinite the computation of R^*_H on $\mathcal{C}$ does not yield a syntactic model representation. However we can compute a finite AR of the model $\mathcal{M}$ directly out of $\mathcal{A}$ itself; such a representation is $\mathcal{B} : \{P(b), P(f(x))\}$ (over $H(\mathcal{C})$). Note that $\mathcal{M}$ does not represent a minimal Herbrand model, in contrast to $POS(R^*_H(\mathcal{C}))$. ♯

It is not hard to show that the complement set of a finite ground representation always possesses a finite AR too; this property even holds for linear representations, i.e. for finite AR's $\mathcal{A}$ such that, for all $A \in \mathcal{A}$, A contains every variable at most once [Fermüller and Leitsch, 1992]. We will show in the next subsection that these representations can be obtained algorithmically.

1.4 Transformation to Finite Models

We have seen in Proposition 1.1 that the hyperresolution operator can produce atomic representations of Herbrand models also for sets of clauses not having finite models. Thus not every ARM can be transformed into a finite model of the corresponding set of clauses. There are, however, syntactic restrictions on ARMs allowing for extractions of finite models; we will see below that the linearity of atoms (i.e. all variables occur at most once) is such a restriction. Let us consider the set of clauses $\mathcal{C}$ defined in Proposition 1.1:

$$\{P(x,x),\ \neg P(f(x), f(y)) \vee P(x,y),\ \neg P(c, f(x))\}.$$

The ARM produced by hyperresolution is $\{P(x,x)\}$, which defines the Herbrand model $\{P(s,s) | s \in H\}$ for $H = \{f^n(c) | n \geq 0\}$. There exist infinitely many Herbrand models of $\mathcal{C}$, but in all of them the ground atoms $P(f^k(c), f^l(c))$ for $k < l$ are set to **false**. This property cannot be "compressed" into a finite domain. Here it is essential that x occurs twice in $P(x,x)$: if we replace the positive clause $P(x,x)$ in $\mathcal{C}$ by $P(x,c)$ we also obtain a satisfiable set of clauses $\mathcal{C}'$, but $\mathcal{C}'$ has finite models too.

Under specific circumstances a Herbrand model $\mathcal{M}$ can be transformed into a finite model $\mathcal{N}$ in a "canonical" way, i.e. $\mathcal{N}$ preserves the truth values of $\mathcal{M}$.

EXAMPLE 3.41

$$\mathcal{C} = \{\neg P(f(x,y),z),\ P(a,f(a,x))\}.$$

$\mathcal{C}$ is satisfiable and, by $R_H(\mathcal{C}) = \mathcal{C}$, $P(a,f(a,x))$ is an ARM of of a Herbrand model $\mathcal{M} : \{P(a,f(a,t)) | t \in H\}$ of $\mathcal{C}$.

Clearly there exists a two-element model $\mathcal{N} : (D, \Phi, I)$ of $\mathcal{C}$ with $\Phi(P) = \pi$, $\Phi(f) = \phi$, $\Phi(a) = 1$ and $D = \{1,2\}$ where

$\phi(d,d') = 2$ for $d, d' \in D$ and

$\pi(1,d) = \mathbf{true}$, $\pi(2,d) = \mathbf{false}$ for $d \in D$.

However $\mathcal{N}$ is not compatible with $\mathcal{M}$: in particular the ground atom $P(a,f(f(a,a),a))$ is false in $\mathcal{M}$, but true in $\mathcal{N}$ (note that $\pi(1,d) = \mathbf{true}$ for all $d \in D$). Indeed $\mathcal{M}$ is also a model of the extended set of clauses $\mathcal{D}$ (with the same ARM), for

$$\mathcal{D} = \{\neg P(f(x,y),z),\ \neg P(a,f(f(x,y),z)),\ P(a,f(a,x))\},$$

while $\mathcal{N}$ falsifies $\mathcal{D}$.

Now let $\mathcal{P} : (D', \Phi', I)$ be a finite interpretation of $\mathcal{C}$ where $D' = \{1,2,3\}$, $\Phi'(a) = 1$, $\Phi'(f) = \psi$ and $\Phi'(P) = \rho$, where

$\psi(1,d) = 2,\ \psi(2,d) = 3,\ \psi(3,d) = 3$ for $d \in D'$ and

$\rho(1,2) = \mathbf{true}$, $\rho(d,d') = \mathbf{false}$ for $(d,d') \neq (1,2)$.

$\mathcal{P}$ is compatible with $\mathcal{M}$ in the sense that

$v_{\mathcal{P}}(A) = v_{\mathcal{M}}(A)$ for all $A \in \mathrm{AT}(\mathcal{C})$.

Clearly $\mathcal{P}$ is also a model of $\mathcal{D}$ and of *any* set of Horn clauses over the signature of $\mathcal{C}$ having a Herbrand model represented by the ARM $\{P(a,f(a,x))\}$! Thus every set of Horn clauses over $\{P,f,a\}$ satisfied by a Herbrand model with ARM $\{P(a,f(a,x))\}$ has a finite model, namely $\mathcal{P}$. ♯

Example 3.41 shows that, in order to guarantee the model property of the finite interpretation, we must preserve the truth values on the atom set.

DEFINITION 3.42 Let $\mathcal{M}$ be a Herbrand model of $\mathcal{C}$. A finite model $\mathcal{N}$ of $\mathcal{C}$ *finitely realizes* $\mathcal{M}$ if for all ground atoms $A \in \mathrm{AT}(\mathcal{C})$ $v_{\mathcal{N}}(A) = v_{\mathcal{M}}(A)$

(if $\mathcal{C}$ does not contain constant symbols then $\mathcal{N}$ is supposed to interpret the constant symbol of the Herbrand universe). If there exists such a finite model $\mathcal{N}$ we call $\mathcal{M}$ *finitely realizable.* $\sharp$

Clearly a Herbrand model $\mathcal{M}$ of a set of clauses $\mathcal{C}$ is not finitely realizable if $\mathcal{C}$ does not have finite models at all. But finite realizability essentially depends on the ARM, not so much on the set of clauses $\mathcal{C}$.

PROPOSITION 1.4 *There exists a set of Horn clauses $\mathcal{C}$ with the following property*

(1) *$\mathcal{C}$ has finite models and*

(2) *the minimal Herbrand model of $\mathcal{C}$ is not finitely realizable.*

Proof: Let $\mathcal{C}$ be the set of clauses $\{P(x,x),\ \neg P(x, f(x))\}$. Obviously there exists a two-element model of $\mathcal{C}$. By $R_H(\mathcal{C}) = \mathcal{C}$ the minimal Herbrand model $\mathcal{M}$ is represented by $\mathcal{A} : \{P(x,x)\}$; therefore $v_{\mathcal{M}}(P(f^{(n)}(a), f^{(m)}(a))) = \mathbf{true}$ iff $n = m$.

We show that $\mathcal{M}$ is not finitely realizable: Let us assume that $\mathcal{N} : (D, \Phi, I)$ is a finite model of $\mathcal{C}$ which realizes $\mathcal{M}$. Then

$$(I)\ v_{\mathcal{N}}(P(f^{(n)}(a), f^{(m)}(a))) = \mathbf{true} \text{ iff } n = m.$$

Let $\phi = \Phi(f)$ and $\pi = \Phi(P)$. Then ϕ maps D to D and, because D is finite, the infinite sequence $(\phi^{(n)}(d))_{n\in\mathbb{N}}$ is cyclic for all $d \in D$; i.e. for every $d \in D$ there exists numbers k, l s.t. $k < l$ and $\phi^{(k)}(d) = \phi^{(l)}(d)$.

Now let $\Phi(a) = d$ and $\phi^{(k)}(d) = \phi^{(l)}(d)$ for some k, l with $k < l$. Then, because of (I),

$$(II)\ \pi(\phi^{(k)}(d), \phi^{(l)}(d)) = v_{\mathcal{N}}(P(f^{(k)}(a), f^{(l)}(a))) = \mathbf{false}.$$

On the other hand there exists a d' with

$d' = \phi^{(k)}(d) = \phi^{(l)}(d)$ and

$t_{\mathcal{N}}(f^{(k)}(a)) = t_{\mathcal{N}}(f^{(l)}(a)) = d'$,

$v_{\mathcal{N}}(P(f^{(k)}(a), f^{(k)}(a))) = \pi(d', d')$.

Because of (I) $\pi(d', d') = v_{\mathcal{N}}(P(f^{(k)}(a), f^{(k)}(a))) = \mathbf{true}$. This, however, contradicts (II) and the definition of d'. Therefore there is no finite realization of $\mathcal{M}$. $\diamond$

The decidable classes introduced in the last section do not only allow for the construction of atomic representation of models but also guarantee a very simple term structure of the representing atoms. The resulting representations are called linear. We will show below that Herbrand models having such representations are finitely realizable.

DEFINITION 3.43 An expression E is called *linear* if each variable in E occurs only once, i.e. $OCC(x, E) = 1$ for all $x \in V(E)$. A set of expressions is said to be linear if all its elements are linear. $\sharp$

We show that every linear atomic representation can algorithmically be transformed into another one representing the same Herbrand model in an "orthogonal" manner and consisting of instances of the original atoms. This new orthogonal representation allows for a natural projection of the Herbrand model into a finite model.

W.l.o.g. we speak of terms only in the following definitions. The generalization to atoms is obvious.

DEFINITION 3.44 Let s, t be expressions and H a Herbrand universe. We define $s \leq_H t$ if there exists a substitution σ over the signature of H with $s\sigma = t$ (i.e. t is an H-instance of s). Let T_1 and T_2 be two sets of expressions. Then we write $T_1 \leq_H T_2$ if for every $t \in T_2$ there exists an $s \in T_1$ with $s \leq_H t$. $\sharp$

According to Definition 3.44, a ground atom B is true in the Herbrand model represented by $\mathcal{A}$ iff $\mathcal{A} \leq_H \{B\}$.

DEFINITION 3.45 Let us denote by $G_H(s)$ the set of all ground instances of a term s over the Herbrand universe H. Two terms s, t are called *orthogonal* (notation $s \perp t$) if they are not unifiable, i.e. if $G_H(s) \cap G_H(t) = \emptyset$ for every Herbrand universe H containing the signature of $\{s, t\}$. If T is a set of terms then $T^\perp$ denotes all terms $s \in H$ with $s \perp t$ for all $t \in T$. $\sharp$

DEFINITION 3.46 Let $T : \{t_1, \ldots, t_n\}$ be a finite set of non-variable linear terms and H a Herbrand universe containing $H(T)$. Then T is called *orthogonal representation* of $G_H(T)$ if for all $i, j \leq n$ and $i \neq j$: $t_i \perp t_j$.

Let T' be an arbitrary representation of $G_H(T)$. Then T is called an *orthogonal extension* of T' if T is an orthogonal representation of $G_H(T)$ and $T' \leq_H T$. $\sharp$

EXAMPLE 3.47 $T = \{a,\ f(x, f(u, v)),\ f(a, y),\ f(x, a)\}$ represents $G_H(T)$, where H is the set of all terms built up from a and f only. We even have $G_H(T) = H$. T is not an orthogonal representation since, e.g., $f(x, f(u, v)) \perp f(a, y)$ does not hold.

$T_1 = \{a, f(x, y)\}$ is an orthogonal representation of H but it is not an orthogonal extension of T since $T \not\leq_H T_1$ ($f(x, y)$ is not instance of a term in T).

However, $T_2 = \{a,\ f(x, f(u, v)),\ f(a, a),\ f(f(x, y), a)\}$ is an orthogonal extension of T w.r.t. H. $\sharp$

Below we present an algorithm that constructs an orthogonal extension of T for any finite set T of linear terms. For this purpose we need some additional terminology:

DEFINITION 3.48 The set of *base terms* BT_H of some Herbrand universe H is the set of all constants in H and all linear terms of the form $f(x_1, \ldots, x_n)$, where f is a function symbol in the signature of H. $\sharp$

Remark: Within a set of terms we do not distinguish terms that only differ in the names of their variables. Therefore BT_H contains exactly one term for each constant and function symbol. It can be considered as the simplest orthogonal representation of H (Remember that a variable cannot be element of a representation by definition.). $\sharp$

We need a linear representation of the set of all terms that are not instances of a certain term. Such representations have been investigated first in [Lassez et al., 1986].

DEFINITION 3.49 Let t be a linear term and H a Herbrand universe. Then the set of terms $co_H(t)$ is inductively defined as follows:

(1) If t is a constant then $co_H(t) = BT_H - \{t\}$.

(2) If $t = f(t_1, \ldots, t_n)$ then $co_H(t) = (BT_H - \{f(x_1, \ldots, x_n)\}) \cup \{f(s_1, \ldots, s_n) \mid s_i = t_i$ or $s_i \in co_H(t_i)$ but at least one $s_i \neq t_i\}$, where $V(s_i) \cap V(s_j) = \emptyset$ for all $i \neq j$. I.e., the variable names are chosen s.t. $f(s_1, \ldots, s_n)$ is linear. (Observe that $co_H(t)$ is undefined for variables; thus if t_i is a variable then $s_i = t_i$.)

The relativation of $co_H(t)$ to a term s, $co(t|s)$, is defined as the subset of $co_H(t)$ that contains all terms in $co_H(t)$ that are instances of s. $\sharp$

Observe that all elements of $co_H(t)$ are linear terms.

EXAMPLE 3.50 Let $t = f(x, a)$, and let H be the set of ground terms built up from a, f, and the unary function symbol g. Then $co_H(a) = \{g(x),\ f(x, y)\}$ and therefore

$$co_H(t) = \{a,\ g(x),\ f(x, g(y)),\ f(x, f(y, z))\} \text{ and}$$

$$co_H(t|f(x, y)) = \{f(x, g(y)),\ f(x, f(y, z))\}.$$

$\sharp$

The following lemmas serve to guarantee the correctness of our orthogonalization algorithm.

LEMMA 3.51 *Let t be a linear term over the signature of a Herbrand universe H. Then $t \cup co_H(t)$ is an orthogonal representation of H.*

Proof: Obviously it is sufficient to prove the following propositions:

(1) $\{t\} \cup co_H(t)$ represents H.

(2) $\{t\} \bot co_H(t)$.

(3) $co_H(t)$ is an orthogonal representation.

We prove (1), (2), and (3) simultaneously by induction on the term depth of t:

Obviously, the propositions hold if t is a constant. We have the following induction hypothesis:

(IH) Propositions (1), (2), and (3) hold for all t where $\tau(t) \leq n$.

Let $t' = f(t_1, \ldots, t_k)$ be any term of depth $n+1$. Since by the induction hypothesis $\{t_i\} \cup co_H(t_i)$ represents H for each $1 \leq i \leq k$, it follows by part (2) of definition 3.49 that $co_H(t')$ represents $H - G_H(t')$. This proves proposition (1).

Clearly, no base term except $f(x_1, \ldots, x_n)$ is unifiable with t'. On the other hand for any term $f(s_1, \ldots, s_n) \in co_H(t')$ there is at least one i with $s_i \in co_H(t_i)$. It follows by part (2) of (IH) that also $f(s_1, \ldots, s_n)$ is not unifiable with t'. This proves proposition (2).

To prove proposition (3) first observe that any two different base terms are not unifiable. Let $s = f(s_1, \ldots, s_n)$ and $s' = f(s'_1, \ldots, s'_n)$ be two different elements of $co_H(t')$. By definition, there must be an i s.t. $s_i \neq s'_i$ and both terms are not variables. Hence, by part (2) of definition 3.49, for some i either both, s_i and s'_i, are in $co_H(t_i)$ or exactly one of the two terms equals t_i. In either case, $s \bot s'$ follows by the induction hypothesis. ◇

LEMMA 3.52 *Let t be a linear term and $t \leq_H s$. Then $co_H(s|t)$ is an orthogonal representation of $G_H(t) - G_H(s)$.*

Proof: $co_H(s|t)$ is defined as the subset of all terms in $co_H(s)$ that are instances of t. We first consider the set $co'_H(s|t)$ of all terms in $co_H(s|t)$ that are unifiable with t, and then show that $co'_H(s|t) = co_H(s|t)$.

Since $co'_H(s|t)$ is a subset of $co_H(s)$ it follows from Lemma 3.51 that the subset $co'_H(s|t)$ is orthogonal and does not represent any instances of s. I.e.,

$$G_H(co'_H(s|t)) \subseteq G_H(t) - G_H(s).$$

To see that also the converse holds, consider any term $r \in G_H(t) - G_H(s)$. Since $r \notin G_H(s)$ there must be some $r' \in co_H(s)$ s.t. $r' \leq_H r$. But r is also an instance of t by definition. Consequently, t and r' are unifiable and thus $r' \in co'_H(s|t)$. This means that $r \in G_H(co'_H(s|t))$. Therefore

$$G_H(co'_H(s|t)) = G_H(t) - G_H(s).$$

It remains to show that any term $r' \in co'_H(s|t)$ is also an instance of t. Suppose to the contrary that there is a position $p \in \mathcal{P}(r') \cap \mathcal{P}(t)$ s.t. r'_p is a variable but t_p is not a variable. (If there is no such p then, since r' and t are unifiable, r' is an instance of t).

By the definition of $co_H(s)$, the fact that r'_p is a variable implies that there is some subposition p' of p in $\mathcal{P}(s)$ s.t. $s_{p'}$ is a variable, too. Since s is an instance of t we conclude that there exists a subposition p'' of p' in $\mathcal{P}(t)$ s.t. $t_{p''}$ is a variable. But this clearly contradicts the assumption that t_p is not a variable. We have thus shown that r' is an instance of t and consequently

$$co'_H(s|t) = co_H(s|t).$$

To sum up we have proved that $co_H(s|t)$ is an orthogonal representation of $G_H(t) - G_H(s)$. ◇

We are now prepared to present an orthogonalization algorithm.

function *Orthogonalize*
input: a finite set T of linear terms, alphabet of H
output: an orthogonal extension of T w.r.t. H
begin
 $T' := T$;
 while $\exists t_1, t_2 \in T'$ s.t. $(t_1 \neq t_2)$ and $\{t_1, t_2\}$ is unifiable **do**
  begin
  $\theta :=$ mgu of t_1 and t_2;
  $T' := (T' - \{t_1, t_2\}) \cup \{t_1\theta(= t_2\theta)\} \cup co_H(t_1\theta|t_1) \cup co_H(t_1\theta|t_2)$
  end;
 return T'
end.

Remark: H is assumed to be based on an alphabet that at least contains the constant and function symbols occurring in T. Moreover, it should be clear that the terms are considered to be variable disjoint when subjected to the unification algorithm. Thus, e.g., $f(x, g(y))$ is unifiable with $f(g(x), y)$ (i.e. with $f(g(u), v)$). ♯

Before proving the correctness of *Orthogonalize* we illustrate its behaviour by an example.

EXAMPLE 3.53 Let

$$T = \{f(x, g(y)),\ g(x),\ g(f(x, a)),\ f(x, a),\ a,\ f(g(x), y)\}$$

and H be based on the alphabet $\{a, g, f\}$. Denoting T as

$$T = \{t_1, t_2, \dots t_6\}$$

where

$$t_1 = f(x, g(y)),\ t_2 = g(x),\ t_3 = g(f(x, a))$$

$$t_4 = f(z, a),\ t_5 = a,\ t_6 = f(g(u), v)$$

one easily sees that (t_1, t_6), (t_2, t_3) and (t_4, t_6) are pairs of unifiable terms.

We apply *Orthogonalize* to T and choose t_1 and t_6 as terms to be replaced in $T = T'^{(0)}$. The substitution $\sigma = \{x \leftarrow g(u),\ v \leftarrow g(y)\}$ is an mgu of (t_1, t_6). Thus $t_1\sigma = t_6\sigma = f(g(u), g(y))$. We have

$$co_H(t_1\sigma|t_1) = \{f(a, g(x))(= t_7),\ f(f(x, y), g(z))(= t_8)\} \text{ and}$$

$$co_H(t_1\sigma(= t_6\sigma)|t_6) = \{f(g(x), a)(= t_9),\ f(g(x), f(y, z))(= t_{10})\}.$$

We arrive at

$$T'^{(1)} = \{t_2, t_3, t_4, t_5, t_7, t_8, t_9, t_{10}, t_{11} = t_1\sigma(= t_6\sigma)\}.$$

In $T'^{(1)}$ $t_4 = f(z, a)$ and $t_9 = f(g(x), a))$ are unifiable with mgu $\sigma = \{z \leftarrow g(x)\}$. $t_4\sigma(= t_9\sigma = t_9) = f(g(x), a)$, thus

$$co_H(t_4\sigma|t_4) = \{f(a, a) = t_{12},\ f(f(x, y), a)(= t_{13})\} \text{ and}$$

$$co_H(t_4\sigma(= t_9)|t_9) = \emptyset.$$

We get

$$T'^{(2)} = \{t_2, t_3, t_5, t_7, t_8, t_9, t_{10}, t_{11}, t_{12}, t_{13}\}$$

$t_2 = g(x)$ and $t_3 = g(f(x, a)) = g(f(y, a))$ are unifiable with mgu $\sigma = \{x \leftarrow f(y, a)\}$.

$$\begin{aligned} co_H(t_2\sigma|t_2) = \{& g(a)(= t_{14}),\ g(g(x))(= t_{15}), g(f(x, g(y))(= t_{16}), \\ & g(f(x, f(y, z))(= t_{17})\} \qquad \text{and} \end{aligned}$$

$$co_H(t_2\sigma(= t_3)|t_3) = \emptyset.$$

We therefore have

$$T'^{(3)} = \{t_3, t_5, t_7, t_8, t_9, t_{10}, t_{11}, t_{12}, t_{13}, t_{14}, t_{15}, t_{16}, t_{17}, \}$$

Since no pair of different terms in $T'^{(3)}$ is unifiable the algorithm terminates and returns

$$T_O = T'^{(3)} = \{a,\ g(f(x,a)),\ g(a),\ g(g(x)),\ g(f(x,g(y))),\ g(f(x,f(y,z)),\ f(a,g(x)),\ f(f(x,y),a),\ f(a,a),\ f(g(u),g(y)),\ f(f(x,y),g(z)),\ f(g(x),a),\ f(g(x),f(y,z))\}.$$

It is not hard to verify that T_O is an orthogonal extension of T. ♯

LEMMA 3.54 *For each finite set T of linear terms the function Orthogonalize computes a finite, orthogonal extension T_O of T.*

Proof: It follows immediately from the definition of $co_H(t)$ and T that T' is finite and contains only linear terms at any stage of the algorithm. Therefore, if *Orthogonalize* terminates then T_O is finite and linear, too.

By Lemma 3.52, if t is an instance of s then $G_H(\{s\}) = G_H(\{t\}) \cup G(co_H(t|s)$. In particular we have

$$G_H(\{t_1, t_2\}) = G_H(\{t_1\theta(= t_2\theta)\}) \cup G_H(co_H(t_1\theta|t_1)) \cup G_H(co_H(t_1\theta|t_2)),$$

where θ is an mgu of t_1 and t_2.

It follows that $G_H(T') = G_H(T)$ is a loop–invariant of *Orthogonalize.* Therefore T_O represents $G_H(T)$. Since, by the **while**–condition, the terms in T_O are pairwise not unifiable the representation is orthogonal. Since for any t_i that is removed from T' the instance $t_i\theta$ is added to T' we conclude that T_O is an extension of T. We have thus proved the correctness of the algorithm.

To prove the termination of *Orthogonalize* we first show that there is only a fixed, finite number of terms that may occur in T_O. In fact the maximal depth of occurrence of a non-variable term (in terms in T') cannot increase. More formally, let

$$\tau_c(t) = \max\{|p| \mid p \in \mathcal{P}(t) \text{ but } t_p \notin V(t)\}.$$

If T is a set of terms then define

$$\tau_c(T) = \max\{\tau_c(t) \mid t \in T\}.$$

By definition of $co_H(t)$ we have $\tau_c(co_H(t)) = \tau_c(t)$. Moreover, for all linear terms t_1, t_2 that are unifiable by an mgu θ we have $\tau_c(t_1\theta(= t_2\theta)) = \tau_c(\{t_1, t_2\})$ (see [Fermüller et al., 1993], chapter 3 for details).

Therefore $\tau_c(T')$ does not increase throughout the execution of the algorithm. But since we identify terms that are equal up to renaming of variables, and since we assume that H is based on a finite signature this means that there are only finitely many different possible values for T'.

We still have to exclude the possibility that *Orthogonalize* enters an infinite loop (i.e. that T' takes the same value infinitely many times): That this cannot happen is easily seen by observing the fact that in each traversal of the **while**–loop at least one term in T' is replaced by *proper* instances of it: If t_1 and t_2 are unifiable by the mgu θ then $t_1\theta = t_2\theta$ is a proper instance either of t_1 or of t_2, unless t_1 and t_2 are equal (up to a renaming of variables). But any two terms in the set T' are assumed to be different. The elements of $co_H(t_1\theta|t_1)$ and $co_H(t_1\theta|t_2)$ consist of proper instances of t_1 and t_2, respectively, by definition. It follows that, throughout the whole execution of the algorithm, T' never takes the same value twice. ◇

As a remark on the complexity of the algorithm, observe that $|co_H(t)|$ may be exponential (in the length of t.)

EXAMPLE 3.55 Let $t = f(a, \ldots, a)$ for some n-ary function symbol f. Then

$$co(t) = \{a\} \cup \{f(t_1, \ldots, t_n) \mid (t_1, \ldots, t_n) \in Q\},$$

where Q is the set of all tuples $(t_1, \ldots, t_n)$ s.t. each t_i is either an f-base term or a, but at least one t_i is not a. Clearly $|Q| = 2^n$ and so $|co(t)| = 2^n$. ♯

The following example shows that not only our algorithm, but the problem of orthogonalization itself is of exponential space (and time) complexity. Still, experiments with implementations of *Orthogonalize* show that for many relevant cases orthogonal extensions can be computed very fast.

EXAMPLE 3.56 Let T be the set of terms

$$\{f(a, x_2, x_3, \ldots, x_n), f(x_1, a, x_3, \ldots, x_n), \ldots, f(x_1, x_2, \ldots x_{n-1}, a)\}.$$

Let H be based on the signature $\{f, a\}$. Clearly, T represents the set of all f-terms where at least one argument equals a. Of course T is not orthogonal since all its elements are pairwise unifiable. *Orthogonalize* constructs the set of terms

$$T_O = \{a\} \cup \{f(t_1, \ldots, t_n)\} \mid (t_1, \ldots, t_n) \in Q\}$$

where Q consists of all term tuples $(t_1, \ldots, t_n)$ s.t. each t_i is an f-base term or a, but not all t_i are a. Therefore $|Q| = 2^n$ and thus $|T_O| = 2^n$ (in contrast to $|T| = n$). However, it is not hard to see that not only T_O (like any other orthogonal extension of T) but *any* set of terms that orthogonally represents $G_H(T)$ contains at least 2^n elements. ♯

Remark: We are now in the possession of the machinery for the *transformation* of linear atomic representations into representations of *finite models*. This procedure can be considered as a *final step of the model building procedure by hyperresolution*. Indeed, on the classes $\mathcal{PVD}$ and $\mathcal{OCCIN}$, hyperresolution terminates and yields linear atomic representations. The method of transforming specifications of Herbrand models into representations of finite models differs from the standard methods of finite model building in principle: the latter essentially consists of finite domain search and requires backtracking, while the former one is *backtracking-free*. Note that also the construction of ARMs on the class $\mathcal{PDC}$ is an iteration procedure, which is free of backtracking! Thus the whole procedure of constructing a finite model of a set of clauses $\mathcal{C} \in \mathcal{PVD} \cup \mathcal{OCCIN}$ can be divided into two steps:

1. the iteration procedure constructing the (linear) ARM of a Herbrand model $\mathcal{M}$ and

2. the computation of the finite realization of $\mathcal{M}$.

The remaining part of this subsection is devoted to the second step defined above. The reader is invited to compare this method here with finite domain search presented in Chapter 6. ♯

In order to specify the domain of our finite models we need the orthogonal extensions of sets of terms as introduced in Definition 3.46.

DEFINITION 3.57 Let $\mathcal{A}$ be a linear atomic representation of a Herbrand model w.r.t. a Herbrand universe H (containing $H(\mathcal{A})$). Let $T(\mathcal{A})$ be the set of all nonvariable terms occurring (as subterms) in some atom in $\mathcal{A}$. An orthogonal extension of the set $BT_H \cup T(A)$ is denoted by $\hat{T}(\mathcal{A}, H)$. ♯

By definition of the set of base terms BT_H, BT_H is an orthogonal representation of the Herbrand universe H. Thus $BT_H \cup T(A)$ is a representation of H too, but it is not orthogonal in general. The orthogonal extension $\hat{T}(\mathcal{A}, H)$ also represents H but is more finegrained than BT_H (i.e. the Herbrand universe is partitioned into more different sets).

EXAMPLE 3.58 Let $\mathcal{A} = \{P(f(f(a))), P(g(x,y))\}$ and H be the Herbrand universe defined over the signature $\{P, f, g, a, b\}$ where $g \in \mathrm{FS}_2$, $f \in \mathrm{FS}_1$, $a, b \in \mathrm{CS}$ and $P \in \mathrm{PS}$. Then $BT_H = \{f(x), g(x,y), a, b\}$ and $T(\mathcal{A}) = \{a, f(a), f(f(a)), g(x,y)\}$. Finally

$$\begin{aligned}\hat{T}(\mathcal{A}, H) \quad = \quad & \{a, b, f(a), f(b), g(x,y), f(f(a)), f(f(b)), f(g(x,y)), \\ & f(f(f(x))), f(f(g(x,y)))\}.\end{aligned}$$

♯

Being an orthogonal representation of the Herbrand universe H, the set of terms $\hat{T}(\mathcal{A}, H)$ defines a partition and thus an equivalence relation on H. We will see later on that these equivalence classes may serve as the domain of a finite model.

DEFINITION 3.59 Let H be a Herbrand universe, $\mathcal{A}$ be a linear atomic representation and $s, t \in H$. We define $s \sim t$ (w.r.t. $\hat{T}(\mathcal{A}, H)$) if there exists a $u \in \hat{T}(\mathcal{A}, H)$ with $u \leq_H s$ and $u \leq_H t$. ♯

LEMMA 3.60 *The relation $\sim$ defined in Definition 3.59 is an equivalence relation on H.*

Proof: Reflexivity:
It suffices to show that for every $s \in H$ there exists a $u \in \hat{T}(\mathcal{A}, H)$ with $u \leq_H s$. But this follows from the fact that $\hat{T}(\mathcal{A}, H)$ is a representation of H.

Symmetry: trivial.
Transitivity:
Let us assume that $s \sim t$ and $t \sim r$ for $s, t, r \in H$. Then there are terms $u, v \in \hat{T}(\mathcal{A}, H)$ with

$$u \leq_H s, u \leq_H t, \text{ and } v \leq_H t, v \leq_H r.$$

In particular we have $u \leq_H t$ and $v \leq_H t$, i.e. t is an instance of u *and* v. As $\hat{T}(\mathcal{A}, H)$ is orthogonal u must be equal to v. This, in turn, implies $u \leq_H s$ and $u \leq_H r$, i.e. $s \sim r$. ◇

EXAMPLE 3.61 The term set $\hat{T}(\mathcal{A}, H)$ in Example 3.58 defines 10 equivalence classes w.r.t. $\sim$. In particular $g(a, b) \sim g(f(a), a)$ but $f(a) \not\sim f(f(a))$. ♯

Because $\hat{T}(\mathcal{A}, H)$ is an orthogonal extension of the set of linear terms $T(\mathcal{A})$ occurring in the atomic representation $\mathcal{A}$ we get a kind of interpolation lemma:

LEMMA 3.62 *Let $\mathcal{A}$ be a linear ARM, $P(r_1, \ldots, r_n) \in \mathcal{A}$ and $P(r_1, \ldots, r_n) \leq_H P(t_1, \ldots, t_n)$ for a ground atom $P(t_1, \ldots, t_n)$ over the Herbrand universe H. Then there exist elements $u_1, \ldots, u_n \in \hat{T}(\mathcal{A}, H)$ s.t.*

$$P(r_1, \ldots, r_n) \leq_H P(u_1, \ldots, u_n) \leq_H P(t_1, \ldots, t_n).$$

Proof: $\hat{T}(\mathcal{A}, H)$ is an extension of $\mathcal{A} \cup BT_H$. Therefore there exist elements $u_1, \ldots, u_n \in \hat{T}(\mathcal{A}, H)$ with

$$r_1 \leq_H u_1 \leq_H t_1, \ldots, r_n \leq_H u_n \leq_H t_n.$$

We may assume without loss of generality that the u_i are pairwise variable disjoint. The r_j are pairwise variable disjoint because $\mathcal{A}$ is linear. Therefore there are substitutions θ_i with disjoint domains with $r_i\theta_i = u_i$ for $i = 1, \ldots, n$. But then

$$P(r_1, \ldots, r_n)(\theta_1 \cup \ldots \cup \theta_n) = P(u_1, \ldots, u_n)$$

and finally

$$P(r_1, \ldots, r_n) \leq_H P(u_1, \ldots, u_n) \leq_H P(t_1, \ldots, t_n).$$

◇

The following lemma shows that the semantics defined by the ARM $\mathcal{A}$ is invariant under $\sim$.

LEMMA 3.63 *Let $\mathcal{A}$ be a linear ARM and $P(s_1, \ldots, s_n)$ be a ground atom over a Herbrand universe H with $\mathcal{A} \leq_H P(s_1, \ldots, s_n)$. If $t_1, \ldots, t_n \in H$ with $s_i \sim t_i$ for $i = 1, \ldots, n$ w.r.t. $\hat{T}(\mathcal{A}, H)$ then $\mathcal{A} \leq_H P(t_1, \ldots, t_n)$.*

Proof: By Lemma 3.62 there exist $u_1, \ldots, u_n \in \hat{T}(\mathcal{A}, H)$ with

$$(*)\ \mathcal{A} \leq_H P(u_1, \ldots, u_n) \leq_H P(s_1, \ldots, s_n).$$

By $s_i \sim t_i$ there exist (variable disjoint) $v_1, \ldots, v_n \in \hat{T}(\mathcal{A}, H)$ with

$$v_1 \leq_H s_1, \ldots, v_n \leq_H s_n,$$
$$v_1 \leq_H t_1, \ldots, v_n \leq_H t_n.$$

Combining these last relations with $(*)$ we obtain $u_i \leq_H s_i$ and $v_i \leq_H s_i$ for $i = 1, \ldots, n$. By the orthogonality of $\hat{T}(\mathcal{A}, H)$ we obtain $u_i = v_i$ for $i = 1, \ldots, n$ and therefore

$$P(u_1, \ldots, u_n) \leq_H P(t_1, \ldots, t_n)$$

and so $\mathcal{A} \leq_H P(t_1, \ldots, t_n)$. ◇

The lemmas above prepare the ground for the definition of a finite interpretation based on a linear ARM $\mathcal{A}$. If s a is ground term in H then we write $[s] : \{t | t \sim s \text{ w.r.t. } \hat{T}(\mathcal{A}, H)\}$ for the equivalence class containing s. The set of equivalence classes is denoted by $H|_\sim$. Note that, as every element of $\hat{T}(\mathcal{A}, H)$ represents exactly one equivalence class, the set $H|_\sim$ is finite.

DEFINITION 3.64 Let $\mathcal{A}$ be a linear ARM w.r.t. a signature Σ and H be the set of ground terms over Σ. We define an interpretation $\mathcal{M}(\mathcal{A}, H) = (D, \Psi)$ in the following way:

1. $D = H|_\sim$,

2. $\Psi(c) = [c]$ for all $c \in \mathrm{CS}(\Sigma)$,

3. $\Psi(f)([t_1], \ldots, [t_n]) = [f(t_1, \ldots, t_n)]$ for all terms $t_1, \ldots, t_n \in H$ and $f \in \mathrm{FS}_n(\Sigma)$ and

4. $\Psi(P)([t_1], \ldots, [t_m]) = \mathbf{true}$ iff $\mathcal{A} \leq_H P(t_1, \ldots, t_m)$ for $P \in \mathrm{PS}_m(\Sigma)$ and terms $t_1, \ldots, t_m \in H$. ♯

Remark: Note that the interpretation $\mathcal{M}(\mathcal{A}, H)$ is well defined; in case of the definition of $\Psi(P)$ for $P \in \mathrm{PS}$ this is not completely trivial, but it follows from Lemma 3.63. ♯

EXAMPLE 3.65 Let

$$\mathcal{A} : \{P(x, f(f(y))), P(x, f(g(y))), Q(a)\}$$

be an ARM; it is evident, that $\mathcal{A}$ is linear. $\mathcal{A}$ defines a Herbrand model $\mathcal{M}$ of the set of clauses

$$\mathcal{C} : \{Q(x) \vee P(x, f(x)),\ \neg P(f(x), a)\}$$

over the Herbrand universe H defined by the set of ground terms over the signature $\Sigma : \{P, Q, a, f, g\}$. Indeed,

$$\mathcal{A} \not\leq_H \{P(f(x), a)\sigma\}$$

for all ground instances σ over Σ. Thus all ground instances $P(f(x), a)\sigma$ are false in $\mathcal{M}$ and so the clause $\neg P(f(x), a)$ is true in $\mathcal{M}$. In order to prove that $Q(x) \vee P(x, f(x))$ is true in $\mathcal{M}$ we consider the ground substitutions $\sigma(a) : \{x \leftarrow a\}$, $\sigma(f(t)) : \{x \leftarrow f(t)\}$, $\sigma(g(t)) : \{x \leftarrow g(t)\}$ for $t \in H$. Clearly $\sigma(a), \sigma(f(t)), \sigma(g(t))$ cover all ground substitution over H with domain $\{x\}$. The following relations are obviously valid for $C = Q(x) \vee P(x, f(x))$:

$$\begin{array}{rcl} \{Q(a)\} & \leq_H & C\sigma(a) \\ \{P(x, f(f(y)))\} & \leq_H & C\sigma(f(t)) \text{ for all } t \in H \\ \{P(x, f(g(y)))\} & \leq_H & C\sigma(g(t)) \text{ for all } t \in H \end{array}$$

This shows that all ground instances of C are true in $\mathcal{M}$.

Now we are going to construct $\mathcal{F} : (D, \Psi)$ for $\mathcal{F} = \mathcal{M}(\mathcal{A}, H)$. The set of all nonvariable subterms occurring in $\mathcal{A}$ is

$$T(\mathcal{A}) : \{a,\ f(y),\ f(f(y)),\ g(y),\ f(g(y))\}.$$

$T(\mathcal{A})$ contains all the base terms, i.e. $BT_H \subseteq T(\mathcal{A})$. Therefore $\hat{T}(\mathcal{A}, H)$ is defined as an orthogonal extension of $T(\mathcal{A})$ and

$$\hat{T}(\mathcal{A}, H) = \{a,\ f(a),\ f(f(y)),\ f(g(y)),\ g(y)\}.$$

By definition of the equivalence relation $\sim$ the terms in $\hat{T}(\mathcal{A}, H)$ represent the equivalence classes. Thus the domain of $\mathcal{F}$ is given by

$$D = \{[a],\ [f(a)],\ [f(f(a))],\ [f(g(a))],\ [g(a)]\}.$$

To save space we write 1 for $[a]$, 2 for $[f(a)]$, 3 for $[f(f(a))]$, 4 for $[f(g(a))]$ and 5 for $[g(a)]$.

By definition of $\mathcal{F}$ we have $\Psi(f)[t] = [f(t)]$ and $\Psi(g)(t) = [g(t)]$ for $t \in H$. This yields the function tables for f and g in $\mathcal{F}$:

D	$\Psi(f)$	$\Psi(g)$
1	2	5
2	3	5
3	3	5
4	3	5
5	4	5

As an example take $\Psi(f)(4)$: $[f(f(g(a)))] = [f(f(a))]$ as both terms match $f(f(y))$ and so $[f(f(g(a)))] = 3$. $\Psi(g)(i) = 5$ for all $i = 1, \ldots, 5$ because $[g(t)] = [g(a)]$ for all $t \in H$.

For the definition of $\psi(P)$ and $\Psi(Q)$ recall that $\Psi(P)([s],[t]) = \mathbf{true}$ iff $\mathcal{A} \leq_H P(s,t)$ for $s, t \in H$ and $\Psi(Q)([s]) = \mathbf{true}$ iff $\mathcal{A} \leq_H Q(s)$ for $s \in H$. This generates the following tables, where t stand for **true** and f stands for **false**:

$\Psi(P)$	1	2	3	4	5
1	f	f	t	t	f
2	f	f	t	t	f
3	f	f	t	t	f
4	f	f	t	t	f
5	f	f	t	t	f

$\Psi(P)$ shows a relatively uniform structure. Indeed, the third and the fourth column are always **true**, the other columns **false**. It is easy to see where this comes from: As $\mathcal{A} \not\leq_H P(t,a)$ for all $f \in H$, $\Psi(P)(i,1) = \mathbf{false}$ for all $i = 1, \ldots 5$. Similarly $\mathcal{A} \leq_H P(t, f(f(s)))$ for all $s, t \in H$ and so $\Psi(P)(i,3) = \mathbf{true}$ for all $i = 1, \ldots, 5$. For the instances with predicate symbol Q we have $\mathcal{A} \leq_H Q(a)$ and $\mathcal{A} \not\leq_H Q(t)$

for $t \in H - \{a\}$. Therefore we obtain the following table for $\Psi(Q)$:

$\Psi(Q)$	
1	t
2	f
3	f
4	f
5	f

It is easy to check that $\mathcal{F}$ is indeed a model of $\mathcal{C}$ (just perform all possible substitutions of the variables by $\{1, \ldots, 5\}$). Moreover, $\mathcal{F}$ is a finite realization of $\mathcal{M}$, i.e. $v_{\mathcal{M}}(A) = v_{\mathcal{F}}(A)$ for all ground atoms over Σ. As an illustration take the atom $A : P(f(a), f(g(f(a))))$. A is true in $\mathcal{M}$ by $\{P(x, f(g(y)))\} \leq_H A$. By definition of $\Psi(P)$ we have

$$\Psi(P)([f(a)], [f(g(f(a)))]) = \Psi(P)([f(a)], [f(g(a))]) = \Psi(P)(2, 4) = \mathbf{true}.$$

♯

It remains to prove in general that, given a linear atomic representation $\mathcal{A}$ of a Herbrand model $\mathcal{M}$ of a set of clauses $\mathcal{C}$, the finite interpretation $\mathcal{M}(\mathcal{A}, H)$ is a finite realization of $\mathcal{M}$. $\mathcal{M}(\mathcal{A}, H)$ is not only a model of $\mathcal{C}$, it even yields the same truth values as $\mathcal{M}$ on all clauses over the signature of $\mathcal{C}$. Moreover the specification of $\mathcal{M}(\mathcal{A}, H)$ is not based on search, but on the computation of $\hat{T}(\mathcal{A}, H)$ and then on the computation of $\Psi(f)$ for $f \in \mathrm{FS}$ and $\Psi(P)$ for $P \in \mathrm{PS}$ via $\sim$. We show first that $\mathcal{M}$ and $\mathcal{M}(\mathcal{A}, H)$ agree on ground atoms:

LEMMA 3.66 *Let $\mathcal{A}$ be a linear ARM w.r.t. a signature Σ and H be the Herbrand universe defined by Σ. Let $\mathcal{M}$ be a Herbrand model of a clause set $\mathcal{C}$ represented by $\mathcal{A}$ and $\mathcal{F} : \mathcal{M}(\mathcal{A}, H)$ be the corresponding finite interpretation. Then the following properties hold:*

(1) $t_{\mathcal{F}}(t) = [t]$ *for all* $t \in H$ *and*

(2) $v_{\mathcal{F}}(A) = v_{\mathcal{M}}(A)$ *for all ground atoms* A *over* Σ.

Proof: Let $\mathcal{M} = (H, \Phi)$ and $\mathcal{F} = (D, \Psi)$ for $D = D(\mathcal{A}, H)$ and $\Psi = \Psi(\mathcal{A}, H)$.

We prove (1) by induction on term depth:
$\tau(t) = 0$:
Then t is a constant symbol c and, by definition of $\mathcal{F}$, $t_{\mathcal{F}}(c) = \Psi(c) = [c]$.
(IH) Assume that (1) holds for all terms t with $\tau(t) \leq n$.
Let $t : f(t_1, \ldots, t_m)$ be a term of depth $n + 1$. Then

$$\begin{array}{rcl} t_{\mathcal{F}}(f(t_1, \ldots, t_n)) & = & \Psi(f)(t_{\mathcal{F}}(t_1), \ldots, t_f(t_m)) =_{(IH)} \\ \Psi(f)([t_1], \ldots, [t_n]) & = & [f(t_1, \ldots, t_n)] \end{array}$$

by definition of $\mathcal{F}$.

For (2) let $A = P(t_1, \ldots, t_k)$. By definition of $\mathcal{M}$ via $\mathcal{A}$, $v_{\mathcal{M}}(A) = \mathbf{true}$ iff $\mathcal{A} \leq_H P(t_1, \ldots, t_k)$. By definition of $\mathcal{F}$ from H and $\mathcal{A}$ the last relation holds iff $\Psi(P)([t_1], \ldots, [t_k]) = \mathbf{true}$.

But

$$\Psi(P)([t_1], \ldots, [t_k]) = \Psi(P)(t_{\mathcal{F}}(t_1), \ldots, t_{\mathcal{F}}(t_k))$$

by (1) and by the semantics of first-order logic

$$\Psi(P)(t_{\mathcal{F}}(t_1), \ldots, t_{\mathcal{F}}(t_k)) = v_{\mathcal{F}}(P(t_1, \ldots, t_k)).$$

This finally yields $v_{\mathcal{M}}(P(t_1, \ldots, t_k)) = v_{\mathcal{F}}(P(t_1, \ldots, t_k))$. ◇

COROLLARY 1.1 *Let $\mathcal{A}$ be a linear ARM w.r.t. a signature Σ and H be the Herbrand universe defined by Σ. Let $\mathcal{M}$ be a Herbrand model of a clause set $\mathcal{C}$ represented by $\mathcal{A}$ and $\mathcal{F} : \mathcal{M}(\mathcal{A}, H)$ be the corresponding finite interpretation. Then*

$$v_{\mathcal{F}}(C) = v_{\mathcal{M}}(C)$$

for all ground clauses over Σ.

Proof: Follows immediately from Lemma 3.66. ◇

Suppose we know that, for every model $\mathcal{M}$ of $\mathcal{C}$ represented by a linear ARM $\mathcal{A}$, $\mathcal{M}(\mathcal{A}, H)$ is also a model of $\mathcal{C}$. Then, by Lemma 3.66, $\mathcal{M}(\mathcal{A}, H)$ is a finite realization of $\mathcal{M}$ (indeed we know that the interpretations coincide on ground atoms!). It remains to show that Corollary 1.1 can be extended to clauses containing variables. To this aim we have to define a mapping between the variable assignments VA($\mathcal{M}$) of $\mathcal{M}$ and VA($\mathcal{M}(\mathcal{A}, H)$) of $\mathcal{M}(\mathcal{A}, H)$:

DEFINITION 3.67 Let $\mathcal{M}_I = (H, \Phi, I)$ be a V-instance of a Herbrand model $\mathcal{M}$ defined by a linear ARM $\mathcal{A}$ w.r.t. a signature Σ and a Herbrand universe H. Let $\mathcal{F} = \mathcal{M}(\mathcal{A}, H)$ and $I = \{(x_1, t_1), \ldots, (x_n, t_n)\}$ for $x_1, \ldots, x_n \in V$ and $t_1, \ldots, t_n \in H$. Let ϕ be a mapping VA($\mathcal{M}$) $\rightarrow$ VA($\mathcal{F}$) with

$$\phi(I) = \{(x_1, [t_1]), \ldots, (x_n, [t_n])\}.$$

Then the V-instance $\mathcal{F}_{\phi(I)} : (D, \Psi, \phi(I))$ of $\mathcal{F}$ is called the *canonical image* of $\mathcal{M}_I$ relative to $\sim$. ♯

LEMMA 3.68 *Let $\mathcal{M}_I = (H, \Phi, I)$ be a V-instance of a Herbrand model $\mathcal{M}$ defined by a linear ARM $\mathcal{A}$ w.r.t. a signature Σ and a Herbrand universe H. Let $\mathcal{F}_J = (D, \Psi, J)$ be the canonical image of $\mathcal{M}_I$ relative to $\sim$. Then*

$$v_{\mathcal{M}_I}(L) = v_{\mathcal{F}_J}(L)$$

for all literals L over Σ.

Proof: It is sufficient to show the lemma for all positive literals (i.e. for atoms). The extension to negative literals is trivial.

Let $A = P(s_1, \ldots, s_n)$ and $I = \{(x_1, t_1), \ldots, (x_m, t_m)\}$. Then

$$(I)\ v_{\mathcal{M}_I}(P(s_1, \ldots, s_n)) = \Phi(P)(t_{\mathcal{M}_I}(s_1), \ldots, t_{\mathcal{M}_I}(s_n))$$

and

$$(II)\ \Phi(P)(t_{\mathcal{M}_I}(s_1), \ldots, t_{\mathcal{M}_I}(s_n)) = \Phi(P)(s_1\eta, \ldots, s_n\eta)$$

for $\eta = \{x_1 \leftarrow t_1, \ldots, x_m \leftarrow t_m\}$.

As $\mathcal{F}_J$ is a canonical image of $\mathcal{M}_I$ we have

$$J = \{(x_1, [t_1]), \ldots, (x_m, [t_m])\}.$$

By Lemma 3.66 we have

$$(III)\ \Phi(P)(s_1\eta, \ldots, s_n\eta) = \Psi(P)([s_1\eta], \ldots, [s_n\eta]).$$

Moreover we will show below that

$$[s_i\eta] = t_{\mathcal{F}_J}(s_i) \text{ for } i = 1, \ldots, n$$

and therefore

$$\Psi(P)([s_1\eta], \ldots, [s_n\eta]) = \Psi(P)(t_{\mathcal{F}_J}(s_1), \ldots, t_{\mathcal{F}_J}(s_n) = $$
$$= v_{\mathcal{F}_J}(P(s_1, \ldots, s_n)).$$

By combining these last equations with (I), (II) and (III) we eventually get

$$v_{\mathcal{M}_I}(P(s_1, \ldots, s_n) = v_{\mathcal{F}_J}(P(s_1, \ldots, s_n)).$$

It remains to show

$$[s_i\eta] = t_{\mathcal{F}_J}(s_i) \text{ for } i = 1, \ldots, n:$$

We prove $t_{\mathcal{F}_J}(s) = [s\eta]$ for all terms s over the variables $\{x_1, \ldots, x_n\}$ by induction on term depth.

$\tau(s) = 0$:

Then $s = c$ for a constant symbol c and $t_{\mathcal{F}_J}(c) = [c] = [c\eta]$. If s is a variable then $s = x_i$ for some $x_i \in dom(\eta)$ and

$$t_{\mathcal{F}_J}(s) = t_{\mathcal{F}_J}(x_i) = J(x_i) = [t_i] = [x_i\eta] = [s\eta].$$

(IH) assume that the assertion holds for s with $\tau(s) \leq k$:

Thus let $\tau(s) = k + 1$. Then $s = f(r_1, \ldots, r_p)$ for terms $r_1, \ldots, r_p$ of depth $\leq k$ and an $f \in \mathrm{FS}$. But then

$$t_{\mathcal{F}_J}(f(r_1, \ldots, r_p)) = \Psi(f)(t_{\mathcal{F}_J}(r_1), \ldots, t_{\mathcal{F}_J}(r_p))$$

and, by (IH)

$$\Psi(f)(t_{\mathcal{F}_J}(r_1), \ldots, t_{\mathcal{F}_J}(r_p)) = \Psi(f)([r_1\eta], \ldots, [r_p\eta]).$$

By definition of Ψ we have

$$\Psi(f)([r_1\eta], \ldots, [r_p\eta]) = [f(r_1\eta, \ldots, r_p\eta)].$$

But

$$[f(r_1\eta, \ldots, r_p\eta)] = [f(r_1, \ldots, r_p)\eta] = [s\eta].$$

◇

COROLLARY 1.2 *Let $\mathcal{M}_I = (H, \Phi, I)$ be a V-instance of a Herbrand model $\mathcal{M}$ defined by a linear ARM $\mathcal{A}$ w.r.t. a signature Σ and a Herbrand universe H. Let $\mathcal{F}_J = (D, \Psi, J)$ be the canonical image of $\mathcal{M}_I$ relative to $\sim$. Then*

$$v_{\mathcal{M}_I}(C) = v_{\mathcal{F}_J}(C)$$

for all clauses C over Σ with $V(C) \subseteq dom(I)$.

Proof: Let $C = L_1 \vee \cdots \vee L_m$. Then

$$\begin{aligned} v_{\mathcal{M}_I}(C) &= or(v_{\mathcal{M}_I}(L_1), \ldots, v_{\mathcal{M}_I}(L_m)) \text{ and} \\ v_{\mathcal{F}_J}(C) &= or(v_{\mathcal{F}_J}(L_1), \ldots, v_{\mathcal{F}_J}(L_m)). \end{aligned}$$

Thus the result follows directly from Lemma 3.68. ◇

PROPOSITION 1.5 *Let $\mathcal{M} = (H, \Phi)$ be a Herbrand interpretation of a set of clauses $\mathcal{C}$ defined by a linear ARM $\mathcal{A}$ w.r.t. a signature Σ and a Herbrand universe H. Let $\mathcal{F} = \mathcal{M}(\mathcal{A}, H)$ the corresponding finite interpretation; then*

$$v_{\mathcal{M}}(\mathcal{C}) = v_{\mathcal{F}}(\mathcal{C}),$$

i.e. $\mathcal{F}$ is a model of $\mathcal{C}$ iff $\mathcal{M}$ is a model of $\mathcal{C}$.

Proof: It is sufficient to show that $v_{\mathcal{M}}(C) = v_{\mathcal{F}}(C)$ for all $C \in \mathcal{C}$.
By definition of the semantics of clause logic,

$v_{\mathcal{M}}(C) = \textbf{true}$ iff $v_{\mathcal{M}_I}(C) = \textbf{true}$

for all V-instances $\mathcal{M}_I$ of $\mathcal{M}$. A similar property holds for $v_{\mathcal{F}}(C)$.

(1) $v_{\mathcal{M}}(C) = \textbf{true} \;\Rightarrow\; v_{\mathcal{F}}(C) = \textbf{true}$.
We may assume that

$(*)$ $v_{\mathcal{M}_I}(C) = \textbf{true}$ for all $I \in \mathrm{VA}(\mathcal{M})$.

Now let $J \in \mathrm{VA}(\mathcal{F})$. Then

$$J = \{(x_1, [t_1]), \ldots, (x_n, [t_n])$$

for some variables x_i and terms $t_1, \ldots, t_n \in H$. Let

$$I = \{(x_1, t_1), \ldots, (x_n, t_n)\}$$

Then $v_{\mathcal{M}_I}(C) = \mathbf{true}$ by $(*)$. But $\mathcal{F}_J$ is the canonical image of $\mathcal{M}_I$ and thus, by Corollary 1.2, $v_{\mathcal{F}_J}(C) = \mathbf{true}$. As J is arbitrary, $v_{\mathcal{F}}(C) = \mathbf{true}$.

(2) $v_{\mathcal{F}}(C) = \mathbf{true} \Rightarrow v_{\mathcal{M}}(C) = \mathbf{true}$.
Assume that

$(**)$ $v_{\mathcal{F}_J}(C) = \mathbf{true}$ for all $J \in \mathrm{VA}(\mathcal{F})$.

Let $I : \{(x_1, t_1), \ldots, (x_n, t_n)\}$ be an assignment in $\mathrm{VA}(\mathcal{M})$. Then, by $(**)$ $v_{\mathcal{F}_J}(C) = \mathbf{true}$ for the canonical image of $\mathcal{M}_I$. But by Corollary 1.2

$$v_{\mathcal{M}_I}(C) = v_{\mathcal{F}_J}(C) = \mathbf{true}.$$

As this last equation holds for all $I \in \mathrm{VA}(\mathcal{M})$ we obtain $v_{\mathcal{M}}(C) = \mathbf{true}$. $\diamond$

By putting things together we eventually obtain

THEOREM 1.8 *Let $\mathcal{A}$ be a linear ARM of a model $\mathcal{M}$ of a set of clauses $\mathcal{C}$. Then $\mathcal{M}(\mathcal{A}, H)$ is a finite realization of $\mathcal{M}$.*

Proof: By Proposition 1.5, $\mathcal{M}(\mathcal{A}, H)$ is a model of $\mathcal{C}$. By Lemma 3.66 $v_{\mathcal{M}}(A) = v_{\mathcal{M}(\mathcal{A},H)}(A)$ for all ground atoms over the signature of $\mathcal{C}$. Therefore $\mathcal{M}(\mathcal{A}, H)$ is a finite realization of $\mathcal{M}$. $\diamond$

2. Model Building in Equational Clause Logic

2.1 Equational Atomic Representations of Models

Perhaps the most important predicate in mathematics is equality. In particular, any approach to automated deduction will be considered poor and incomplete, unless there is a way to handle equality. From the point of view of derivability equality is not a problematic predicate; in fact we may translate every theory (over a finite signature) containing equality into ordinary predicate logic just by adding a finite set of axioms, the so-called equality axioms. However, in automated deduction, such a translation is not really useful. First of all, deduction becomes highly inefficient, due to the blind handling of symmetry, transitivity and substitutivity. Second, termination of theorem provers is severely deteriorated by those additional axioms, making decision procedures and model building a hopeless task. In the next subsection we will present (ordered) paramodulation as the appropriate inference principle, which handles equational clause logic without equational axioms with the exception of reflexivity. But before we investigate the problems of inference, we look at the semantic expressivity of atomic representations in equational clause logic. In particular we are interested in the expressive power of ARMs when equational atoms are admitted.

DEFINITION 3.69 Let $\mathcal{C}$ be a set of clauses in equational clause logic. An interpretation $\mathcal{M} : (D, \Phi)$ of $\mathcal{C}$ is called an *equality interpretation* (E-interpretation) if $\Phi(\doteq)$ is the equality predicate on D. ♯

As we are mainly interested in Herbrand interpretations, we may ask whether the special interpretation of $\doteq$ interferes with the specific domain definition of Herbrand interpretations. The following example shows that this is indeed the case.

EXAMPLE 3.70 Let

$$\mathcal{C} = \{P(a, x),\ \neg P(x, y) \vee P(x, f(y)),\ a \doteq f(a)\}$$

and $\mathcal{M} : (H, \Phi)$ be a Herbrand model and an E-model of $\mathcal{C}$ over the signature of $\mathcal{C}$. Then, by definition of E-models, $\Phi(\doteq)$ is the equality predicate on H. In particular $\mathcal{M}$ must verify every clause in $\mathcal{C}$ and so we obtain $a = f(a)$ on H and, eventually, $a = f^n(a)$ for all $n \in \mathbb{N}$. Because the terms $f^n(a)$ define H, H consists of one element only. This, however, is incompatible with the definition of H as the infinite set $\{a, f(a), f(f(a)), \ldots\}$. Thus we see that no Herbrand model of $\mathcal{C}$ is also an E-model of $\mathcal{C}$. ♯

Example 3.70 illustrates that Herbrand domains and E-interpretations are incompatible concepts. As Herbrand domains are the most natural ones in the syntactic environment of automated deduction we have to look for a new type of equality interpretation slightly weaker than E-interpretations and admitting Herbrand domains.

DEFINITION 3.71 Let $\mathcal{C}$ be a set of equational clauses. An interpretation $\mathcal{M} : (D, \Phi)$ of $\mathcal{C}$ is called an *EH-interpretation* if the following conditions are fulfilled:

(1) $D = H(\mathcal{C})$,

(2) $\Phi(c) = c$ for all $c \in \mathrm{CS}(\mathcal{C})$
$\Phi(f)(t_1, \ldots, t_n) = f(t_1, \ldots, t_n)$ for all $f \in \mathrm{FS}_n(\mathcal{C})$ and $n \geq 1$.

(3) $\Phi(\doteq)$ is a reflexive, symmetric and transitive relation on $H \times H$.

(4) If $t_1 = f(s_1, \dots, s_n)$ and $t_2 = f(r_1, \dots, r_n)$ for two terms $t_1, t_2 \in H(\mathcal{C})$ and $s_i \doteq r_i$ is true in $\mathcal{M}$ for $i = 1, \dots, n$ then $t_1 \doteq t_2$ is true in $\mathcal{M}$.

(5) If $s_i \doteq r_i$ is true in $\mathcal{M}$ for terms $s_i, r_i \in H(\mathcal{C})$ and $i = 1, \dots, n$ and P is an n-place predicate symbol then $v_{\mathcal{M}}(P(s_1, \dots, s_n)) = v_{\mathcal{M}}(P(r_1, \dots, r_n))$.

The properties (4) and (5) are called the substitution properties of an EH-interpretation. An EH-interpretation which is a model of $\mathcal{C}$ is called an *EH-model* of $\mathcal{C}$. ♯

EXAMPLE 3.72 Let

$$\mathcal{C} = \{P(a, x),\ \neg P(x, y) \vee P(x, f(y)),\ a \doteq f(a)\}$$

as in Example 3.70. Then the interpretation $\mathcal{M} : (H, \Phi)$ defined below is an EH-model of $\mathcal{C}$:
$H = \{f^n(a) | n \in \mathbb{N}\}$, $\Phi(P)(s, t) =$ **true** for all $s, t, \in H$ and $\Phi(\doteq)(s, t) =$ **true** for all $s, t \in H$. It is easy to see that, under these settings, $\mathcal{M}$ is indeed an EH-model and even is the only one. Indeed, as every EH-model $\mathcal{N}$ must verify $a \doteq f(a)$ it also verifies $f(a) \doteq f(f(a))$ by the substitution property (4) and $a \doteq f(f(a))$ by transitivity. By continuing along this line we obtain $v_{\mathcal{N}}(a \doteq f^n(a)) =$ **true** for all $n \geq 0$ and, by use of symmetry and transitivity, $v_{\mathcal{N}}(f^m(a) \doteq f^n(a)) =$ **true** for all $n, m \in \mathbb{N}$. $\mathcal{N}$ must also verify the first clause, i.e. $v_{\mathcal{N}}(P(a, s)) =$ **true** for all $s \in H$. But then the substitution property (5) yields $v_{\mathcal{N}}(P(s, t)) =$ **true** for all $s, t \in H$. This finally gives $\mathcal{N} = \mathcal{M}$. ♯

Obviously E-interpretations and EH-interpretations are closely related. Thus it is not surprising that there exist a simple transformation mapping EH-interpretations into E-interpretations. The only thing we have to do is to factorize the interpretation w.r.t. equivalence classes defined by $\Phi(\doteq)$ (note that, by property (3) in Definition 3.71, $\doteq$ is interpreted as equivalence relation).

DEFINITION 3.73 Let $\mathcal{C}$ be a set of equational clauses and let $\mathcal{M} : (H, \Phi)$ be an EH-interpretation of $\mathcal{C}$. We define $s \sim t$ for terms $s, t \in H$ if $v_{\mathcal{M}}(s \doteq t) =$ **true**; then $\sim$ is an equivalence relation and we denote the equivalence class containing s by $[s]$. Moreover we define an equality interpretation $\mathcal{M}^* = (D, \Phi^*)$ in the following way:

(1) $D = H|_{\sim}$ (i.e. $D = \{[s] | s \in H\}$).

(2) $\Phi^*(f)([t_1], \dots, [t_n]) = [f(t_1, \dots, t_n)]$ for $f \in \mathrm{FS}_n(\mathcal{C})$ and $t_1, \dots, t_n \in H$, and $\Phi^*(c) = [c]$ for constant symbols $c \in \mathrm{CS}(\mathcal{C})$.

(3) $\Phi^*(P)([t_1],\ldots,[t_n]) = \Phi(P)(t_1,\ldots,t_n)$ for $P \in \mathrm{PS}_n(\mathcal{C})$ and $t_1,\ldots,t_n \in H$.

$\mathcal{M}^*$ is called the E-interpretation corresponding to $\mathcal{M}$. ♯

Note that there might be an ambiguity in point (3) of Definition 3.73. Indeed, $\Phi(P)$ for a predicate symbol P is only well-defined if the values of $\Phi(P)$ are invariant on equivalence classes. Thus we have to show:

LEMMA 3.74 *Let $\mathcal{M} : (H, \Phi)$ be an EH-interpretation of a set of clauses $\mathcal{C}$ and let P be an n-place predicate symbol in the signature of $\mathcal{C}$. If $s_i \sim t_i$ for $s_i, t_i \in H$ and $i = 1,\ldots,n$ then $\Phi(P)(s_1,\ldots,s_n) = \Phi(P)(t_1,\ldots,t_n)$.*

Proof: By definition of $\sim$ we have $v_{\mathcal{M}}(s_i \doteq t_i) = \mathbf{true}$ for $i = 1,\ldots,n$. But then, by (5) in Definition 3.71

$$v_{\mathcal{M}}(P(s_1,\ldots,s_n)) = v_{\mathcal{M}}(P(t_1,\ldots,t_n)).$$

But

$$\begin{aligned} v_{\mathcal{M}}(P(s_1,\ldots,s_n)) &= \Phi(P)(s_1,\ldots,s_n) \text{ and} \\ v_{\mathcal{M}}(P(t_1,\ldots,t_n)) &= \Phi(P)(t_1,\ldots,t_n) \end{aligned}$$

by $t_{\mathcal{M}}(s) = s$ for all $s \in H$ (which, in turn, follows from (2) in Definition 3.71). ◇

COROLLARY 2.1 *$\Phi^*(P)$ for predicate symbols P in Definition 3.73 is well-defined.*

Proof: Trivial by Lemma 3.74. ◇

EH-interpretations are elementarily equivalent to their corresponding E-interpretations, i.e. they yield the same truth values on clauses over the signature of the corresponding set of clauses.

PROPOSITION 2.1 *Let $\mathcal{C}$ be a set of equational clauses, $\mathcal{M}$ be an EH-interpretation of $\mathcal{C}$ and $\mathcal{M}^*$ be the corresponding E-interpretation. Then $v_{\mathcal{M}}(C) = v_{\mathcal{M}^*}(C)$ for all clauses C over the signature of $\mathcal{C}$.*

Proof: Let $P(t_1,\ldots,t_n)$ be a ground atom over $H(\mathcal{C})$. Then by Definition 3.73

$$\Phi(P)(t_1,\ldots,t_n) = \Phi^*(P)([t_1],\ldots,[t_n])$$

and, particularly, $t_{\mathcal{M}^*}(t_i) = [t_i]$ by (2) in Definition 3.73. This gives us

$$v_{\mathcal{M}}(P(t_1,\ldots,t_n)) = v_{\mathcal{M}^*}(P(t_1,\ldots,t_n)).$$

That $v_{\mathcal{M}}(C) = v_{\mathcal{M}^*}(C)$ for ground clauses C is thus straightforward. The extension to clauses containing variables can be achieved by canonic images (see Definition 3.67). $\diamond$

We are now in the position to extend atomic representations of models to the equational case.

DEFINITION 3.75 Let $\mathcal{A}$ be a finite set of atoms (some of them may be equational atoms) and let H bc a Herbrand universe containing $H(\mathcal{A})$. Then $\mathcal{A}$ is called equational atomic representation (abbreviated by EARM) of an EH-interpretation $\mathcal{M}(\mathcal{A}) : (H, \Phi)$ defined by

$$\Phi(P)(t_1, \ldots, t_n) = \textbf{true} \text{ iff } \mathcal{A} \models P(t_1, \ldots, t_n)$$

for ground atoms $P(t_1, \ldots, t_n)$, i.e. iff $P(t_1, \ldots, t_n)$ is true in all E-models of $\mathcal{A}$. $\sharp$

Although it is trivial that $\mathcal{M}(\mathcal{A})$ defined above is indeed an EH-interpretation, it is much less trivial, however, to define an algorithm for evaluating $\Phi(P)$ over H. In fact EARMs are too strong to admit even decision procedures for the truth values of ground atoms. This is based on the undecidability of the word problem in general equational theories. Thus unrestricted EARMS are too expressive to be of computational value. For this reason we have to find some reasonable restrictions, which admit algorithmic clause evaluation and equivalence tests, but are sufficiently expressive too.

DEFINITION 3.76 An EARM $\mathcal{A}$ containing only ground atoms is called a ground equational atomic representation (or, for short, *GEARM*). $\sharp$

GEARMs admit algorithmic clause evaluation and equivalence tests (see Chapter 2). But, in the first moment, GEARMs appear too weak a representation tool to be of major importance. In absence of equational atoms this would be indeed the case; but we will see below that the strong expressivity of equality adds considerable weight – even to ground representations.

PROPOSITION 2.2 *There are Herbrand models representable by GEARMs which are not representable by ARMs.*

Proof: Let

$$\mathcal{C} = \{P(x) \vee P(f(x)),\ \neg P(x) \vee \neg P(f(x))\}.$$

$\mathcal{C}$ has two Herbrand models $\mathcal{M}_1$ and $\mathcal{M}_2$; in $\mathcal{M}_1$ $P(f^n(a))$ is true iff n is even, and in $\mathcal{M}_2$ $P(f^n(a))$ is true iff n is odd.

Neither $\mathcal{M}_1$ nor $\mathcal{M}_2$ are representable by ARMs, as is easily shown below:
consider an ARM $\mathcal{A}$ over the signature of $\mathcal{C}$. Then $\mathcal{A} = \{P(t_1), \dots, P(t_n)\}$ for terms $t_1, \dots, t_n$ over the signature $\{f, a\}$ for some constant symbol a. If all terms t_i are ground and $n = \max\{m | P(f^m(a)) \in \mathcal{A}\}$ then, clearly, for all $k > n$ $P(f^k(a))$ is false in the Herbrand model $\mathcal{N}$ represented by $\mathcal{A}$. Thus $\mathcal{N}$ is different from $\mathcal{M}_1$ and $\mathcal{M}_2$ and falsifies the first clause in $\mathcal{C}$.

Now let us assume that $P(t) \in \mathcal{A}$ where t is a nonground term. Then $t = f^n(x)$ for some $n \geq 0$. If $\mathcal{N}$ is the model represented by $\mathcal{A}$ then, according to the definition of ARM-representation, $P(f^m(a))$ is true in $\mathcal{N}$ for all $m \geq n$. But then $\mathcal{N}$ falsifies the second clause in $\mathcal{C}$ and thus is not a model of $\mathcal{C}$. Putting things together we see that all models defined by ARMs falsify $\mathcal{C}$.

We are going to show that $\mathcal{M}_1$ and $\mathcal{M}_2$ are representable by GEARMs. To this aim consider the GEARM $\mathcal{A} : \{P(a), f(f(a)) \doteq a\}$ and let $\mathcal{M}(\mathcal{A})$ be the corresponding EH-interpretation according to Definition 3.75. Then, by the substitution axioms of equality, $f^{2n}(a) \doteq a$ is true in $\mathcal{M}(\mathcal{A})$ and so $\mathcal{A} \models f^{2n}(a) \doteq a$. By the substitution property (5) in Definition 3.71 and by $\mathcal{A} \models P(a)$ we also get $\mathcal{A} \models P(f^{2n}(a))$ and the atoms $P(f^{2n}(a))$ are true in $\mathcal{M}(\mathcal{A})$ for $n \in \mathbb{N}$. It remains to show that $\mathcal{A} \not\models P(f^{2n+1}(a))$, i.e. that the ground atoms $P(f^{2n+1}(a))$ are false in $\mathcal{M}(\mathcal{A})$. For this purpose we define an E-interpretation $\mathcal{N} : (D, \Phi)$ with $D = \{\lambda, \beta\}$, $\Phi(f) = \phi$, where $\phi(\lambda) = \beta$ and $\phi(\beta) = \lambda$; moreover we define $\Phi(a) = \lambda$, $\Phi(P)(\lambda) = \mathbf{true}$ and $\Phi(P)(\beta) = \mathbf{false}$. Then $\mathcal{N}$ verifies $\mathcal{A}$ but falsifies the atoms $P(f^{2n+1}(a))$ for all $n \in \mathbb{N}$. But this means that $\mathcal{A} \not\models P(f^{2n+1}(a))$ and $v_{\mathcal{M}(\mathcal{A})}(P(f^{2n+1}(a)) = \mathbf{false}$ for all $n \in \mathbb{N}$. Eventually we see that $\mathcal{A}$ represents $\mathcal{M}_1$.

It can be shown in the same way that the GEARM $\mathcal{A}' : \{P(f(a)), f(f(a)) \doteq a\}$ represents the model $\mathcal{M}_2$. ◇

Now let us determine the E-models corresponding to the EH-models represented by $\mathcal{A}$ and by $\mathcal{A}'$ in the proof of Proposition 2.2. By the equation $f(f(a)) \doteq a$ there are only two equivalence classes induced by $\sim$, namely $[a]$ and $[f(a)]$. Therefore the equational model $\mathcal{M}^*(\mathcal{A})$ is defined by

$$\Phi^*(P)([a]) = \mathbf{true}, \qquad \Phi^*(P)([f(a)]) = \mathbf{false},$$
$$\Phi^*(f)([a]) = [f(a)] \qquad \Phi^*(f)([f(a)]) = [a],$$

and $\Phi^*(a) = [a]$. So $\mathcal{M}^*(\mathcal{A})$ is a model with two elements only. Note that $\mathcal{M}^*(\mathcal{A})$ is isomorphic to the model $\mathcal{N}$ constructed in the proof of

Proposition 2.2 (where we proved $\mathcal{A} \not\models P(f^{2n+1}(a)))$; indeed λ corresponds to $[a]$ and β to $[f(a)]$.

We will demonstrate later that all sets of clauses representable by GEARMs also have finite models. By Proposition 1.1 there exist clause sets with only infinite models, some of them representable by ARMs. Thus there are also models representable by ARMs, but not representable by GEARMs. The two different representation formalisms are in fact incomparable.

Even for a GEARM $\mathcal{A}$, the problem to decide whether $\mathcal{A} \models B$ for ground atoms B is not completely trivial. Note that, in contrast to ground ARMs, $\mathcal{A} \models B$ may hold for infinitely many atoms B (like in the example in the proof of Proposition 2.2) and similarly for $\mathcal{A} \not\models B$. We will defer this problem to the next section where $\models$ is formalized by equational inference systems.

2.2 An Inference System for Equational Model Building

Because of the central importance of equality in mathematics and computer science, specific equational inference techniques (demodulation and paramodulation) were developed already in the sixties [Robinson and Wos, 1969]. Perhaps the most significant feature of equality is the law of substitutivity

$$(x_1 \doteq y_1 \land \ldots \land x_n \doteq y_n) \rightarrow f(x_1, \ldots, x_n) \doteq f(y_1, \ldots, y_n)$$

for n-place function symbols f. This law admits replacement of terms by others on *any* position of a term within an atom. The law can be "simulated" by a rule of the form

$$\frac{s \doteq t \quad E[s]_\alpha}{E[t]_\alpha}$$

where E is an arbitrary expression containing the term s on position α. We obtain *paramodulation* by combining this rule with the principle of most general unification. Because paramodulation can be applied to any position in a literal it is a very fertile rule and creates a huge search space; therefore the development of refinements became a task of major importance. One of the oldest refinements is hyperparamodulation, a combination of hyperresolution and paramodulation. Below we present a slightly weaker version, namely positive paramodulation.

DEFINITION 3.77 Let $C' : C_1 \lor s \doteq t \lor C_2$ be a factor of a positive clause C and let $D' : D_1 \lor L[u]_\alpha \lor D_2$ be a variant of an arbitrary clause

D, where L is a (possibly equational) literal and C' and D' are variable disjoint. Moreover let σ be a most general unifier of $\{s, u\}$. Then the clause

$$(C_1 \vee C_2 \vee D_1 \vee L[t]_\alpha \vee D_2)\sigma$$

is called *P-paramodulant* of C and D ("P" stands for "positive"). C is called the *active* and D the *passive* clause of the paramodulation. ♯

In case of Horn logic P-paramodulation is of a specific simple form: the active clause is always an equation $s \doteq t$, and the passive clause $D_1 \vee L[u]_\alpha \vee D_2$ transforms to $(D_1 \vee L[t]_\alpha \vee D_2)\sigma$.

Definition 3.77 only leads to a complete inference system if we observe the symmetry of equality. That means we have also to consider the atom $t \doteq s$ and not only $s \doteq t$. This leads to a further increase of the search space and typically also to nontermination of the inference system. Beginning with the paper of Knuth and Bendix [Knuth and Bendix, 1970] orientation of equality became a major issue in equational reasoning, leading to a discipline on its own best known as term rewriting. The following example shows the importance of orienting equality. Indeed, P-paramodulation does not even terminate on equational ground clause logic, making it an inadequate tool for automated model building.

EXAMPLE 3.78 Let $\mathcal{C} = \{a \doteq f(a), P(a), \neg P(b)\}$.

Clearly $\mathcal{C}$ is E-satisfiable and $\{a \doteq f(a), P(a)\}$ is a GEARM of an EH-model of $\mathcal{C}$. However this fact cannot be "detected" by paramodulation: note that all literals $P(f^n(a))$ (for $n \in \mathbb{N}$) are derivable by P-paramodulation! Indeed $P(f(a))$ is a P-paramodulant of $a \doteq f(a)$ and $P(a)$, $P(f(f(a)))$ of $a \doteq f(a)$ and $P(f(a))$ and so on.... Moreover, by applying $a \doteq f(a)$ to itself, we can derive all equations $a \doteq f^n(a)$ for $n \geq 1$. Thus P-paramodulation does not terminate on $\mathcal{C}$, a set of ground unit clauses! Note that resolution is not applicable to $\mathcal{C}$. ♯

Example 3.78 shows that, unlike resolution on ordinary ground clause logic, P-paramodulation does not terminate on equational ground clause logic. But note that, by the law of substitutivity, equational ground clause logic does not correspond to propositional logic! In order to achieve termination on ground clause logic we have to get rid of symmetry in equality. In Example 3.78 we obtain infinitely many P-paramodulants by replacing the term a by the more complex term $f(a)$ again and again. Thus the goal must be to *simplify* expressions rather than extending them. This idea of simplification was already the driving force behind the demodulation principle developed by L. Wos in the sixties; demodulation, though quite useful in automated deduction, was used as heuristic principle only and questions of completess and termination remained largely unsolved.

In 1970 Knuth and Bendix presented the method of completion for pure equational logic, a major breakthrough in equational reasoning. In this method equations are ordered according to a term ordering and the loss of symmetry is compensated by systematically adding new oriented equations and using them as rewrite rules on the whole system. This method, however, is not directly applicable to more complex logical formalisms like equational clause logic; moreover it only works if all present and newly generated equations are indeed orientable via a term ordering. A full inference system for equational clause logic based on orderings (for resolution and paramodulation) was given by Hsiang and Rusinowitch [Hsiang and Rusinowitch, 1991]); stronger refinements, based on the superposition calculus, were defined by Bachmair and Ganzinger [Bachmair and Ganzinger, 1994]. For the purpose of model building, the refinement of Hsiang and Rusinowitch is even a little too strict (we will see later that ordered resolution may be destructive to model building procedures, while ordered paramodulation is absolutely necessary). The key concept in the definition of ordered paramodulation is that of complete simplification ordering.

DEFINITION 3.79 A partial ordering $\prec$ on the set of expressions over a finite signature is called a *complete simplification ordering* (CSO) if it satisfies the following conditions:

(O1) $\prec$ is well-founded,

(O2) $\prec$ is total on ground expressions,

(O3) For all expressions M, N and substitutions θ, $M \prec N$ implies $M\theta \prec N\theta$ (*lifting property* of $\prec$).

(O4) For all terms s, t and expressions M, $s \prec t$ implies $M[s]_\alpha \prec M[t]_\alpha$ for any position α in M.

(O5) For all terms s, t, r, q with $t \preceq s$ (where $\preceq$ stands for $t = s$ or $t \prec s$), for all expressions M and all atoms A we have

1. If s is a proper subterm of M then $s \prec M$.
2. If s is a subterm of A, where A is not an equality atom, then $(s \doteq t) \prec A$.
3. If s is a proper subterm of r or of q then $(s \doteq t) \prec (r \doteq q)$. ♯

EXAMPLE 3.80 Let $\Sigma = \{P, f, a\}$ where $P \in \mathrm{PS}_1$, $f \in \mathrm{FS}_1$ and $a \in \mathrm{CS}$. We define $\prec$ as follows:

(1) For all $s, t \in T(\Sigma)$: $s \prec t$ iff s is a proper subterm of t.

(2) $P(s) \prec P(t)$ iff $s \prec t$ (for $s, t \in T(\Sigma)$).

(3) $s \prec P(t)$ iff $s \preceq t$ (for $s, t \in T(\Sigma)$).

(4) $s \doteq t \prec P(r)$ iff $s \prec r$ (for $r, s, t \in T(\Sigma)$).

(5) If $s \prec r$ or $s \prec q$ then $s \doteq t \prec r \doteq q$ (for $s, t, r, q \in T(\Sigma)$).

(6) $\prec$ is invariant under negation of atoms.

By (1) $\prec$ is well-founded and thus (O1) is fulfilled. Moreover $\prec$ is total on ground terms in $T(\Sigma)$: for $f^n(a), f^m(a)$ and $n \neq m$ either $f^m(a)$ is a proper subterm of $f^n(a)$ or vice versa. This gives (O2).

For (O3) consider two terms expressions M, N with $M \prec N$. We have to show $M\theta \prec N\theta$ for all substitutions θ. By (2) to (6) it is sufficient to show the lifting properties for terms, i.e. whenever $s \prec t$ for terms s and t then $s\theta \prec t\theta$ for all substitutions θ. If s and t are both ground then the lifting property trivially holds. If s and t are not both ground then both are nonground (note that, as the only function symbol is unary, s cannot be a subterm of t if one of the terms is ground and the other nonground!). Then s, t must be of the form $s = f^n(x)$ and $t = f^m(x)$ for $x \in V$ and $n < m$; clearly $s\theta \prec t\theta$ for all substitutions θ.

(O4) follows from (1), (2) and (6). (O5)-1 is fulfilled because of (3) and (6). (4) and (5) directly correspond to (O5)-2 and (O5)-3.

Note that $\prec$, as it is defined above, cannot be carried over to the signature $\Sigma' = \{P, f, a, b\}$ for an additional constant symbol b; indeed the ordering obtained this way would not be total on ground expressions. Note that a and b are not in a subterm relation, thus neither $a \prec b$ nor $b \prec a$ and (O2) is violated. In order to obtain a CSO, the definition of $\prec$ has to be extended to a relation $\ll$.

To this aim we first define $a \ll b$. Then (O4) enforces $f(a) \ll f(b)$ and generally $f^n(a) \ll f^n(b)$. By $a \ll f(a)$ (we already have $a \prec f(a)$) we also get $f^m(a) \ll f^n(b)$ for $m \leq n$. Thus, for defining a CSO $\ll$ we have to replace (1) by the following condition (1)':

$s \ll t$ iff one of the following conditions hold

(a) $s \prec t$ or

(b) $s = f^m(a), t = f^n(b)$ for $m \leq n$ or

(c) $s = f^m(b), t = f^n(a)$ for $m < n$.

♯

We are ready now to define the rule of PO-paramodulation, i.e. positive paramodulation under an ordering restriction based on CSOs. From now on we write $\prec$ for an arbitrary CSO, which we do not specify in more detail.

DEFINITION 3.81 Let $C' = C_1 \vee s \doteq t \vee C_2$ be a factor of a positive clause C and $D' = D_1 \vee L[u]_\alpha \vee D_2$ be a variant of a clause D s.t. C' and D' are variable disjoint. Moreover assume that the term u (on position α in the literal L) is not a variable. Let σ be a most general unifier of $\{s, u\}$. If $s\sigma \not\preceq t\sigma$ (i.e. neither $s\sigma \prec t\sigma$ nor $s\sigma = t\sigma$) then the P-paramodulant

$$(C_1 \vee C_2 \vee L[t]_\alpha \vee D_2)\sigma$$

is called a PO-paramodulant of C and D (w.r.t. $\prec$). ♯

Our basic inference system is defined by operators in the same way as for hyperresolution in Definition 3.1. This time we have to combine two inference principles, resolution and paramodulation.

DEFINITION 3.82 Let $\mathcal{C}$ be a set of equational clauses and $\prec$ be a complete simplification ordering. We define $OP_\prec(\mathcal{C})$ as the set of all condensed paramodulants definable from $\mathcal{C}$ and $\varrho_H(\mathcal{C})$ as the set of all condensed resolvents definable by hyperresolution from $\mathcal{C}$. Let

$$X(\mathcal{C}) = OP_\prec(\mathcal{C}) \cup \varrho_H(\mathcal{C}) \cup \mathcal{C} \cup \{x \doteq x\}.$$

Then X is called a *PROP-operator* (**p**ositive hyper-**r**esolution and **o**rdered **p**aramodulation) based on $\prec$. We define the deductive closure of X by

$$X^*(\mathcal{C}) = \bigcup_{i \in \mathbb{N}} X^i(\mathcal{C}).$$

♯

Remark: The addition of the reflexivity clause $\{x \doteq x\}$ in the definition of a PROP-operator is necessary in order to obtain completeness. In fact the empty clause can only be derived by resolution! ♯

EXAMPLE 3.83 Let

$$\mathcal{C} = \{\neg f(a) \doteq f(b), a \doteq b\}.$$

Clearly $\mathcal{C}$ is E-unsatisfiable. Assume that $a \prec b$ in a CSO $\prec$. Then the only ordered paramodulants definable from $\mathcal{C}$ are $\neg f(a) \doteq f(a)$ and $a \doteq a$; note that the second paramodulant comes from paramodulation of $a \doteq b$ on itself (we have no ordering restrictions on the passive clauses!).

Clearly the new equational clause is unsuited for ordered paramodulation and no further paramodulation rules are applicable to the new negative clause. Resolution is neither applicable to $\mathcal{C}$ nor to $\mathcal{D}$ for

$$\mathcal{D} = \mathcal{C} \cup \{a \doteq a, \neg f(a) \doteq f(a)\}.$$

Therefore, unless we add the clause $x \doteq x$, the deductice closure is $\mathcal{D}$ and our inference system is incomplete. Clearly we obtain the empty clause by resolving $x \doteq x$ and $\neg f(a) \doteq f(a)$ and $\Box \in X^*(\mathcal{C})$ for the PROP-operator X based on $\prec$. ♯

As in Section 1 subsumption (in form of replacement) will play a major role in the model building process. Here we give a slightly different definition of the replacement operator which is based on prior computation of the deduction levels $X^i(\mathcal{C})$. This is surely not optimal in implementations but it makes the theoretical analysis more transparent.

DEFINITION 3.84 Let X be a PROP-operator as in Definition 3.82. Then we define $X_s^i(\mathcal{C}) = sub(X^i(\mathcal{C}))$ for $i \geq 0$. The *replacement closure* of $\mathcal{C}$ under X_s is defined by

$$X_s^*(\mathcal{C}) = \bigcup_{i \geq 0} \bigcap_{j \geq i} X_s^j(\mathcal{C}).$$

♯

Although the computation of X_s^* is problematic in general, it is easy if the deductive closure under X is finite (i.e. if X converges on a set of clauses).

PROPOSITION 2.3 *Let $\mathcal{C}$ be a set of equational clauses and X be a PROP-operator s.t. $X^*(\mathcal{C})$ is finite. Then $X_s^*(\mathcal{C}) = sub(X^*(\mathcal{C}))$.*

Proof: By definition of the deductive closure of X and by finiteness of $X^*(\mathcal{C})$ there exists a number i with $X^*(\mathcal{C}) = X^i(\mathcal{C})$ and $X^j(\mathcal{C}) = X^i(\mathcal{C})$ for $j \geq i$. But then, by Definition 3.84,

$$X_s^i(\mathcal{C}) = sub(X^i(\mathcal{C})) \text{ and } X_s^j(\mathcal{C}) = X_s^i(\mathcal{C})$$

for all $j \geq i$. In particular $\bigcap_{j \geq k} X_s^j(\mathcal{C}) = X_s^i(\mathcal{C})$ for all $k \geq i$. Therefore, by definition of X_s^*,

$$X_s^*(\mathcal{C}) = X_s^i(\mathcal{C}) = sub(X^i(\mathcal{C})) = sub(X^*(\mathcal{C})).$$

◇

EXAMPLE 3.85 Let

$$\mathcal{C} = \{a \doteq f(a),\ P(a),\ \neg P(b)\}$$

as in Example 3.78. We have seen that there are infinitely many paramodulants definable by the first two clauses. Now let $\prec$ be an arbitrary CSO based on the signature of $\mathcal{C}$ and let X be the PROP-operator based on $\prec$. Then, by Definition 3.82,

$$X(\mathcal{C}) = OP_{\prec}(\mathcal{C}) \cup \varrho_H(\mathcal{C}) \cup \mathcal{C} \cup \{x \doteq x\}.$$

Clearly $\varrho_H(\mathcal{C}) = \emptyset$ (there are no resolvents at all definable from $\mathcal{C}$). But, as $a \prec f(a)$, the paramodulant $P(a)$ is blocked by the ordering condition and so $OP_{\prec}(\mathcal{C}) = \emptyset$. This gives

$$X(\mathcal{C}) = \mathcal{C} \cup \{x \doteq x\}.$$

But the clause $\{x \doteq x\}$ can neither be used for ordered paramodulation (by principle) and does not yield any resolvents either. Therefore $X^*(\mathcal{C}) = X(\mathcal{C})$ and also $X_s^*(\mathcal{C}) = X(\mathcal{C})$. So we obtain termination of the operators X and X_s on $\mathcal{C}$. ♯

The clause set $\mathcal{C}$ in the last example is E-satisfiable and $X_s^*(\mathcal{C}) = \mathcal{C} \cup \{x \doteq x\}$. But note that the computation of X_s^* can only be considered as a proof of satisfiability of $\mathcal{C}$ if we already know that X and Xs are complete operators. But this is indeed the case.

On one hand, the PROP-operators define a less restricted deduction principle than this in the paper of Hsiang and Rusinowitch [Hsiang and Rusinowitch, 1991], where ordering is used in resolution too. On the other hand, the hyperresolution rule is used instead of positive resolution, which may reduce the number of potential inferences. As we shall see, using positive resolution instead of hyperresolution may lead to non-termination in some cases.

Since PROP-operators are not comparable with existing resolution refinements, we have to provide a new completeness proof for them.

PROPOSITION 2.4 *Let X be a PROP operator. Then X and Xs are complete, i.e. for any E-unsatisfiable set of equational clauses $\mathcal{C}$ we have $\square \in X^*(\mathcal{C})$ and $X_s^*(\mathcal{C}) = \{\square\}$.*

Proof:

We first prove the refutational completeness of X.

Let S be a set of clauses such that $X(S) = S$. Assume that S does not contain the empty clause. We show that S is satisfiable.

Let $\mathcal{B}$ be the set of ground atoms built on the signature Σ. Note that $(\mathcal{B}, \prec)$ is isomorphic to an ordinal λ. For any ordinal $\alpha \in \lambda$, we denote by A_α the atom corresponding to α and by $\mathcal{B}_\alpha$ the set of atoms that are smaller than A_α.

We construct - by transfinite induction - a sequence of functions $(\mathcal{I}_\alpha)_{\alpha<\lambda}$ from $\{L \mid L < \mathcal{A}_\alpha\}$ into $\{\mathbf{true}, \mathbf{false}\}$ as follows.

- **Base case.** If $\alpha = 0$, then $\mathcal{B}_\alpha = \emptyset$ hence the domain of $\mathcal{I}_\alpha$ is empty.
- **Limit ordinal.** If α is a limit ordinal, then we define the function $\mathcal{I}_\alpha$ as follows: for any $\beta < \alpha$, $\mathcal{I}_\alpha(A_\beta) = \mathcal{I}_{\beta+1}(A_\beta)$.
- **Successor ordinal.** If $\alpha = \alpha' + 1$, then $\mathcal{I}_\alpha$ is defined as follows: For all $\beta < \alpha'$, we define $\mathcal{I}_\alpha(A_\beta) = \mathcal{I}_{\alpha'}(A_\beta)$. Moreover, $\mathcal{I}_\alpha(A_{\alpha'})$ is defined as follows.
  - If A_α is of the form $(t \doteq t)$ then $\mathcal{I}_\alpha(A_{\alpha'}) = \mathbf{true}$.
  - If A_α is of the form $L[t]_p$ where there exists an equation $(t \doteq s)$ such that $s \prec t$, $(t \doteq s) \prec L[t]_p$ and $\mathcal{I}_\alpha(t \doteq s) = \mathbf{true}$ then $\mathcal{I}_\alpha(A_{\alpha'}) = \mathcal{I}_{\alpha'}(L[s]_p)$ (remark: we must have $L[s]_p \prec L[t]_p$ hence $L[s]_p$ must be in the domain of $\mathcal{I}_{\alpha'}$).
  - If there exists a positive clause $L_1 \vee \ldots \vee L_n \vee R \in S$ and a ground substitution σ of the variables in $L_1 \vee \ldots \vee L_n \vee R$, unifying $\{L_1, \ldots, L_n\}$, such that $R\sigma \prec L_1\sigma$, and $A_\alpha \equiv L_1\sigma$ and for all atoms A in R, $\mathcal{I}_{\alpha'}(A) = \mathbf{false}$, then $\mathcal{I}_\alpha(A_\alpha) = \mathbf{true}$.

The reader can easily check that for all $\alpha < \beta$, $\mathcal{I}_\alpha$ is a restriction of $\mathcal{I}_\beta$. Moreover, by construction of the sequence $\mathcal{I}$, $\mathcal{I}_\lambda$ must satisfy the following properties:

- $\mathcal{I}_\lambda$ is a *total* function from $\mathcal{B}$ to $\{\mathbf{true}, \mathbf{false}\}$.
- For any term t, $\mathcal{I}_\lambda(t \doteq t) = \mathbf{true}$.
- If $L \equiv L[t]_p$ and there exists $(t \doteq s) \in D$ such that $\mathcal{I}(t \doteq s) = \mathbf{true}$ then $\mathcal{I}(L[t]_p) = \mathcal{I}(L[s]_p)$.

Therefore, $\mathcal{I}_\lambda$ is a total interpretation on the domain $H(\Sigma)$. We are going to show that $\mathcal{I}_\lambda \models S$ (which proves that S is satisfiable).

Let S_g be the set of ground instances of clauses in S.

We introduce the following definition.

DEFINITION 3.86 *An atom A is said to be* reducible *iff there exists a clause of the form $(t \doteq s) \vee R \in S_g$ satisfying the following conditions:*

- $(t \doteq s) \prec A$.

- $R \prec (t \doteq s)$.
- $s \prec t$.
- t *occurs in* A.
- $\mathcal{I}_\lambda \models (t \doteq s)$.
- $\mathcal{I}_\lambda \not\models R$.

A clause C *is said to be* reducible, *iff there exists an atom* A *occurring in* C *such that* A *is reducible.* ♯

Let C_{min} be the minimal clause (according to the multiset extension of the ordering $\prec$) in S_g such that $\mathcal{I}_\lambda \not\models C_{min}$. By definition of S_g, $C_{\min}$ is equivalent to $C\sigma$ where $C \in S$ and σ is a ground substitution of the variables in C.

First we show that C_{min} is irreducible. Assume that there exists a clause of the form $(t \doteq s) \vee R \in S$ and a substitution θ of the variables in $(t \doteq s) \vee R$ and an atom A occurring in C such that

- $(t \doteq s)\theta \prec A\sigma$.
- $R\theta \prec (t \doteq s)\theta$.
- $s\theta \prec t\theta$.
- $t\theta$ occurs at a position p in $A\sigma$.
- $\mathcal{I}_\lambda \not\models R\theta$.
- $\mathcal{I}_\lambda \models (t \doteq s)\theta$.

Then, two cases must be distinguished.

1 If p is not a position in A or is a position of a variable in A, then this means that p must be of the form $p_1.p_2$ where p_1 is a position in A, $A_{|p_1}$ is a variable x, and p_2 is a position in $x\sigma$ (with possibly $p_2 = \epsilon$). Let σ' be the substitution of the variables in C, defined as follows: for all variables $y \neq x$, $y\sigma' = y\sigma$, and $x\sigma' = (x\sigma)[s\theta]_{p_2}$. Since $s \prec t$, we must have $(x\sigma)[s\theta]_{p_2} \prec x\sigma$ hence $x\sigma' \prec x\sigma$. Therefore, we have $C\sigma' \prec C\sigma$. Moreover, since $\mathcal{I}_\lambda \models (t \doteq s)\theta$ and $\mathcal{I}_\lambda \not\models C\sigma$, we must have $\mathcal{I}_\lambda \not\models C\sigma'$. But this is impossible, since C_{min} is the smallest clause in S_g not true in $\mathcal{I}_\lambda$.

2 If p is a non-variable position in C, then the ordered positive paramodulation rule can be applied between the clause C and $(t \doteq s) \vee R$, yielding a clause of the form $(C[s]_q \vee R)\theta'$ where θ' is a

m.g.u. of t and $A_{|p}$. Since S is saturated w.r.t. the ordered positive paramodulation rule, $(C[s]_q \vee R)\theta'$ must occur in S. Let $\gamma = \sigma \cup \theta$. Since γ is a unifier of t and A_p, θ' must be more general than γ. Hence $(C[s]_q \vee R)\gamma$ is an instance of $(C[s]_q \vee R)\theta'$, hence occurs in S_g. We have $R\theta \prec (t \doteq s)\theta$ and $(t \doteq s)\theta \prec A\sigma$, hence $R\theta \prec A\sigma$. Moreover, $C[s]_q \prec C[t]_q$, hence $(C[s]_q \vee R)\gamma \prec C_{min}$. Moreover, since $\mathcal{I}_\lambda \models (t \doteq s)\theta$ and $\mathcal{I}_\lambda \models \neg R\theta$, $(C[s]_q \vee R)\gamma$ must have the same truth value as C_{min} in $\mathcal{I}_\lambda$. This is impossible by definition of C_{min}.

Therefore, C_{min} must be irreducible. We distinguish two cases.

- C_{min} **is positive.** Let $L_1, \ldots, L_n$ be the set of literals such that $L_1\sigma = \cdots = L_n\sigma$ and $L_1\sigma$ is maximal in C_{min}. C is of the form $L_1 \vee \ldots \vee L_n \vee R$ where for all $L' \in R$, $L'\sigma \prec L_1\sigma$. For simplicity we write L instead of L_1 below.

  Let α be the ordinal such that $A_\alpha = L\sigma$. Since $R_\sigma \prec L\sigma$, any atom A in $R\sigma$ must occur in the domain of $\mathcal{I}_{\alpha+1}$. Since $\mathcal{I}_\lambda$ is an extension of $\mathcal{I}_{\alpha+1}$ we must have, for all atom $A \in R$, $\mathcal{I}_\lambda(L) =$ **false**. Therefore, by construction of $\mathcal{I}_{\alpha+1}$, since we have $\mathcal{I}_\lambda(L\sigma) =$ **false**, there exists an equation $(t \doteq s) \prec L\sigma$ such that $s \prec t$, $t = (L\sigma)_p$, $\mathcal{I}_\alpha(t \doteq s) =$ **true** and $\mathcal{I}_\alpha \not\models L\sigma[s]_p$. W.l.o.g. we assume that $t \doteq s$ is the smallest (according to $\prec$) equation having this property.

  Since $\mathcal{I}_\alpha(t \doteq s) =$ **true**, by construction of $\mathcal{I}$, one of the following conditions hold.

  - Either $t \equiv s$. This is impossible since $s \prec t$.
  - Or there exists an equation $(u \doteq v) \prec (t \doteq s)$ such that u occurs in $(t \doteq s)$, $v \prec u$ and $\mathcal{I}_\alpha(u \doteq v) =$ **true**. If u occurs in t, u also occurs in $L\sigma$; but this is impossible, by definition of $(t \doteq s)$. Hence u occurs at a position q in s. Then, we must have $\mathcal{I}_\alpha(t \doteq s[v]_q) =$ **true**, $s[v]_q \prec t$, and $\mathcal{I}_\alpha \not\models L\sigma[s[v]_q]_p$ which is impossible by definition of $t \doteq s$ since $(t \doteq s[v]_q) \prec (t \doteq s)$.
  - Or there exists a clause $t \doteq s \vee R \in S_g$ such that $\mathcal{I}_\alpha \models \neg R$. In this case, $L\sigma$ is reducible, which is impossible.

- C is of the form $\bigvee_{i=1}^{n} \neg L_i \vee R$ where $n \geq 1$, $L_1, \ldots, L_n$ are atoms, R is a positive clause.

  Since $\mathcal{I}_\lambda \not\models C\sigma$, we have for all $i \in \{1, \ldots, n\}$, $\mathcal{I}_\lambda \models L_i\sigma$. By construction of $\mathcal{I}$, this means that for all $i \in \{1, \ldots, n\}$, one of the following conditions holds:

  - Either $L_i\sigma \equiv (t \doteq t)$. In this case, $L_i\sigma$ is an instance of a clause in S, since the clause $x \doteq x$ must occur in S. Therefore there exists a clause $L'_i \vee R'_i$ and a substitution σ_i such that $R_i\sigma_i \prec L'_i\sigma_i$, $L'_i\sigma_i = L_i\sigma$ and $\mathcal{I}_\alpha \not\models R$ (it suffices to consider $L'_i \equiv (x \doteq x)$, $R_i = \Box$, $\sigma_i = \{x \to t\}$).

- Or there exists an equation $(t \doteq s)$ such that $(t \doteq s) \prec L_i\sigma$, $s \prec t$, $t = (L_i\sigma)_p$, $\mathcal{I}_\alpha(t \doteq s) = \mathbf{true}$ and $\mathcal{I}_\alpha \not\models L_i\sigma[s]_p$. Then, we show exactly as above that C_{min} is reducible, which is impossible.
- Or there exists a clause $L'_i \vee R_i$ and a substitution σ_i such that $R_i\sigma_i \prec L'_i\sigma_i$, $L'_i\sigma_i = L_i\sigma$ and $\mathcal{I}_\alpha \not\models R$.

Therefore, the hyperresolution rule may be applied on the clause C with the electrons $\{L'_i \vee R_i \mid i \in \{1,\dots,n\}\}$. We obtain a clause of the form $(R \vee \bigvee_{i=1}^n R_i)\theta$ where θ is a m.g.u of L'_i, L_i $(\forall i \in \{1,\dots,n\})$. Since $\sigma_1 \cup \ldots \cup \sigma_n \cup \sigma$ is a unifier of L'_i, L_i $(\forall i \in \{1,\dots,n\})$, the clause $R\sigma \vee \bigvee_{i=1}^n R_i\sigma_i$ is an instance of $(R \vee \bigvee_{i=1}^n R_i)\theta$, and thus must occur in S_g. But we have $\mathcal{I}_\lambda \not\models R\sigma$, and for all $i \in \{1,\dots,n\}$, $\mathcal{I}_\lambda \not\models R_i\sigma_i$. However, we have $R\sigma \vee \bigvee_{i=1}^n R_i\sigma_i \prec C\sigma$ (since for all $i \in \{1,\dots,n\}$, $R_i\sigma \prec L_i\sigma$). Hence this contradicts the definition of C_{min}.

For the completeness of X_s consider $X^*(\mathcal{C})$: if $\Box \in X^*(\mathcal{C})$ then there exists an $i \in \mathbb{N}$ with $\Box \in X^i(\mathcal{C})$. By definition of X_s we have $X_s^j(\mathcal{C}) = sub(X^j(\mathcal{C}))$ for all $j \in \mathbb{N}$. For all j with $j \geq i$ we have $\Box \in X^j(\mathcal{C})$ and, as $\Box$ subsumes every other clause, $X_s^j(\mathcal{C}) = \{\Box\}$ for all j with $j \geq i$. But this results in $X_s^*(\mathcal{C}) = \{\Box\}$. $\diamond$

Remark: In Proposition 2.4, the completeness of X_s relative to that of X was quite easy to prove. The key was the property $X_s^j(\mathcal{C}) = sub(X^j(\mathcal{C}))$ for all $j \in \mathbb{N}$. In particular $X_s^j(\mathcal{C}) \leq_{sub} X^j(\mathcal{C})$ for all equational clause sets $\mathcal{C}$ and $j \geq 0$. Let us define a replacement operator X_r in a different, intuitively more attractive way:

$$X_r^0(\mathcal{C}) = sub(\mathcal{C}),\ X_r^{i+1}(\mathcal{C}) = sub(X_r^i(\mathcal{C}) \cup X(X_r^i(\mathcal{C}))).$$

X_r defines the following principle. Take the set of clauses generated so far, compute X on it and reduce the result under subsumption. But in this definition we have the problem that $X_r^i(\mathcal{C})$ need not subsume $X^i(\mathcal{C})$. Just take

$$\mathcal{C} = \{f(x) \doteq g(x),\ P(x) \vee Q(x),\ P(f(a)) \vee Q(f(a))\}$$

As, for any PROP-operator X, paramodulation into variables is forbidden; there are no paramodulants from the first two clauses. Assuming $g(a) \prec f(a)$ there exist paramodulants between the first and the third clause; one of them is

$$P(g(a)) \vee Q(f(a))$$

and so $P(g(a)) \vee Q(f(a)) \in X^1(\mathcal{C})$.

On the other hand, $X_r^1(\mathcal{C}) = \mathcal{C} - \{P(f(a)) \vee Q(f(a))\}$:
first of all the third clause is deleted in the computation of X_r^0, i.e. $X_r^0(\mathcal{C}) = \mathcal{C} - \{P(f(a)) \vee Q(f(a))\}$. But, as mentioned above, no paramodulants are allowed between the first and the second clause and so $X_r^1(\mathcal{C}) = X_r^0(\mathcal{C})$. Obviously there is no clause in $X_r^1(\mathcal{C}))$ which subsumes the clause $P(g(a)) \vee Q(f(a))$ in $X^1(\mathcal{C})$. Therefore we obtain

$$X_r^1(\mathcal{C}) \not\leq_{sub} X^1(\mathcal{C}) \text{ and } X_r^*(\mathcal{C}) \not\leq_{sub} X^*(\mathcal{C}).$$

This does not mean that the operator X_r is incomplete, but only that it cannot be proved complete in the way above. Moreover the relations among the deduction levels of X^i and X_r^i are more complex than among X^i and X_s^i. ♯

DEFINITION 3.87 Let X be a PROP-operator and $X_s(\mathcal{C}) = \mathcal{C} \cup \{x \doteq x\}$. Then $\mathcal{C}$ is called (X_s-) stable. ♯

The following proposition generalizes Lemma 3.24 from Section 1 to equational clause logic. Here the fixed points of PROP-operators represent equational Herbrand models.

PROPOSITION 2.5 *Let $\mathcal{C}$ be a finite set of nonpositive equational clauses and let $\mathcal{A}$ be a finite set of positive (equational) unit clauses s.t. $\mathcal{C} \cup \mathcal{A}$ is satisfiable and X_s-stable. Then $\mathcal{A}$ is an EARM of a model of $\mathcal{C} \cup \mathcal{A}$ (over the Herbrand universe of $\mathcal{C} \cup \mathcal{A}$).*

Proof: We give a proof by contradiction:
Assume that the interpretation $\mathcal{M}$ induced by $\mathcal{A}$ falsifies $\mathcal{C} \cup \mathcal{A}$. Then the set $\mathcal{C} \cup \mathcal{A} \cup \{\neg B | \mathcal{A} \not\models B, B \text{ ground }\}$ is E-unsatisfiable. By the compactness theorem of first-order logic there exists a finite subset $\mathcal{F}$ of of $\{\neg B | \mathcal{A} \not\models B, B \text{ ground }\}$ s.t. $\mathcal{C} \cup \mathcal{A} \cup \mathcal{F}$ is E-unsatisfiable. By the completeness of X_s we get $\Box \in X_s^*(\mathcal{C} \cup \mathcal{A} \cup \mathcal{F})$. But as $\mathcal{C} \cup \mathcal{A}$ is satisfiable and $\Box \notin \mathcal{F}$ there exists an $i > 0$ with

$$\Box \in X_s^i(\mathcal{C} \cup \mathcal{A} \cup \mathcal{F}) - X_s^{i-1}(\mathcal{C} \cup \mathcal{A} \cup \mathcal{F}).$$

By assumption, $\mathcal{C} \cup \mathcal{A}$ is X_s-stable and so $X_s^*(\mathcal{C} \cup \mathcal{A}) = \mathcal{C} \cup \mathcal{A} \cup \{x \doteq x\}$. By definition of X no inferences between clauses of $\mathcal{F}$ and of $\mathcal{C}$ are possible (inference among nonpositive clauses is forbidden); the only potentially applicable rule is subsumption which may yield a proper subset of $\mathcal{F}$ or of $\mathcal{C}$. By $\Box \notin X_s^{i-1}(\mathcal{C} \cup \mathcal{A} \cup \mathcal{F})$ all clauses in $X_s^{i-1}(\mathcal{C} \cup \mathcal{A} \cup \mathcal{F}) - X_s^*(\mathcal{C} \cup \mathcal{A})$ must be ordered paramodulants from $\mathcal{A} \cup \mathcal{F}$. In particular there exists a finite set of negative unit clauses $\mathcal{F}'$ (some of them possibly nonground) s.t.

$$\begin{aligned} \mathcal{F}' &\subseteq X^*(\mathcal{A} \cup \mathcal{F}) \text{ and} \\ X_s^{i-1}(\mathcal{C} \cup \mathcal{A} \cup \mathcal{F}) &= \mathcal{C} \cup \mathcal{A} \cup \mathcal{F}'. \end{aligned}$$

Note that $\mathcal{A}$ itself is stable under ordered paramodulation, i.e. $OP_{\prec}(\mathcal{A}) = \mathcal{A}$.

By $\Box \in X_s^i(\mathcal{C} \cup \mathcal{A} \cup \mathcal{F})$ there exists an $A \in \mathcal{A}$ and a $\neg B \in \mathcal{F}'$ s.t. A and $\neg B$ resolve to $\Box$. Then, in particular, there exists a ground substitution θ (a substitution instance of the most general unifier) s.t. $A\theta = B\theta$. By definition of $OP_{\prec}$ and of $\mathcal{F}$ (consisting of negative literals only) there exists an atom C with $\neg B \in OP^*_{\prec}(\mathcal{A} \cup \{\neg C\})$ and $\neg C \in \mathcal{F}$. By the correctness of $OP_{\prec}$ we have $\mathcal{A} \cup \{\neg C\} \models \neg B\theta$ and (as C is ground) also $\mathcal{A} \cup \{B\theta\} \models C$. By definition of $\mathcal{F}$ C is false in $\mathcal{M}$, i.e. $\mathcal{A} \not\models C$. This, in turn, implies $\mathcal{A} \not\models B\theta$ (for otherwise we obtain $\mathcal{A} \models C$). On the other hand $\mathcal{A} \models A\theta$ because $A \in \mathcal{A}$ contradicting $A\theta = B\theta$.

So we conclude that $\mathcal{C} \cup \mathcal{A} \cup \mathcal{F}$ is satisfiable which contradicts the choice of $\mathcal{F}$. Finally we realize that $\mathcal{M}$, represented by the EARM $\mathcal{A}$, is indeed a model of $\mathcal{C} \cup \mathcal{A}$. $\diamond$

The problem remains to determine a syntactic characterization of sets of clauses $\mathcal{C}$ having a finite replacement closure $X_s^*(\mathcal{C})$. Only on sets of clauses of this type we can guarantee termination and thus success of model building procedures.

It follows from Proposition 2.4 that (the E-satisfiability problem of) a class Γ of equational clause sets is decidable if for all $\mathcal{C} \in \Gamma$ the closure $X^*(\mathcal{C})$ is finite. A natural candidate is the class $\mathcal{PVD}$ (see Definition 3.18) presented in Section 1, but extended by equality. Unfortunately the class extended in this way is undecidable, a frequent phenomenon appearing in equational extensions of syntax classes (see [Börger et al., 1997]).

DEFINITION 3.88 $\mathcal{PVD}^=$ is the class of all finite equational clause sets $\mathcal{C}$ s.t. for all $C \in \mathcal{C}$:

(V) $V(C_+) \subseteq V(C_-)$ and

(D) for all $x \in V(C_+) : \tau_{\max}(x, C_+) \leq \tau_{\max}(x, C_-)$.

$\sharp$

PROPOSITION 2.6 *$\mathcal{PVD}^=$ is undecidable.*

Proof: It is well-known that the word problem for general equational theories (e.g. for the theory of combinators, [Barendregt, 1984]) is undecidable. Formally this is the problem

$$(*)\ \{s_1 \doteq t_1, \ldots, s_n \doteq t_n\} \models s \doteq t.$$

for equations $s_i \doteq t_i$ and $s \doteq t$.

Clearly the problem $(*)$ is equivalent to the problem of deciding whether

$$\mathcal{C} : \{s_1 \doteq t_1, \ldots, s_n \doteq t_n, \neg s' \doteq t'\}$$

is E-unsatisfiable, where $\neg s' \doteq t'$ is the skolemization of $\neg\forall(s \doteq t)$.

The set of clauses $\mathcal{C}$ as defined above need not be in $\mathcal{PVD}^=$, as there may be equations $s_i \doteq t_i$ containing variables and thus violating the condition (V) in Definition 3.88. However, if we replace $\mathcal{C}$ by a set of clauses $\mathcal{D}$ of the form

$$\{s_1 \doteq t_1 \vee \neg s_1 \doteq s_1 \vee \neg t_1 \doteq t_1, \ldots, s_n \doteq t_n \vee \neg s_n \doteq s_n \vee \neg t_n \doteq t_n, \neg s' \doteq t'\}$$

then $\mathcal{D} \in \mathcal{PVD}^=$ and $\mathcal{D} \sim_E \mathcal{C}$. Note that, in any E-interpretation $\mathcal{M}$, the truth values of $s_i \doteq t_i \vee \neg s_i \doteq s_i \vee \neg t_i \doteq t_i$ and of $s_i \doteq t_i$ are equal because $r \doteq r$ is true in $\mathcal{M}$ for all terms r. So we obtain a reduction of the general word problem for equational theories to the E-satisfiability problem for the class $\mathcal{PVD}^=$. Therefore $\mathcal{PVD}^=$ is undecidable. ◇

Proposition 2.6 tells us that a decidable version of $\mathcal{PVD}$ in equational clause logic can be obtained only by restricting the appearence of equality. Note that, in some sense, the PVD-restriction (V) is always violated by a PROP-operator X. Indeed by

$$X(\mathcal{C}) = \rho_H(\mathcal{C}) \cup OP_{\prec}(\mathcal{C}) \cup \{x \doteq x\}$$

we always have to add a positive clause containing variables. Like in the restriction of EARMs to GEARMs we demand that all equality literals occurring in a set of clauses must be ground. The class below has been introduced in [Fermüller and Leitsch, 1998].

DEFINITION 3.89 $\mathcal{PVD}_g^=$ is the subclass of $\mathcal{PVD}^=$ defined by the restriction that all equality literals occurring in a set of clauses $\mathcal{C}$ for $\mathcal{C} \in \mathcal{PVD}^=$ are ground. ♯

Note that $\mathcal{PVD}_g^=$ also generalizes $\mathcal{PVD}$ because, in absence of equality literals, it collapses to $\mathcal{PVD}$. Before we prove that the class $\mathcal{PVD}_g^=$ is decidable we show that the property of being in $\mathcal{PVD}_g^=$ is closed under PROP-operators.

LEMMA 3.90 *Let X be a PROP-operator and $\mathcal{C} \in \mathcal{PVD}_g^=$, then*

$$X(\mathcal{C} \cup \{x \doteq x\}) - \{x \doteq x\} \in \mathcal{PVD}_g^=.$$

Proof: Let $\mathcal{C}$ be a set of clauses in $\mathcal{PVD}_g^=$. First observe that the condition

$$(D) \quad \text{for all } x \in V(C_+): \ \tau_{\max}(x, C_+) \leq \tau_{\max}(x, C_-)$$

is invariant under the application of substitutions. In particular, if $\{C\} \in \mathcal{PVD}_g^=$ then $\{C\theta\} \in \mathcal{PVD}_g^=$ for all factors $C\theta$ of C. Since every positive clause $C \in \mathcal{C}$ is ground, the most general unifiers defined in $\varrho_H(\mathcal{C})$ and in $OP_{\prec}(\mathcal{C})$ are ground. Thus neither (D) nor (V) are violated by application of X, with the exception of the clause $\{x \doteq x\}$.

It remains to show that resolution between a clause C in $\mathcal{C}$ and $\{x \doteq x\}$ only yields a clause in $\mathcal{PVD}_g^=$; note that ordered paramodulation with $\{x \doteq x\}$ is impossible anyway. Let L be the resolved literal in C. Then, by $\mathcal{C} \in \mathcal{PVD}_g^=$, L must be of the form $\neg t \doteq t$ for a ground term t. But then the resolvent of C and $\{x \doteq x\}$ is a subclause of C and thus in $\mathcal{PVD}_g^=$ too.

Therefore any clause obtained by resolution from $\mathcal{C} \cup \{x \doteq x\}$ is in $\mathcal{PVD}_g^=$. Hence $\varrho_H(\mathcal{C} \cup \{x \doteq x\}) \in \mathcal{PVD}_g^=$.

But

$$X(\mathcal{C} \cup \{x \doteq x\}) = \mathcal{C} \cup \{x \doteq x\} \cup \varrho_H(\mathcal{C} \cup \{x \doteq x\}) \cup OP_{\prec}(\mathcal{C})$$

and so

$$X(\mathcal{C} \cup \{x \doteq x\}) - \{x \doteq x\} \in \mathcal{PVD}_g^=.$$

◇

COROLLARY 2.2 *Let X be a PROP-operator and $\mathcal{C}$ be a set of clauses in $\mathcal{PVD}_g^=$. Then for all $i \in \mathbb{N}$:*

(1) $X^i(\mathcal{C}) - \{x \doteq x\} \in \mathcal{PVD}_g^=$,

(2) $X_s^i(\mathcal{C}) - \{x \doteq x\} \in \mathcal{PVD}_g^=$.

Proof: (1) follows from a straightforward induction argument from Lemma 3.90. For (2) note that, for all i, $X_s^i(\mathcal{C}) \subseteq X^i(\mathcal{C})$ by definition of X_s. But subsets of clause sets in $\mathcal{PVD}_g^=$ are in $\mathcal{PVD}_g^=$ too. ◇

In Corollary 2.2 we did not state the property $X^*(\mathcal{C}) - \{x \doteq x\} \in \mathcal{PVD}_g^=$. The simple reason is that $\mathcal{PVD}_g^=$ consists of *finite* sets of clauses only and we do not know yet that $X^*(\mathcal{C})$ is finite. Indeed, it is just the finiteness property of X^* on $\mathcal{PVD}_g^=$ we are looking for.

THEOREM 2.1 *Let X be a PROP-operator and $\mathcal{C}$ be a set of clauses in $\mathcal{PVD}_g^=$; then $X^*(\mathcal{C})$ is finite.*

Proof: We first show that neither $\tau_v(C)$ (the maximal depth of a variable in C) nor $|V(C)|$ (the number of variables occurring in C) can increase by *binary* positive resolution (hence by hyperresolution) or by ordered paramodulation:

So let E be a binary resolvent of a clause C and a positive clause D with $\{C, D\} \in \mathcal{PVD}_g^=$; then, by definition of $\mathcal{PVD}_g^=$, D is ground. Let σ be the m.g.u. of the resolution; then σ is a ground substitution and therefore $|V(E)| \leq |V(C)|$. By condition (D) we also have $\tau_{\max}(x, C_+) \leq \tau_{\max}(x'C_-)$ for all $x \in V(C_+)$. Because $x\sigma$ is ground for all $x \in dom(\sigma)$ we have

$$\begin{aligned} \tau_{\max}(y, E_+) &\leq \tau_{\max}(y, E_-) \text{ for } y \in V(E_+), \\ \tau_{\max}(y, E_-) &\leq \tau_{\max}(y, C_-) \text{ for } y \in V(E_-). \end{aligned}$$

This gives $\tau_v(E) \leq \tau_v(C)$.

Another possibility for a binary resolution step within X is the inference between a clause $C \in \mathcal{PVD}_g^=$ and $x \doteq x$. But then C contains a ground literal $t \doteq t$ and the binary resolvent E is of the form $C \backslash \{\neg t \doteq t\}$. Thus E is a subclause of C and $|V(E)| \leq |V(C)|$, $\tau_v(E) \leq \tau_v(C)$.

If $\mathcal{D}$ is a set of clauses in $\mathcal{PVD}_g^=$ and $E \in OP_{\prec}(\mathcal{D} \cup \{x \doteq x\})$ then E is a paramodulant of a positive ground clause $D \in \mathcal{D}$ and a clause $C \in \mathcal{D}$. Again the equation in the active clause D and the corresponding m.g.u. are ground and so $|V(E)| \leq |V(C)|$, $\tau_v(E) \leq \tau_v(C)$.

Let us define

$$\begin{aligned} \parallel \mathcal{D} \parallel &= \max\{|V(C)| \mid C \in \mathcal{D}\} \text{ and} \\ \tau_v(\mathcal{D}) &= \max\{\tau_v(C) \mid C \in \mathcal{D}\}. \end{aligned}$$

Our next goal is to show that for any $\mathcal{C} \in \mathcal{PVD}_g^=$ there are constants c_1, c_2 with

$$(I) \ \parallel X^*(\mathcal{C}) \parallel \leq c_1 \text{ and } \tau_v(X^*(\mathcal{C})) \leq c_2.$$

Afterwards it remains to prove that also the depth of ground terms in $X^*(\mathcal{C})$ is bounded. This eventually yields the finiteness of $X^*(\mathcal{C})$.

Note that we cannot derive $\tau_v(X^*(\mathcal{C})) \leq \tau_v(\mathcal{C})$ as the depth of variables can increase by factoring.

Now let us assume that $C\theta$ is a factor of C. It is easy to see that the following inequality holds in general:

$$(F) \ \tau_v(C\theta) \leq 2^{|V(C)|} \cdot \tau_v(C).$$

We now define a function ϕ on clauses by

$$\phi(C) = 2^{2^{|V(C)|}} \cdot \tau_v(C).$$

Let $\mathcal{C} \in \mathcal{PVD}_g^=$, $k > \parallel \mathcal{C} \parallel$ and $l > \tau_v(\mathcal{D})$. Our next step consists in proving that

$$(II) \ \ \phi(C) \leq 2^{2^k} \cdot l \text{ for all } C \in X^*(\mathcal{C}).$$

Obviously (II) yields constants c_1, c_2 fulfilling (I). We prove (II) by induction on the deduction levels defined by the operator X, i.e. we have to prove $A(i)$ for $i \in \mathbb{N}$ where

$$A(i): \ \phi(C) \leq 2^{2^k} \cdot l \text{ for all } C \in X^i(\mathcal{C}).$$

Induction basis $i = 0$: trivial by definition of k and l.

Induction hypothesis (IH): Assume that $A(i)$ holds.

case $i + 1$:
By Lemma 3.90 $X^i(\mathcal{C}) \setminus \{x \doteq x\} \in \mathcal{PVD}_g^=$. So let $E \in X^{i+1}(\mathcal{C})$. Then by definition of a PROP-operator

$$E \in \varrho_H(X^i(\mathcal{C})) \cup OP_{\prec}(X^i(\mathcal{C})) \cup X^i(\mathcal{C}) \cup \{x \doteq x\}.$$

If $E \in OP_{\prec}(X^i(\mathcal{C}))$ then E is paramodulant of a positive ground clause D in $\mathcal{PVD}_g^=$ and of an arbitrary clause C in $\mathcal{PVD}_g^=$. Thus, as we have shown above, $|V(E)| \leq |V(C)|$ and $\tau_v(E) \leq \tau_v(C)$. But then, by $\phi(E) \leq \phi(C)$, the induction hypothsis implies $\phi(E) \leq 2^{2^k} \cdot l$.

If E is an hyperresolvent then, again, $\phi(E)$ is bounded by $\phi(C)$ where C is one of the resolving clauses, the same bound as above applies to E.

It remains to handle the case that E is an hyperresolvent of $C\theta$ and $D_1, \ldots, D_m$ where $C\theta$ is a factor of a clause $C \in X^i(\mathcal{C})$ and $D_1, \ldots, D_m$ are positive clauses in $X^i(\mathcal{C})$).

If $|V(C\theta)| \leq |V(C)|$ and $\tau_v(C\theta) \leq \tau_v(C)$ then the argument is the same as for hyperresolvents.

So let us assume $\tau_v(C\theta) > \tau_v(C)$. Then θ has a nonempty domain and, by being a factoring substitution, $|V(C\theta)| < |V(C)|$. Moreover inequality (F) holds, i.e.

$$\tau_v(C\theta) \leq 2^{|V(C)|} \cdot \tau_v(C).$$

For the hyperresolvent E of $C\theta$ and $D_1, \ldots, D_m$ we have $\tau_v(E) \leq \tau_v(C\theta)$ and $|V(E)| \leq |V(C\theta)|$. By combining these inequalities we obtain

$$\phi(E) = 2^{2^{|V(E)|}} \cdot \tau_v(E) \leq 2^{2^{|V(C)|-1}} \cdot 2^{|V(C)|} \cdot \tau_v(C).$$

Moreover

$$2^{2^{|V(C)|-1}} \cdot 2^{|V(C)|} \cdot \tau_v(C) \leq 2^{2^{|V(C)|}} \cdot \tau_v(C) = \phi(C).$$

by $|V(C)| \geq 1$ and by the inequality

$$n + 2^{n-1} \leq 2^n \text{ for } n \geq 1.$$

But, by induction hypothesis, $\phi(C) \leq 2^{2^k} \cdot l$ and therefore

$$\phi(E) \leq 2^{2^k} \cdot l.$$

This completes the induction proof and we have established the uniform bound (I).

What is still missing is a uniform bound on term depth. Note that $X^*(\mathcal{C})$ might be infinite by containing arbitrary complex ground terms.

Let T be the set of ground terms inductively defined as follows.

- If t is a ground term occurring in a clause in S then $t \in S$.
- If t occurs in T and s is a subterm of t then $s \in T$.
- If $t[u]_p$ is a term in T, and if $(u \doteq v)$ is an equation occurring in a clause in S with $v \prec u$, then $t[v]_p$ is a term in T.

T is clearly finite, since the value of the term decreases strictly according to the ordering $\prec$ at each inductive step.

First we show, by induction on n, that for all equations $(t \doteq s)$ occurring in a clause in $X^n(S)$, s and t must occur in T.

- $n = 0$. The proof is immediate, since $t \doteq s$ must be ground (by definition of the $\mathcal{PVD}_g^=$ class).
- $n > 0$. Since any equation occurring in $X(S)^{n-1} \setminus \{x \doteq x\}$ is ground, no new equation may be generated by the hyperresolution rule. Therefore, we only have to consider the case of positive paramodulation. Any new equation occurring in a clause generated by positive paramodulation must be of the form $t[v]_p = s$ where $t \doteq s$ and $u = v$ are ground equations occurring in a clause in $X^{n-1}(S)$ and where $v \prec u$. By induction hypothesis, we have $t, u, v \in T$. By definition of T, we deduce that $t[v]_p \in T$.

Let $d = \max\{\tau_v(C) \mid C \in X^*(S)\}$ and C be a clause obtained from S by a derivation using the positive ordered paramodulation and hyperresolution rules. We are going to prove, by induction on the length of the derivation n, that any term occurring at a depth strictly greater than d in C must occur in T.

- **Base case.** $n = 0$. We have $X^0(S) = S$, hence the proof is immediate, by definition of d.

- **Hyper-resolution.** Let $C = R\theta \vee \bigvee_{i=1}^{n} R_i$ be a clause deduced from a clause $R \vee \bigvee_{i=1}^{n} \neg L_i$ and n positive clauses $L'_i \vee R_i$ ($1 \leq i \leq n$) by applying the hyperresolution rule. Since $L'_i \vee R_i$ is positive (for all $i \in \{1, \ldots, n\}$), $L'_i \vee R_i$ must be ground. Let t be a term occurring at a depth strictly greater than d in C. If t occurs at a depth strictly greater than d in $R \vee \bigvee_{i=1}^{n} \neg L_i$ then $t \in T$ hence the proof is completed.

  Else, since the depth of the variables cannot increase, t must occur at a position p in a term $x\theta$, where x is a variable occurring at a position q in R. Then, since $R \vee \bigvee_{i=1}^{n} \neg L_i$ is in $\mathcal{PVD}_g^=$, x must occur at a position q' such that $|q'| > |q|$ in a literal L_j (for some $j \in \{1, \ldots, n\}$). But since $L'_j = L_j\sigma$, this implies that t occurs at position $q'.p$ in L'_j. Since $|q'| \geq |q|$ and $|q.p| > d$, we deduce that $|q'.p| > d$, hence by induction hypothesis, that $t \in T$.

- **Positive ordered paramodulation.** Let $C = (L[s]_p \vee D)\theta \vee R$ be a clause deduced from two clauses $L[t']_p \vee D$ and $(t \doteq s) \vee R$ by ordered positive paramodulation. Then $s \prec t$ and θ is the m.g.u. of t and t'.

  Let u be a term in C occurring at a depth strictly greater than d. We distinguish several cases.

  - If u occurs in D or in R, then by induction hypothesis, t must occur in T.
  - If u occurs in $x\theta$ for some variable x, then u must occur in t hence in T.
  - If u occurs in s then we must have $u \in T$.
  - Otherwise, u must occur at a prefix q of p in $L[s]_p$. Since $|q| > d$, $L[t']_q$ must occur in T. Hence by definition of T, since $t, s \in T$ and $s \prec t$, we have $L[s]_q \in T$.

Therefore $\max\{\tau(C) \mid C \in X^*(S)\}$ is defined (and thus finite). Since the maximal number of variables occurring in the clauses is not increasing, we conclude that the number of distinct clauses occurring in $X^*(S)$ is finite.

Hence there exists an i with $X^*(\mathcal{C}) = X^i(\mathcal{C})$. ◇

COROLLARY 2.3 *$\mathcal{PVD}_g^=$ is decidable.*

Proof: X is correct and complete by Proposition 2.4. By Theorem 2.1 X^* is computable and terminating on $\mathcal{PVD}_g^=$. Therefore X^* is a decision procedure of the satisfiability problem of $\mathcal{PVD}_g^=$. ◇

COROLLARY 2.4 *Let X be a PROP-operator and $\mathcal{C} \in \mathcal{PVD}_g^=$. Then $X_s^*(\mathcal{C})$ is finite.*

Proof: By Theorem 2.1 $X^*(\mathcal{C})$ is finite for $\mathcal{C} \in \mathcal{PVD}_g^=$. Thus, by Proposition 2.3 $X_s^*(\mathcal{C}) = sub(X^*(\mathcal{C}))$ and the finiteness of $X_s^*(\mathcal{C})$ follows from $sub(X^*(\mathcal{C})) \subseteq X^*(\mathcal{C})$. ◇

Remark: It is interesting that Theorem 2.1 does not hold any more if positive resolution is considered instead of hyperresolution. Indeed, allowing the application of the paramodulation rule on "intermediate" resolvents produced by positive resolution may lead to non-termination, as shown in the following example.

We consider the following set of clauses S (P, Q denote predicate symbols, f denotes a function symbol, a denotes a constant symbol and x denotes a variables).

1	$f(a) \doteq a$
2	$\neg P(f(x)) \vee \neg Q(x) \vee Q(f(x))$
3	$P(a)$
4	$Q(a)$

The reader can easily check that S is in $\mathcal{PVD}_g^=$.

Positive resolution and positive ordered paramodulation produce the following clauses:

5	$\neg P(f(a)) \vee Q(f(a))$	$(resolution, 2, 4)$
6	$\neg P(a) \vee Q(f(a))$	$(paramodulation, 1, 5)$
7	$Q(f(a))$	$(resolution, 3, 6)$
8	$\neg P(f(f(a))) \vee Q(f(f(a)))$	$(resolution, 7, 2)$
9	$\neg P(f(a)) \vee Q(f(f(a)))$	$(paramodulation, 8, 1)$
10	$\neg P(a) \vee Q(f(f(a)))$	$(paramodulation, 9, 1)$
11	$Q(f(f(a)))$	$(resolution, 10, 3)$
...		

It is straightforward to see that an infinite number of positive ground clauses of the form $Q(f^n(a))$, for all $n \in \mathbb{N}$ may be deduced. ♯

Remark: Some decidable extensions of the $\mathcal{PVD}_g^=$ class have been considered.

For example, $\mathrm{PVD}_b^=$ is an extension of $\mathcal{PVD}_g^=$, that was defined and proven to be decidable in [Rudlof, 2000]. This class is obtained by weakening the condition on the negative equational literals, i.e. clauses in

$\mathrm{PVD}_b^=$ may contain literals of the form $t \neq s$ where t *and* s are not ground, provided that some conditions are satisfied to make sure that no unifier of t and s modulo the set of equations in S may exist. It is shown in [Rudlof, 2000] that the SHR tableaux calculus (a tableau-based procedure that can be seen as an extension of SATCHMO [Manthey and Bry, 1988]) is a a decision procedure for $\mathrm{PVD}_b^=$.

In [Peltier, 2001b], another extension PVD_E of $\mathcal{PVD}_g^=$ is considered by allowing the occurrence of non ground positive equational literals in the clauses, provided at these literals are of the form $f(t_1, \ldots, t_n) \doteq s$ where s is ground and strictly lower than $f(t_1, \ldots, t_n)$ and where the symbol f occurs only at *root* position in the clause set. A special refinement of the paramodulation rule, the so-called simultaneous ordered positive paramodulation is defined, and proven to be a decision procedure for PVD_E. In [Peltier, 2001b], a model building procedure is defined for PVD_E.

It is worthwhile to note that $\mathrm{PVD}_b^=$ class is incomparable to PVD_E. Hence these two classes may be considered as "orthogonal" extensions of $\mathcal{PVD}_g^=$. ♯

2.3 Model Building by MBEQ

In Section 1 we have defined a model building method based on iterations of unit reducts and closure under $R_{H,s}$. A similar method can be established on equational clause logic when the hyperresolution operator is replaced by a PROP-operator. Many steps like unit reduction are independent of equality can be carried over directly, leaving the overall structure of the model building procedure unchanged. We only have to take care about ordered paramodulation in the new operators. The method below is derived from this in [Fermüller and Leitsch, 1998], but positive resolution has been replaced by hyperresolution. Again the key lemma is the following one:

LEMMA 3.91 *Let X be a PROP-operator and $\mathcal{C}$ be a finite set of equational clauses s.t. $\mathcal{C}$ is satisfiable and X_s-stable, and $X^*(\mathcal{C})$ is finite. If $\mathcal{D}$ is a unit reduct of $\mathcal{C}$ then $\mathcal{D}$ is satisfiable and $\mathcal{D}$ implies $\mathcal{C}$.*

Proof: $\mathcal{D}$ implies $\mathcal{C}$ because $\mathcal{D}$ is a split problem of $\mathcal{C}$. It remains to prove that $\mathcal{D}$ is satisfiable.

Let $\mathcal{D} : (\mathcal{C} - \{D\}) \cup \{P\}$ be a nontrivial unit reduct for some positive (nonunit) clause $D \in \mathcal{D}$ and let $D' = D \setminus P$.

We use proof by contradiction and assume that $\mathcal{D}$ is unsatisfiable. Then $\Box \in X^*(\mathcal{D})$.

As in Lemma 3.28 we show that the clauses in $X^*(\mathcal{D})$ differ from those in $X^*(\mathcal{C})$ "at most by the clause D'". So let $\leq_{D'}$ be the relation defined in Definition 3.27. Our first step consists in proving

$$\text{(I)} \quad X^*(\mathcal{D}) \leq_{D'} X^*(\mathcal{C}).$$

We prove (I) by induction on the iteration number of X, i.e. we have to show

$$\text{(II)} \quad X^i(\mathcal{D}) \leq_{D'} X^*(\mathcal{C})$$

for all $i \in \mathbb{N}$.

(IB) $i = 0$: trivial by $\mathcal{D} \leq_{D'} \mathcal{C}$ and $\mathcal{C} \subseteq X^*(\mathcal{C})$.

(IH) Let us assume $X^i(\mathcal{D}) \leq_{D'} X^*(\mathcal{C})$.

Now let $E \in X^{i+1}(\mathcal{D})$. If $E \in X^i(\mathcal{D})$ there is nothing to prove; thus it remains consider the case $E \in X^{i+1}(\mathcal{D}) - X^i(\mathcal{D})$.

By definition of a PROP-operator we have

$$X(\mathcal{E}) = \varrho_H(\mathcal{E}) \cup OP_{\prec}(\mathcal{E}) \cup \mathcal{E} \cup \{x \doteq x\}$$

for all equational clause sets $\mathcal{E}$. Therefore we have to investigate the cases $E \in \varrho_H(X^i(\mathcal{D}))$ and $E \in OP_{\prec}(X^i(\mathcal{D}))$.

a. $E \in \varrho_H(X^i(\mathcal{D}))$.
We have to show that $\{E\} \leq_{D'} X^*(\mathcal{C})$.
By definition of ϱ_H the clause E is a clash resolvent of a clause $G : R \vee \neg A_1 \vee \cdots \vee \neg A_m$ and positive clauses $H_1, \ldots, H_m$ in $X^i(\mathcal{C})$.

By (IH) there are clauses $G', H'_1, \ldots, H'_m \in X^*(\mathcal{C})$ with

$$G \leq_{D'} G', \ H_j \leq_{D'} H'_j \text{ for } j = 1, \ldots m.$$

By definition of $\leq_{D'}$ there are renaming substitutions $\eta, \eta_1, \ldots, \eta_m$ with

$$\begin{array}{ccc} N_c(G') = N_c(G \vee D'\eta) & \text{or} & N_c(G') = N_c(G) \\ N_c(H_j) = N_c(H_j \vee D'\eta_j) & \text{or} & N_c(H'_j) = N_c(H_j) \end{array}$$

for $j = 1, \ldots, m$.

Now let $G_i : Q'_i \vee \neg A'_i \vee \ldots \neg A'_m$ be the i-th intermediary resolvent of the clash. Then G_i is obtained by successively resolving G with $H_1, \ldots, H_{i-1}$.

For $i = 0$ $G_i = G$ and we have $G_i \leq_{D'} G'_i = G'$.

Now let us assume that for G_i there exists an intermediary resolvent G'_i of the clash $(G'; H'_1, \ldots, H'_m)$ with $G_i \leq_{D'} G'_i$. If $i = m$ then we have resolved through the whole clash; it remains to consider the case $i < m$.

By definition of $\leq_{D'}$ we have $N_c(G'_i) = N_c(G_i)$ or $N_c(G'_i) = N_c(G_i \vee D'\xi)$ for some renaming substitution ξ. We only treat the more interesting case $N_c(G'_i) = N_c(G_i \vee D'\xi)$, the other case is even simpler.

Now consider G_{i+1}, the $(i+1)$-th intermediary resolvent of G_i and H_i; then G_{i+1} is a binary resolvent of a factor $H_i\sigma :\ A_i\sigma \vee R_i\sigma$ and of G_i. Let θ be the m.g.u. of $\{A_i\sigma, A'_i\}$. Then

$$G_{i+1} = Q'_i\theta \vee R_i\sigma\theta \vee \neg A'_{i+1}\theta \vee \cdots \vee \neg A'_m\theta.$$

We have to construct an intermediary $(i+1)$-th clash resolvent G'_{i+1} of $(G'; H'_1, \ldots, H'_m)$ with $G_{i+1} \leq_{D'} G'_{i+1}$.

By $H_i \leq_{D'} H'_i$ we either have $H_i = H'_i$ or $N_c(H'_i) = N_c(H_i \vee D'\rho_i)$ for some renaming substitution ρ_i with $V(D'\rho_i) \cap V(H_i) = \emptyset$. Now let us assume that H_i and H'_i "differ" by D'. Then there exists the factor

$$H'_i\sigma :\ A_i\sigma \vee R_i\sigma \vee D'\rho_i$$

of H'_i. Moreover

$$N_c(G'_i) = Q'_i \vee D'\xi \vee \neg A'_i \vee \cdots \vee A_m.$$

So there exists a a binary resolvent G'_{i+1} of $H'_i\sigma$ and G'_i such that

$$G'_{i+1} = Q'_i\theta \vee R_i\sigma\theta \vee D'\xi \vee D'\rho_i \vee \neg A'_{i+1}\theta \cdots \vee \neg A'_m\theta.$$

By definition of $\leq_{D'}$ we get $G_{i+1} \leq_{D'} G'_{i+1}$.

This way we obtain a clash resolvent E' of $(G', H_1, \ldots, H'_m)$ with $E \leq_{D'} E'$. But E' is a clash resolvent of clauses in $X^*(\mathcal{C})$ and $X^*(\mathcal{C})$ is closed under hyperresolution (by definition of a PROP-operator); so we have $E' \in X^*(\mathcal{C})$ and eventually

$$\{E\} \leq_{D'} X^*(\mathcal{C}).$$

b. $E \in OP_{\prec}(X^i(\mathcal{D}))$.

By definition of $OP_{\prec}$ E is a ordered paramodulant of a positive clause G_1 in $X^i(\mathcal{D})$ and a clause G_2 in $X^i(\mathcal{D})$. Again, by (IH), there are clauses $H_1, H_2 \in X^*(\mathcal{C})$ with $G_1 \leq H_1$ and $G_2 \leq H_2$. Once more we consider the most interesting case $G_1 \neq H_1$ and $G_2 \neq H_2$. Then there are renaming substitutions η_1, η_2 with

$$H_1 = N_c(G_1 \vee D'\eta_1) \text{ and } H_2 = N_c(G_2 \vee D'\eta_2).$$

Now let

$$G_1 = N_c(s \doteq t \vee B_1),\ G_2 = N_c(L[u]_\lambda \vee B_2)$$

and

$E = N_c(B_1\sigma \vee L[t]_\lambda\sigma \vee B_2\sigma)$.

Like in case a we can simulate the paramodulation of G_1 and G_2 on H_1, H_2 and obtain an ordered paramodulant F of H_1, H_2 where

$F = E \vee D'\eta_1 \vee D'\eta_2$.

Clearly $E \leq_{D'} F$ and, as $X^*(\mathcal{C})$ is closed under ordered paramodulation, $\{E\} \leq_{D'} X^*(\mathcal{C})$.

This proves the case $i+1$ and thus also (I).

By assumption $\mathcal{D}$ is unsatisfiable and, by the completeness of X, $\Box \in X^*(\mathcal{D})$. Therefore we obtain a clause $H \in X^*(\mathcal{C})$ where H is the condensed form of D' (note that $\Box \notin \mathcal{C}$ as $\mathcal{C}$ is satisfiable!). By assumption $X^*(\mathcal{C})$ is finite and thus, by Proposition 2.3, $X_s^*(\mathcal{C}) = sub(X^*(\mathcal{C}))$. In particular $X_s^*(\mathcal{C})$ subsumes $X^*(\mathcal{C})$ and since $\mathcal{C}$ is X_s-stable there must be a clause $G \in \mathcal{C}$ s.t. $G \leq_{sub} H$. By definition of D and D' we obtain $D' \leq_{sub} D$ and thus $H \leq_{sub} D$; moreover $H \neq D$ – otherwise D is not condensed. Moreover D does not subsume H and thus does not subsume G (note that $\leq_{sub}$ is transitive). Therefore we obtain two clauses D and G in $\mathcal{C}$ with $G \leq_{sub} D$ and $D \not\leq_{sub} G$. This, however, contradicts the assumption that $\mathcal{C}$ is reduced under subsumption. We conclude that $\mathcal{D}$ is satisfiable. ◇

Lemma 3.91 differs from Lemma 3.28 in the requirement that $X^*(\mathcal{C})$ (i.e. the closure under the subsumption-free deduction operator) is finite; note that in Lemma 3.28 we did not need the finiteness of $R_H^*(\mathcal{C})$. But here we face the problem that we can not garantee that $X_s^*(\mathcal{C})$ subsumes $X^*(\mathcal{C})$ unless the latter one is finite. Indeed we have shown, in contrast to hyperresolution, that in some sense paramodulation and replacement are not compatible (see the remark after Proposition 2.4). Lemma 3.91 also shows that ordering of literals and further ordering requirements for paramodulation (concerning literals outside the active equation) is destructive. In fact those ordering conditions could block the simulation of resolution and paramodulation on the larger clauses containg variants of D'. This indicates an interesting feature of the role calculi are playing in model building: If too much clauses are generated then the procedure does not terminate and symbolic model construction becomes impossible; if on the other hand too few clauses are generated we do not have enough information for constructing a model (at least without backtracking). Thus, in model building, we have to keep a delicate balance between weak and strong deduction refinements.

As in Lemma 3.28 stability plays a central role in the validity of Lemma 3.91. As an example we may simply take Example 3.29 ot Section 1. We give another example below involving paramodulation.

EXAMPLE 3.92 Let $\mathcal{C} = \{P(a) \vee P(b), \neg P(c), a \doteq c\}$.
Trivially $\mathcal{C}$ is satisfiable. But if we replace $P(a) \vee P(b)$ by $P(a)$ we obtain the unit reduct

$$\mathcal{C}_1 : \{P(a), \neg P(c), a \doteq c\}$$

which is unsatisfiable. Now let X be a PROP-operator based on a C.S.O. $\prec$ with $c \prec a$. Then, clearly, $\mathcal{C}$ is not X_s-stable; instead we obtain

$$\begin{aligned} X^*(\mathcal{C}) &= \mathcal{C} \cup \{P(c) \vee P(b),\ P(b),\ x \doteq x\}, \\ X_s^*(\mathcal{C}) &= \{P(b),\ \neg P(c),\ a \doteq c,\ x \doteq x\}. \end{aligned}$$

$X_s^*(\mathcal{C})$ does not have nontrivial unit reducts and the subset of its positive clauses $\{P(b), a \doteq c,\ x \doteq x\}$ defines an EARM of $\mathcal{C}$. ♯

Note that, in Example 3.92, unit reduction becomes incorrect even when we replace $\mathcal{C}$ by $X^*(\mathcal{C})$. Thus it is stability under *subsumption* which plays the key role.

The transformation of $\mathcal{C}$ into a unit reduct $\mathcal{D}$ can be described by atom selection functions, which we denote by α. α does not depend on the occurrence of the equality predicate and we refer to Definition 3.30. Like for hyperresolution in Section 1 $\alpha(\mathcal{C})$ need not be X_s-stable, no matter whether $\mathcal{C}$ is X_s-stable. Therefore we have to apply X_s in order to obtain a new stable set.

Let us assume that $X_s^*(\mathcal{C})$ is finite. Then atom selection and closure can be combined in a single operator.

DEFINITION 3.93 (EURC-OPERATOR) Let α be an atom selection function and X be a PROP-operator. We define an operator T by

$$T(\mathcal{C}) = X_s^*(\alpha(\mathcal{C})).$$

Then T is called a EURC-operator (equational unit reduction + closure). ♯

Example 3.92 shows that EURC-operators may be incorrect on non-stable sets. Thus the input to T has to be a finite X_s-stable set. Moreover Lemma 3.91 holds under the condition that the monotone closure $X^*(\mathcal{C})$ is finite too. Thus we have to reformulate the procedure MB, defined in Section 1 accordingly. The role of the operator $R_{H,s}$ is played by X_s for any PROP-operator X. T is a EURC-operator based on X_s and some atom selection function α.

function *MBEQ*
input: A finite set of equational clauses $\mathcal{C}$.
ouput: *satisfiable* or *unsatisfiable* or a set of atoms.
begin
    $\mathcal{C} \leftarrow X^*(\mathcal{C})$; $\mathcal{C} \leftarrow sub(\mathcal{C})$;
    if $\mathcal{C} = \{\Box\}$ **then return** *unsatisfiable*
        else begin
            while $\alpha(\mathcal{C}) \neq \mathcal{C}$ **do** $\mathcal{C} \leftarrow T(\mathcal{C})$
            if unit$(POS(\mathcal{C}))$ **then return** $POS(\mathcal{C})$
              else return *satisfiable*
        end
end

Remark: MBEQ differs from MB in computing the monotone closure $X^*(\mathcal{C})$ in the first step. Only if the computation of $X^*(\mathcal{C})$ terminates we come to the second step $\mathcal{C} \leftarrow sub(\mathcal{C})$; but, for finite $X^*(\mathcal{C})$, $X^*_s(\mathcal{C})$ is just $sub(X^*(\mathcal{C}))$ and we obtain an X^*_s-stable set $\mathcal{D}$. If $\mathcal{D}$ is unsatisfiable then, by the completeness of X_s, we have $\mathcal{D} = \{\Box\}$. Otherwise we enter the while-loop. In contrast to MB the correctness of the while-loop is not so easy to show: we have to guarantee that the set $\mathcal{D}_n$ obtained after n executions of the while loop has a finite X-closure, i.e. $X^*(\mathcal{D}_n)$ is finite. Only then we may safely apply Lemma 3.91. ♯

EXAMPLE 3.94 Let $\mathcal{C}$ be the equational clause set

$$\{P(a) \vee P(f(a)), f(a) \doteq b, f(b) \doteq a,$$

$$\neg P(f(x)) \vee \neg P(x), \neg P(f(f(x))) \vee P(x)\}.$$

Let $\prec$ be a C.S.O. for ordered paramodulation fulfilling $a \prec b$ and $f^n(b) \prec f^m(a)$ for $n < m$ and $n, m \in \mathbb{N}$. Then

$$OP_{\prec}(\mathcal{C}) = \{P(a) \vee P(b),\ \neg P(a) \vee \neg P(b),$$

$$\neg P(f(b)) \vee P(a), \neg P(f(a)) \vee P(b)\}.$$

and

$$\varrho_H(\mathcal{C}) = \{P(a) \vee P(f(a))\}.$$

Consequently we obtain

$$\begin{aligned} X(\mathcal{C}) \;\; = \;\; & \{P(a) \vee P(b), \neg P(a) \vee \neg P(b), \neg P(f(b)) \vee P(a), \\ & \neg P(f(a)) \vee P(b), x \doteq x\} \cup \mathcal{C}. \end{aligned}$$

In the next deduction level we get

$$\begin{aligned} OP_{\prec}(X(\mathcal{C})) &= OP_{\prec}(\mathcal{C}) \cup \{\neg P(a) \vee P(a), \neg P(b) \vee P(b)\}, \\ \varrho_H(X(\mathcal{C})) &= \varrho_H(\mathcal{C}) \cup \{P(a) \vee P(b)\} \text{ and therefore} \\ X^*(\mathcal{C}) &= X^2(\mathcal{C}). \end{aligned}$$

Furthermore it is easy to see that $X^*(\mathcal{C}) = X_s^*(\mathcal{C})$. Thus the deductive closure is

$$\begin{aligned} X_s^*(\mathcal{C}) = \; & \{P(a) \vee P(f(a)), f(a) \doteq b, f(b) \doteq a, P(a) \vee P(b), \neg P(a) \vee \neg P(b), \\ & \neg P(f(b)) \vee P(a), \neg P(f(a)) \vee P(b), \neg P(a) \vee P(a), \neg P(b) \vee P(b), \\ & \neg P(f(x)) \vee \neg P(x), \neg P(f(f(x))) \vee P(x), x \doteq x\}. \end{aligned}$$

Note that we did not delete the tautologies $\neg P(a) \vee P(a)$ and $\neg P(b) \vee P(b)$. However this clauses are useless and the deduction operator X could be extended by tautology deletion without affecting the properties shown in the previous lemmas.

Let us assume that the selection function α selects the atom $P(a)$ from the clause $P(a) \vee P(b)$. Then by $T(X_s^*(\mathcal{C})) = X_s^*(\alpha(X_s^*(\mathcal{C})))$ we obtain

$$\begin{aligned} T(X_s^*(\mathcal{C})) = \; & \{f(a) \doteq b,\ f(b) \doteq a,\ P(a),\ \neg P(b),\ \neg P(f(a)), \\ & \neg P(f(x)) \vee \neg P(x),\ \neg P(f(f(x))) \vee P(x),\ x \doteq x\}. \end{aligned}$$

Now all positive clauses are unit, i.e. unit$(POS(\mathcal{C}))$ evaluates to **true**, and MBEQ return the set of unit clauses

$$\{f(a) \doteq b, f(b) \doteq a, P(a), x \doteq x\}.$$

Clearly $x \doteq x$ is redundant in specification of EARMs and we can replace the former set of unit clauses by the GEARM $\mathcal{A} : \{f(a) \doteq b, f(b) \doteq a, P(a)\}$. Indeed $\mathcal{A}$ represents an EH-model of $\mathcal{C}$.

Note that by selecting $P(b)$ in $P(a) \vee P(b)$ we obtain a GEARM $\mathcal{B} : \{f(a) \doteq b, f(b) \doteq a, P(b)\}$ which specifies another EH-model of $\mathcal{C}$. ♯

In Section 1 we have shown termination of MB on all sets of clauses $\mathcal{C}$ where $R_H^*(\mathcal{C})$ is finite. Here we prove a similar property for the PROP-operators X and X_s. Again we use an ordering $\prec$ (see Definition 3.35) and show that, for any EURC-operator T, the sequence $(T^i(\mathcal{C}))_{i \in \mathbb{N}}$ defines a strictly descending chain w.r.t. $\prec$. This eventually entails the convergence of the sequence $(T^i(\mathcal{C}))_{i \in \mathbb{N}}$ on $\mathcal{C}$. However we will see below that we cannot use the same ordering as in Section 1. We now adapt the terminology to PROP-operators:

DEFINITION 3.95 Let T be a EURC-operator defined by a PROP-operator X and an atom selection function α. Let $\mathcal{C}$ be a satisfiable, finite and X_s-stable set of equational clauses. Then we say that the sequence $(T^i(\mathcal{C}))_{i\in\mathbb{N}}$ *converges* to a set of clauses $\mathcal{D}$ if there exists a number k such that $T^k(\mathcal{C}) = T^{k+1}(\mathcal{C})$ and $T^k(\mathcal{C}) = \mathcal{D}$. ♯

The behavior of subsumption is more delicate for PROP-operators than for hyperresolution (recall the remark after Proposition 2.4). Indeed we cannot reuse the termination proof of Lemma 3.36: an important property in the proof of this lemma was $R_H^*(\alpha(\mathcal{C})) \leq_{sub} R_H^*(\mathcal{C})$. This corresponds to the property $X^*(\alpha(\mathcal{C})) \leq_{sub} X^*(\mathcal{C})$ for PROP-operators X, which unfortunately does not hold:

PROPOSITION 2.7 *Let X be a PROP-operator. Then there exists a set of equational clauses $\mathcal{C}$ s.t. $X^*(\alpha(\mathcal{C})) \not\leq_{sub} X^*(\mathcal{C})$.*

Proof: Let $\mathcal{C}$ be the set of clauses

$$\{P(x, f(y)) \vee P(f(u), v),\ \neg P(w, z) \vee Q(w, w),\ f(a) \doteq a\}.$$

Then $Q(a, f(a)) \in X^*(\mathcal{C})$. This can be seen in the following way: first resolve the first and the second clause by factoring the first clause to $P(f(u), f(y))$. Then the resolvent is $Q(f(u), f(u))$. In any C.S.O. $\prec$ we have $a \prec f(a)$. Thus ordered paramodulation between $Q(f(u), f(u))$ and $f(a) \doteq a$ is admitted and gives the clause $Q(a, f(a))$; so $Q(a, f(a)) \in X^*(\mathcal{C})$.

Now the set $\alpha(\mathcal{C})$ is

$$\{P(x, f(y)),\ \neg P(w, z) \vee Q(w, w),\ f(a) \doteq a\}.$$

We prove that $X^*(\alpha(\mathcal{C})) \not\leq_{sub} Q(a, f(a))$ and therefore $X^*(\alpha(\mathcal{C})) \not\leq_{sub} X^*(\mathcal{C})$. We simply compute $X^*(\alpha(\mathcal{C}))$:

$$\begin{aligned} X^1(\alpha(\mathcal{C})) &= \alpha(\mathcal{C}) \cup \{Q(w, w),\ P(x, a),\ x \doteq x\}, \\ X^*(\alpha(\mathcal{C})) &= X^1(\alpha(\mathcal{C})). \end{aligned}$$

Indeed, none of the clauses in $X^*(\alpha(\mathcal{C}))$ subsumes $Q(a, f(a))$ (note that $Q(w, w) \not\leq_{sub} Q(a, f(a))$!). ◇

Remark: The effect is Proposition 2.7 is based on the restriction that paramodulation into variables is forbidden. In particular no paramodulant between $Q(w, w)$ and $f(a) \doteq a$ can be generated. In $\mathcal{C}$ an appropriate instantiation is generated by factoring which cannot be obtained in $\alpha(\mathcal{C})$. A possible solution to this problem is to allow paramodulation into variables; this however would render the inference operator X

hopelessy inefficient making an implementation of the method pointless. $\sharp$

COROLLARY 2.5 *For every PROP-operator X there are sets of equational clauses $\mathcal{D}, \mathcal{C}$ s.t. $\mathcal{D} \leq_{sub} \mathcal{C}$, but $X^*(\mathcal{D}) \not\leq_{sub} X^*(\mathcal{C})$.*

Proof: Choose $\mathcal{D} = \alpha(\mathcal{C})$ and apply Proposition 2.7. $\diamond$

In order to prove termination of MBEQ (on sets of clauses $\mathcal{C}$ with $X^*(\mathcal{C})$ finite) we introduce an ordering $\leq_c$ which is corresponds to $\leq_D$ (see Definition 3.27) for arbitrary D.

DEFINITION 3.96 Let C and D be clauses in condensation normal form (i.e. $N_c(C) = C$ and $N_c(D) = D$). We define $C \leq_c D$ if a variant of C is a subclause of D. We extend $\leq_c$ to sets of clauses $\mathcal{C}, \mathcal{D}$ in condensation normal form in the following way:

$\mathcal{C} \leq_c \mathcal{D}$ if for all $C \in \mathcal{C}$ there exists a $D \in \mathcal{D}$ with $C \leq_c D$.

The strict (irreflexive) relation $<_c$ is defined by: $\mathcal{C} <_c \mathcal{D}$ if $\mathcal{C} \leq_c \mathcal{D}$, but not $\mathcal{D} \leq_c \mathcal{C}$. $\sharp$

EXAMPLE 3.97 Let $C = P(x_1) \vee P(f(x_1, x_2)) \vee Q(x_2, x_3) \vee R(x_3)$ and $D = P(x_1) \vee R(x_2)$. Then C and D are in condensation normal form and $D \leq_c C$; indeed a variant of D $(P(x_1) \vee R(x_3))$ is a subclause of C.

Let

$$
\begin{aligned}
\mathcal{C} &= \{C\} \text{ and} \\
\mathcal{D} &= \{P(x_1) \vee R(x_2), Q(x_1, x_2) \vee R(x_2), P(f(x_1, x_2)) \vee R(x_3)\}.
\end{aligned}
$$

Then $\mathcal{D} \leq_c \mathcal{C}$ as for every clause $D \in \mathcal{D}$ we have $D \leq_c C$. $\sharp$

It is easy to see that $C \leq_D E$ implies $C \leq_c E$: If $N_c(C \vee D\eta) = N_c(E)$ and $N_c(C) = C$ then, modulo variable-renaming, C is a subclause of E. On the other hand $C \leq_c E$ clearly does not imply $C \leq_D E$ for arbitrary D. Rather $C \leq_c E$ only implies the existence of a clause D with $N_c(C \vee D) = N_c(E)$.

PROPOSITION 2.8 *Let $\mathcal{C}$ be a finite set of clauses in condensation normal form. Then $\leq_c$ fulfils the properties below:*

(1) *If $\mathcal{D} \leq_c \mathcal{C}$ then $\mathcal{D}$ is finite,*

(2) *$\{\mathcal{D} \mid \mathcal{D} \leq_c \mathcal{C}\}$ is finite.*

(3) *$\leq_c$ is reflexive on sets of clauses in condensation normal form.*

(4) $\leq_c$ *is transitive on sets of clauses in condensation normal form.*

Proof: (1): If $\mathcal{D} \leq_c \mathcal{C}$ then for every $D \in \mathcal{D}$ there exists a $C \in \mathcal{C}$ s.t. $D \leq_c C$. That means C contains a subclause E with $N_c(E) = D$. But the number of all subclauses of clauses in $\mathcal{C}$ is finite and so is $\mathcal{D}$.

(2): Let $\mathcal{C}^*$ be the set of all subclauses of $\mathcal{C}$ in condensation normal form. Then for every $\mathcal{D}$ with $\mathcal{D} \leq_c \mathcal{C}$ we have $\mathcal{D} \subseteq \mathcal{C}^*$. But $\mathcal{C}^*$ is finite and thus the number of subsets is finite too.

(3): trivial.

(4): Assume $\mathcal{C}_1 \leq_c \mathcal{C}_2$ and $\mathcal{C}_2 \leq_c \mathcal{C}_3$. Then for every $C \in \mathcal{C}_1$ there exists a $D \in \mathcal{C}_2$ with $C \leq_c D$. For this D, in turn, there exists an $E \in \mathcal{C}_3$ s.t. $D \leq_c E$. But the subclause relation among clauses is transitive and so $C \leq_c E$, thus $\mathcal{C}_1 \leq_c \mathcal{C}_3$. ◇

DEFINITION 3.98 $\mathcal{FC}$ is the class of all sets of clauses $\mathcal{C}$ s.t. $\mathcal{C}$ contains only clauses in condensation normal form and $\mathcal{C}$ is finite. ♯

Our next goal is to prove that, for any EURC-operator T, the sequence $(T^i(\mathcal{C}))_{i \in \mathbb{N}}$ converges on stable sets of clauses $\mathcal{C}$ with $X^*(\mathcal{C}) \in \mathcal{FC}$. In particular we will show that the $T^i(\mathcal{C})$ are finite and the set $\{T^i(\mathcal{C}) \mid i \in \mathbb{N}\}$ is finite too; finally we have to show that the $T^i(\mathcal{C})$ cannot "loop", i.e. there exists no iteration step i s.t. $T^i(\mathcal{C}) \neq T^{i+1}(\mathcal{C})$ and $T^j(\mathcal{C}) = T^i(\mathcal{C})$ for some $j > i + 1$.

LEMMA 3.99 *Let $\mathcal{C}$ be a satisfiable and stable set of clauses s.t. $\mathcal{C}, X^*(\mathcal{C}) \in \mathcal{FC}$ and let T be a EURC-operator based on X and on some atom selection function α. Then*

(a) all $X^*(T^i(\mathcal{C}))$ are finite and

(b) $\{X^*(T^i(\mathcal{C})) \mid i \in \mathbb{N}\}$ is finite.

Proof: By assumption $X^*(\mathcal{C})$ is finite. Thus by Proposition 2.8 it suffices to prove $X^*(T^i(\mathcal{C})) \leq_c X^*(\mathcal{C})$ for all $i \in \mathbb{N}$.
We proceed by induction on i:
$i = 0$:
Trivial as $\leq_c$ is reflexive.

(IH) Assume that $X^*(T^i(\mathcal{C})) \leq_c X^*(\mathcal{C})$.
In order to prove that $X^*(T^{i+1}(\mathcal{C})) \leq_c X^*(\mathcal{C})$ it suffices to show

$$X^*(T^{i+1}(\mathcal{C})) \ \leq_c \ X^*(T^i(\mathcal{C}))$$

and then apply the transitivity of $\leq_c$.

Let $\mathcal{D} = T^i(\mathcal{C})$. Then, by (IH) and Proposition 2.8, $X^*(\mathcal{D})$ is finite; by definition of an EURC-operator $\mathcal{D}$ is X_s-stable. If $\alpha(\mathcal{D}) = \mathcal{D}$ then $T(\mathcal{D}) = \mathcal{D}$ and thus $T^{i+1}(\mathcal{C}) = T^i(\mathcal{C})$ and so the conclusion for $i+1$ follows directly from (IH). Therefore it remains to consider the case where $\alpha(\mathcal{D}) \neq \mathcal{D}$. Then $\alpha(\mathcal{D})$ is a nontrivial unit reduct of a stable set $\mathcal{D}$ in $\mathcal{FC}$. In the first part of the proof of Lemma 3.91 we have shown that $X^*(\alpha(\mathcal{D})) \leq_E X^*(\mathcal{D})$ for some clause E. By definition of $\leq_c$ we also get $X^*(\alpha(\mathcal{D})) \leq_c X^*(\mathcal{D})$. Thus, by transitivity of $\leq_c$, we may reduce the problem to $X^*(T(\mathcal{D})) \leq_c X^*(\alpha(\mathcal{D}))$. But

$$X^*(T(\mathcal{D})) = X^*(X_s^*(\alpha(\mathcal{D}))) = X^*(sub(X^*(\alpha(\mathcal{D})))) \subseteq X^*(X^*(\alpha(\mathcal{D})))$$

and, by idempotence of X^*, $X^*(T(\mathcal{D})) \subseteq X^*(\alpha(\mathcal{D}))$ and so $X^*(T(\mathcal{D})) \leq_c X^*(\alpha(\mathcal{D}))$. ◇

COROLLARY 2.6 *Let $\mathcal{C}$ be a satisfiable and stable set of clauses s.t. $\mathcal{C}, X^*(\mathcal{C}) \in \mathcal{FC}$ and let T be a EURC-operator based on X and on some atom selection function α. Then*

(a) all $T^i(\mathcal{C})$ are finite and

(b) $\{T^i(\mathcal{C}) \mid i \in \mathbb{N}\}$ is finite.

Proof: For all i we have $T^i(\mathcal{C}) \subseteq X^*(T^i(\mathcal{C}))$. Thus the finiteness conditions for $T^i(\mathcal{C})$ directly follows from Lemma 3.99. ◇

The following lemma shows that, in any generation of the sets $T^i(\mathcal{C})$, a substantially new unit clause is generated which is not subsumed by the sets of former generations.

LEMMA 3.100 *Let T be an EURC-operator based on an atom selection function α and on a PROP-operator X; furthermore let i be a natural number with $\alpha(T^i(\mathcal{C})) \neq T^i(\mathcal{C})$ for an X_s-stable set $\mathcal{C}$ in $\mathcal{FC}$. Then there exists a unit clause $P_{i+1} \in T^{i+1}(\mathcal{C})$ s.t.*

$$\bigcup_{j=0}^{i} T^j(\mathcal{C}) \not\leq_{sub} P_{i+1}.$$

Proof: By induction on i:
$i = 0$:
Let $\alpha(\mathcal{C}) = (\mathcal{C} - \{C\}) \cup \{P_1\}$. Then, as C is nonunit and condensed and P_1 is a literal in C, we have $P_1 \leq_{sub} C$ and $C \not\leq_{sub} P_1$. Because $\mathcal{C}$ is X_s-stable, $\mathcal{C}$ does not subsume P_1 (otherwise C would be properly subsumed by another clause in $\mathcal{C}$).

(IH): $\bigcup_{j=0}^{k} T^j(\mathcal{C}) \not\leq_{sub} P_{k+1}$ for $P_{k+1} \in T^{k+1}(\mathcal{C})$ and for k with $k < i$. Now let us assume that $\alpha(T^i(\mathcal{C})) \neq T^i(\mathcal{C})$. Then

$$\alpha(T^i(\mathcal{C})) = (T^i(\mathcal{C}) - \{C_i\}) \cup \{P_{i+1}\}$$

for some nonunit clause $C_i \in T^i(\mathcal{C})$ and P_{i+1} in C_i. Again we have $P_{i+1} \leq_{sub} C_i$ and $C_i \not\leq_{sub} P_{i+1}$. By definition of the operator T, $T^i(\mathcal{C})$ is X_s-stable and thus $T^i(\mathcal{C}) \not\leq_{sub} P_{i+1}$.

We assume that $\bigcup_{j=0}^{i} T^j(\mathcal{C}) \leq_{sub} P_{i+1}$ and derive a contradiction. So let k be the maximal number with $k \leq i$ s.t. $T^k(\mathcal{C}) \leq_{sub} P_{i+1}$. By the argument above k must be smaller than i. Then $\alpha(T^k(\mathcal{C})) \neq T^k(\mathcal{C})$, for otherwise $T^k(\mathcal{C}) = T^{k+1}(\mathcal{C}) = \cdots = T^i(\mathcal{C})$ – which is impossible by $\alpha(T^i(\mathcal{C})) \neq T^i(\mathcal{C})$. But then

$$\alpha(T^k(\mathcal{C})) = (T^k(\mathcal{C}) - \{C_k\}) \cup \{P_{k+1}\}$$

for some clause $C_k \in T^k(\mathcal{C})$ and some atom P_{k+1} in C_k. By assumption $T^k(\mathcal{C}) \leq_{sub} P_{i+1}$ and, by definition of α, $\alpha(T^k(\mathcal{C})) \leq_{sub} T^k(\mathcal{C})$. Thus, by transitivity of subsumption, $\alpha(T^k(\mathcal{C})) \leq_{sub} P_{i+1}$. But $T^{k+1}(\mathcal{C}) = X_s^*(\alpha(T^k(\mathcal{C}))$ and by Proposition 2.8 the sets $X^*(T^j(\mathcal{C}))$ are finite and thus $X^*(\alpha(T^k(\mathcal{C}))$ is finite. Therefore

$$X_s^*(\alpha(T^k(\mathcal{C})) = sub(X^*(\alpha(T^k(\mathcal{C}))) \leq_{sub} \alpha(T^k(\mathcal{C})).$$

Note that $\alpha(T^k(\mathcal{C}))$ is a subset of $X^*(\alpha(T^k(\mathcal{C}))$ and thus is subsumed by it. Putting things together we obtain $T^{k+1}(\mathcal{C}) \leq_{sub} P_{i+1}$. This contradicts the choice of k as maximal number with this property. ◇

PROPOSITION 2.9 *Let $\mathcal{C} \in \mathcal{FC}$ s.t. $\mathcal{C}$ is X_s-stable and $X^*(\mathcal{C}) \in \mathcal{FC}$. Then $(T^i(\mathcal{C}))_{i\in\mathbb{N}}$ converges for any EURC-operator T based on X.*

Proof: By Corollary 2.6 all $T^i(\mathcal{C})$ are finite and there are only finitely many different sets $T^i(\mathcal{C})$. However this does not immediately imply convergence, i.e. the existence of a number i s.t. $T^{i+1}(\mathcal{C}) = T^i(\mathcal{C})$. Indeed it might be the case that there are numbers i, j with $i + 1 < j$ s.t. $T^i(\mathcal{C}) \neq T^{i+1}(\mathcal{C})$ and $T^i(\mathcal{C}) = T^j(\mathcal{C})$ (looping of the sequence $T^i(\mathcal{C})$). But then $i < j$ and $\alpha(T^j(\mathcal{C})) \neq T^j(\mathcal{C})$ (if $\alpha(T^j(\mathcal{C}) = \mathcal{C}$ then, by $T^i(\mathcal{C}) = T^j(\mathcal{C})$, $T^{i+1}(\mathcal{C}) = T^i(\mathcal{C})$). But then, by Lemma 3.100, there exists a unit clause $P_{j+1} \in T^{j+1}(\mathcal{C})$ with $T^{i+1}(\mathcal{C}) \not\leq_{sub} P_{j+1}$. In particular it follows $T^{i+1}(\mathcal{C}) \neq T^{j+1}(\mathcal{C})$, in contradiction to the assumption $T^i(\mathcal{C}) = T^j(\mathcal{C})$. Therefore $(T^i(\mathcal{C}))_{i\in\mathbb{N}}$ is indeed convergent. ◇

DEFINITION 3.101 Let X be a PROP-operator. Then $\mathcal{FC}_X$ is the class of all sets of clauses $\mathcal{C}$ s.t. $\mathcal{C} \in \mathcal{FC}$ and $X^*(\mathcal{C}) \in \mathcal{FC}$ (i.e. $X^*(\mathcal{C})$ is finite). ♯

The following theorem shows that MBEQ always converges on $\mathcal{C}$ if the monotone closure $X^*(\mathcal{C})$ of the PROP-operator X is finite.

THEOREM 2.2 *Let X be the PROP-operator in the definition of MBEQ. Then*

- *MBEQ is sound,*
- *MBEQ is terminating on $\mathcal{FC}_X$.*

Proof: soundness:
MBEQ first computes $X^*(\mathcal{C})$. If $X^*(\mathcal{C})$ is infinite then MBEQ is non-terminating and does not return any answer. Thus we may assume that $X^*(\mathcal{C})$ is finite. If $\mathcal{C}$ is unsatisfiable then, by the completeness of X, $\Box \in X^*(\mathcal{C})$ and $sub(X^*(\mathcal{C})) = \{\Box\}$. In this case MBEQ returns *unsatisfiable.* If $\mathcal{C}$ is satisfiable then, by the soundness of X, $\Box \notin X^*(\mathcal{C})$ and also $\Box \notin sub(X^*(\mathcal{C}))$. Note that $X_s^*(\mathcal{C}) = sub(X^*(\mathcal{C}))$. Thus we enter the while-loop where the sequence $T^i(\mathcal{C}))$ is computed. If the while-loop terminates then there exists a k s.t. $\alpha(T^k(\mathcal{C})) = T^k(\mathcal{C})$. By definition MBEQ returns $POS(T^k(\mathcal{C}))$ if all positive clauses in the set are unit and *satisfiable* otherwise. By Lemma 3.99 all sets $X^*(T^i(\mathcal{C}))$ are finite. Therefore we may apply Lemma 3.91 repeatedly and obtain that $T^k(\mathcal{C})$ is satisfiable and $T^k(\mathcal{C})$ implies $\mathcal{C}$. Thus, in particular, every model of $T^k(\mathcal{C})$ is also a model of $\mathcal{C}$ itself. So, if $POS(T^k(\mathcal{C}))$ contains a nonunit clause, the answer *satisfiable* is correct. If $POS(T^k(\mathcal{C}))$ consists of unit clauses only then, as $T^k(\mathcal{C})$ is X_s-stable, $POS(T^k(\mathcal{C}))$ is an EARM of $T^k(\mathcal{C})$ by Proposition 2.5. But $T^k(\mathcal{C})$ implies $\mathcal{C}$ and thus $\mathcal{D} : POS(T^k(\mathcal{C}))$ is an EARM of $\mathcal{C}$. But in this case MBEQ returns $\mathcal{D}$.

Termination on $\mathcal{FC}_X$:
If $\mathcal{C} \in \mathcal{FC}_X$ then $X^*(\mathcal{C})$ is finite. In this case MBEQ computes a set $\mathcal{D}$ s.t. $\mathcal{D} = X_s^*(\mathcal{C})$. Then $\mathcal{D}$ is X_s-stable, $\mathcal{D} \in \mathcal{FC}$ and $X^*(\mathcal{D}) \subseteq X^*(\mathcal{C})$; in particular $X^*(\mathcal{D})$ is finite. Therefore $(T^i(\mathcal{D}))_{i\in\mathbb{N}}$ is convergent by Proposition 2.9; thus there exists a k with $\alpha(T^k(\mathcal{C})) = T^k(\mathcal{C})$ and the while-loop in MBEQ terminates. $\diamond$

Theorem 2.2 guarantees termination and soundness of the model building procedure, but does not tell whether we actually obtain an EARM of a satisfiable set of clauses $\mathcal{C}$. Indeed MBEQ might terminate with *satisfiable* without giving an EARM. There is some syntactic criterion for the deductive closure $X^*(\mathcal{C})$ which may guarantee full success of MBEQ. Similarly as for MB (see Definition 3.37 and Theorem 1.7) the variable structure in the positive clauses play a crucial role.

DEFINITION 3.102 Let X be a PROP-operator. The class $\mathcal{PDC}_X$ (positively disconnected w.r.t. X) is the subclass of all $\mathcal{C} \in \mathcal{FC}_X$ where all positive clauses in $X^*(\mathcal{C})$ are disconnected. ♯

Although the syntax condition for $\mathcal{C} \in \mathcal{PDC}_X$ is defined via the deductive closure $X^*(\mathcal{C})$ the test is not as expensive as it might appear in the first moment. In any case we are interested to refute $\mathcal{C}$ or to find out whether it is satisfiable. Once we have produced $X^*(\mathcal{C})$ we either have $\Box \in X^*(\mathcal{C})$ and model building is obsolete, or $\Box \notin X^*(\mathcal{C})$ and we easily test whether all positive clauses in $X^*(\mathcal{C})$ are disconnected.

If $X^*(\mathcal{C})$ is finite for $\mathcal{C} \in \mathcal{PVD}^=$ then $\mathcal{C} \in \mathcal{PDC}_X$. In fact it is easy to show that ordered paramodulation and positive resolution preserve the property that all positive clauses are ground (and therefore disconnected). The following lemma shows that $\mathcal{PDC}_X$ is invariant under any EURC-operator based on X.

LEMMA 3.103 *Let $\mathcal{C} \in \mathcal{PDC}_X$ and T be a EURC-operator based on X. Then, for all $i \in \mathbb{N}$, $T^i(\mathcal{C}) \in \mathcal{PDC}_X$.*

Proof: We prove by induction on i that, for all positive clauses C in $X^*(T^i(\mathcal{C}))$, there exists a clause $D \in X^*(\mathcal{C})$ s.t. $C \leq_c D$. By assumption all positive clauses in $X^*(\mathcal{C})$ are disconnected. But if $C \leq_c D$ and D is disconnected then C is disconnected too; this clearly gives $T^i(\mathcal{C}) \in \mathcal{PDC}_X$.

It remains to show

$A(i)$: For all positive clauses in $X^*(T^i(\mathcal{C}))$ there exists a positive clause $D \in X^*(\mathcal{C})$ with $C \leq_c D$.

Proof of $A(0)$: $X^*(T^0(\mathcal{C})) = X^*(\mathcal{C}) \in \mathcal{PDC}_X$.

(IH) Let us assume that $A(i)$ holds.

case $i + 1$:
By definition of T $T^{i+1}(\mathcal{C}) = X^*_s(\alpha(T^i(\mathcal{C})))$. If $\alpha(T^i(\mathcal{C})) = T^i(\mathcal{C})$ then $T^{i+1}(\mathcal{C}) = T^i(\mathcal{C})$ and $A(i+1)$ directly follows from $A(i)$.

Now let us assume $\alpha(T^i(\mathcal{C})) \neq T^i(\mathcal{C})$. Then $\alpha(T^i(\mathcal{C}))$ is a nontrivial unit reduct of $T^i(\mathcal{C})$. By the proof of Lemma 3.91 there exists a positive clause E s.t.

$$X^*(\alpha(T^i(\mathcal{C}))) \leq_E X^*(T^i(\mathcal{C})).$$

In particular, for every positive clause $C \in X^*(\alpha(T^i(\mathcal{C})))$, there exists a clause D in $X^*(T^i(\mathcal{C}))$ with $C \leq_E D$. But then, by Definition 3.27, either $N_c(C) = N_c(D)$ or $N_c(C \vee E\eta) = N_c(D)$ for some renaming substitution η; thus, in both cases, D is positive. For this clause D there exists by

(IH) a positive clause $G \in X^*(\mathcal{C})$ with $D \leq_c G$. But $C \leq_E D$ implies $C \leq_c D$ and $\leq_c$ is transitive; this gives $C \leq_c G$.

Therfore, for all $C \in X^*(\alpha(T^i(\mathcal{C})))$ there exists a $G \in X^*(\mathcal{C})$ s.t. G is positive and $C \leq_c G$. Now $X^*(T^{i+1}(\mathcal{C})) = X^*(X_s^*(\alpha(T^i(\mathcal{C})))) \subseteq X^*(X^*(\alpha(T^i(\mathcal{C}))) = X^*(\alpha(T^i(\mathcal{C})))$. This eventually yields $A(i+1)$. $\diamond$

We are now in the position to give a sharper version of Theorem 2.2 on the class $\mathcal{PDC}_X$:

THEOREM 2.3 *Let $\mathcal{C}$ be a set of clauses in $\mathcal{PDC}_X$. Then MBEQ terminates on $\mathcal{C}$; if $\mathcal{C}$ is unsatisfiable then MBEQ returns* unsatisfiable, *otherwise MBEQ returns an EARM of $\mathcal{C}$.*

Proof: By Theorem 2.2 and $\mathcal{PDC}_X \subseteq \mathcal{FC}_X$ MBEQ is sound and terminating on $\mathcal{PDC}_X$. It suffices to investigate the case where $\mathcal{C}$ is satisfiable. Then MBEQ enters the while-loop and, because it is terminating, leaves it with $\alpha(T^i(\mathcal{C})) = T^i(\mathcal{C})$ for some $i \in \mathbb{N}$. But then either

(a) all positive clauses are unit or

(b) there exists a positive clause C with $|C| \geq 2$ and for no literal L in C $V(L) \cap V(C \setminus L) = \emptyset$.

By Lemma 3.103 $X^*(T^i(\mathcal{C})) \in \mathcal{PDC}_X$ and all positive clauses in $T^i(\mathcal{C})$ (which is a subset of $X^*(T^i(\mathcal{C}))$) are disconnected. Therefore case (b) cannot occur. So MBEQ terminates with (a) and, by the soundness of MBEQ, it returns an EARM of $\mathcal{C}$. $\diamond$

3. Extraction of Finite Models

In Section 1.4 we have shown that Herbrand models representable by linear atomic representation can be transformed into finite models, even under preservation of the truth values for ground atoms. A consequence of this result was the finite model property of the class $\mathcal{PVD}$. In Section 2.3 we have presented an algorithmic procedure for constructing a ground equational representation of models on satisfiable clause sets in $\mathcal{PVD}^=$. So it is quite natural to ask, whether a similar extraction of finite models is possible also in the equational case. We will show in this section that the finite model property for $\mathcal{PVD}_g^=$ holds; on the other hand we will see that the extraction of finite models (as finite realization of Herbrand models) is more problematic in equational clause logic. We will restrict our investigation to GEARMs, where all atoms (equational and nonequational ones) are ground.

DEFINITION 3.104 Let $\mathcal{A}$ be a GEARM. Then we denote by $EQ(\mathcal{A})$ the set of equational atoms in $\mathcal{A}$, and by $PR(\mathcal{A})$ the set of nonequational atoms in $\mathcal{A}$. ♯

The equality of ground terms relative to the set $EQ(\mathcal{A})$ establishes the basic relation for the construction of finite models. Roughly speaking, the Herbrand universe under this congruence relation will yield the domain of the finite models we are going to construct.

DEFINITION 3.105 Let $\mathcal{A}$ be a GEARM w.r.t. a Herbrand universe H with $H(\mathcal{A}) \subseteq H$. We define an equivalence relation $\sim$ on H by

$$s \sim t \text{ iff } EQ(\mathcal{A}) \models s \doteq t.$$

We write $[s]$ for the equivalence class containing s, i.e. for the set $\{t | t \sim s\}$; the set $H_{\sim}$ of all equivalence classes of H under $\sim$ is denoted by $\hat{H}$. ♯

Remark: If $EQ(\mathcal{A}) = \emptyset$, i.e. if $\mathcal{A}$ is just a ground atomic representation, then $\hat{H}$ consists of the one-element sets $\{t\}$ for $t \in H$ and so is isomorphic to H itself. If $EQ(\mathcal{A}) \neq \emptyset$ then the predicate $EQ(\mathcal{A}) \models s \doteq t$ is decidable as well (this is just the property of ground congruence). A simple way to decide $s \sim t$ consists in constructing a canonic term rewrite system $\mathcal{R}$ from $EQ(\mathcal{A})$; then $s \sim t$ iff the normal forms of s and t under $\mathcal{R}$ are identical [Knuth and Bendix, 1970]. ♯

As we are interested in constructing a finite domain from the Herbrand universe H using the congruence relation $\sim$ we might hope that $\hat{H}$ turns out to be finite. However $\hat{H}$ is infinite in general and we have to extend $\sim$ to a new congruence relation making the universe finite. This is goal achieved by extending $EQ(\mathcal{A})$ to a set $EQ^*(\mathcal{A})$ in the definition below.

DEFINITION 3.106 Let $\mathcal{A}$ be a GEARM and $T(\mathcal{A})$ be set of all subterms occurring in $\mathcal{A}$. By $\hat{T}(\mathcal{A})$ we denote the set $\{[t] | t \in T(\mathcal{A})\}$ (i.e. the subset of $\hat{H}$ belonging to elements of $T(\mathcal{A})$). We define the extended set of equations $EQ^*(\mathcal{A})$ by cases:

Let $\mathrm{FS}(H) = \{f_1, \ldots, f_n\}$ and $a(i)$ be the arity of f_i for $i = 1, \ldots, n$.

(1) $\hat{T}(\mathcal{A}) = \hat{H}$: in this case we do not extend the set of equations and define $EQ^*(\mathcal{A}) = EQ(\mathcal{A})$.

(2) $\hat{T}(\mathcal{A}) \subset \hat{H}$:

(2a) There exists $c \in \mathrm{CS}(H)$ with $[c] \notin \hat{T}(\mathcal{A})$: select such a c from H; then perform the following steps:

$\mathcal{E} \leftarrow EQ(\mathcal{A})$;
for all $d \in \mathrm{CS}(H)$ **do if** $[d] \notin \hat{T}(\mathcal{A}) \cup \{[c]\}$ **then** $\mathcal{E} \leftarrow \mathcal{E} \cup \{d \doteq c\}$;
for $i \leftarrow 1$ to n **do**
for $[d_1], \ldots, [d_{a(i)}] \in \hat{T}(\mathcal{A}) \cup \{[c]\}$ do
if $[f_i(d_1, \ldots, d_{a(i)})] \notin \hat{T}(\mathcal{A})$
then $\mathcal{E} \leftarrow \mathcal{E} \cup \{f_i(d_1, \ldots, d_{a(i)}) \doteq c\}$;
Note that d_k stands for an arbitrary element of $[d_k]$.
Finally set $EQ^*(\mathcal{A}) \leftarrow \mathcal{E}$, $\mathcal{A}^* \leftarrow PR(\mathcal{A}) \cup EQ^*(\mathcal{A})$.

(2b) For all $c \in \mathrm{CS}(H)$: $[c] \in \hat{T}(\mathcal{A})$:
Let t be a term in H s.t. $[t] \notin \hat{T}(\mathcal{A})$ but for all proper subterms s of t: $[s] \in \hat{T}(\mathcal{A})$. Clearly t exists and is a functional term. We proceed in a way similar to case (2a):

$\mathcal{E} \leftarrow EQ(\mathcal{A})$;
for $i \leftarrow 1$ to n **do**
for $[d_1], \ldots, [d_{a(i)}] \in \hat{T}(\mathcal{A}) \cup \{[t]\}$ **do**
if $[f_i(d_1, \ldots, d_{a(i)})] \notin \hat{T}(\mathcal{A}) \cup \{[t]\}$
then $\mathcal{E} \leftarrow \mathcal{E} \cup \{f_i(d_1, \ldots, d_{a(i)}) \doteq t\}$;
$EQ^*(\mathcal{A}) \leftarrow \mathcal{E}$, $\mathcal{A}^* \leftarrow PR(\mathcal{A}) \cup EQ^*(\mathcal{A})$. ♯

EXAMPLE 3.107 Let $\mathcal{A}$ be the GEARM

$$\{P(a),\ P(g(a)),\ f(a) \doteq b, f(b) \doteq a\}$$

and let $H = H(\mathcal{A})$. Then $\hat{H}$, the set of all equivalence classes of H under $\{f(a) \doteq b,\ f(b) \doteq a\}$, is infinite: let $t_0 = a$ and $t_{n+1} = g(t_n)$ for all $n \in \mathbb{N}$; then for all m, n with $m \neq n$ we have $[t_n] \neq [t_m]$, i.e.

$$\{f(a) \doteq b, f(b) \doteq a\} \not\models t_n \doteq t_m$$

for $m \neq n$.

$\hat{T}(\mathcal{A}) = \{[a],\ [b],\ [g(a)]\}$.
Note that $[f(a)] = [b]$ and $[f(b)] = [a]$. But then, clearly, $\hat{T}(\mathcal{A}) \subset \hat{H}$ and part (1) of Definition 3.106 does not apply. So we have to look for (2):

There exists no constant symbol c in H s.t. $[c] \notin \hat{T}(\mathcal{A})$ and thus point (2a) does not apply.

In (2b) we have to find a term t with $[t] \notin \hat{T}(\mathcal{A})$, but $[s] \in \hat{T}(\mathcal{A})$ for all subterms s of t. The simplest candidate for this role is $t = g(b)$.

Before we enter the do-loop we have $\mathcal{E} = \{f(a) \doteq a, f(b) \doteq b\}$.
We set $f_1 = f$ and $f_2 = g$.

Thus for $n = 1$ we take care of the f-terms:

$$[f(a)] = [b],\ [f(b)] = [a],\ \text{so } [f(a)], [f(b)] \in \hat{T}(\mathcal{A}) \cup \{[g(b)],$$

$$\text{but } [f(g(a))], [f(g(b))] \notin \{[a],\ [b],\ [g(a)],\ [g(b)]\}.$$

Thus we have to add two equations and set

$$\mathcal{E} = \{f(a) \doteq b,\ f(b) \doteq a,\ f(g(a)) \doteq g(b),\ f(g(b)) \doteq g(b)\}.$$

It remains to take care of the g-terms, i.e. to perform the loop for $n = 2$:

$$[g(a)], [g(b)] \in \hat{T}(\mathcal{A}) \cup \{t\},\ \text{ but } [g(g(a))], [g(g(b))] \notin \hat{T}(\mathcal{A}) \cup \{t\}.$$

So we have to add the equations $\{g(g(a)) \doteq g(b), g(g(b)) \doteq g(b)\}$ and obtain the final result

$$\begin{aligned} EQ^*(\mathcal{A}) \quad = \quad & \{f(a) \doteq b,\ f(b) \doteq a,\ f(g(a)) \doteq g(b),\ f(g(b)) \doteq g(b), \\ & g(g(a)) \doteq g(b), g(g(b)) \doteq g(b)\}. \end{aligned}$$

♯

The first lemma shows that, like $T(\mathcal{A})$, $\hat{T}(\mathcal{A})$ is closed under the subterm relation.

LEMMA 3.108 *Let $\mathcal{A}$ be a GEARM and let t be a term in $H(\mathcal{A})$ with $[t] \in \hat{T}(\mathcal{A})$. Then $[s] \in \hat{T}(\mathcal{A})$ for every subterm s of t.*

Proof: By induction on the term depth $\tau(t)$ of t:

$\tau(t) = 0$:
In this case t is a constant and thus has no proper subterms; so the property trivially holds.

(IH) Let us assume that the lemma holds for t with $\tau(t) \leq n$.

Let $t \in H$ and $\tau(t) = n + 1$.
Then there exists an $m \geq 1$, an $f \in \mathrm{FS}_m(H)$ and $s_1, \ldots, s_m \in H$ with $t = f(s_1, \ldots, s_m)$. By assumption we have $[t] \in \hat{T}(\mathcal{A})$. We have to show that $[s_1], \ldots [s_m] \in \hat{T}(\mathcal{A})$. This is sufficient, as for a proper subterm r of a $[s_i]$ we may apply the induction hypothesis and obtain $[r] \in \hat{T}(\mathcal{A})$.

Now assume, for contradiction, that $[s_i] \notin \hat{T}(\mathcal{A})$ for some $i \in \{1, \ldots, m\}$. As $[t] \in \hat{T}(\mathcal{A})$ there exists a term $r \in T(\mathcal{A})$ with $[r] = [t]$; in particular $EQ(\mathcal{A}) \models r \doteq t$.

(a) $r = f(q_1, \ldots, q_m)$ and $EQ(\mathcal{A}) \models q_i \doteq s_i$:
This is impossible by $q_i \in T(\mathcal{A})$ and $[s_i] = [q_i]$ giving $[s_i] \in \hat{T}(\mathcal{A})$, in contradiction to the choice of s_i.

(b) $r = g(w_1, \ldots, w_p)$ for $g \neq f$.
Here $[r] = [t]$ is only possible if $EQ(\mathcal{A})$ contains an equation of the form

$$g(u_1, \dots, u_p) \doteq f(v_1, \dots, v_m)$$

with $EQ(\mathcal{A}) \models u_j \doteq w_j$, $EQ(\mathcal{A}) \models v_k \doteq s_k$. By definition of $T(\mathcal{A})$ we have $v_1, \dots v_k \in T(\mathcal{A})$ and thus also $[v_1], \dots [v_k] \in \hat{T}(\mathcal{A})$. But then, by $[s_i] = [v_i]$, we obtain $[s_i] \in \hat{T}(\mathcal{A})$, contradicting the choice of s_i. ◇

The second lemma shows that the extension of the GEARM $\mathcal{A}$ to $\mathcal{A}^*$ is sound, i.e. no new equality between terms in $T(\mathcal{A})$ can be derived. For proof-technical reasons we need the concept of P-derivation.

DEFINITION 3.109 A sequence π of equality atoms $e_1, \dots, e_n$ is called a *P-derivation* of e_n from a set of equational atoms $\mathcal{E}$ if for all $i = 1, \dots, n$ either $e_i \in \mathcal{E}$ or there exist $j, k < i$ s.t. e_i is a paramodulant of e_j and e_k. The *length* $l(\pi)$ of π is defined as n. ♯

Remark: It is well known that, for a set of ground equations $\mathcal{E}$, $\mathcal{E} \models s \doteq t$ iff there is a P-derivation of $s \doteq t$ from $\mathcal{E} \cup \mathcal{I}$, where $\mathcal{I} = \{r \doteq r | r \in H(\mathcal{E})\}$ [Chang and Lee, 1973]. ♯

LEMMA 3.110 *Let $\mathcal{A}$ be a GEARM over a signature Σ with $H(\mathcal{A}) \subseteq H(\Sigma)$ and let $EQ^*(\mathcal{A})$ be the set of equations defined in Definition 3.106. Assume that $EQ^*(\mathcal{A}) \models s \doteq t$ for some terms $s, t \in H(\Sigma)$. Then either $EQ(\mathcal{A}) \models s \doteq t$ or $[s], [t] \notin \hat{T}(\mathcal{A})$.*

Proof: By induction on the length $l(\eta)$ of a P-derivation η of $s \doteq t$ from $EQ^*(\mathcal{A}) \cup \mathcal{I}$, where $\mathcal{I} = \{r \doteq r | r \in H(\Sigma)\}$.

$l(\eta) = 1$:
Then $\eta = s \doteq t$. If $s \doteq t \in \mathcal{I} \cup EQ(\mathcal{A})$ then, clearly, $EQ(\mathcal{A}) \models s \doteq t$. If $s \doteq t$ is in $EQ^*(\mathcal{A})$ and $\mathcal{I} \cup EQ(\mathcal{A}) \not\models s \doteq t$ then, by definition of $EQ^*(\mathcal{A})$, $[s], [t] \notin \hat{T}(\mathcal{A})$.

(IH) Assume that for all P-derivations η of $s \doteq t$ from $EQ^*(\mathcal{A}) \cup \mathcal{I}$ with $l(\eta) \leq n$ we either have $EQ(\mathcal{A}) \models s \doteq t$ or $[s], [t] \notin \hat{T}(\mathcal{A})$.

Now let $\eta : \eta', s \doteq t$ be a derivation of length $n + 1$ of $s \doteq t$ from $EQ^*(\mathcal{A}) \cup \mathcal{I}$. If $s \doteq t$ is in $EQ^*(\mathcal{A}) \cup \mathcal{I}$ then the argument of the induction base can be applied again. So the only interesting case is this, where $s \doteq t$ is a paramodulant of equations $s_1 \doteq t_1$ and $s_2 \doteq t_2$ occurring in η'.

If $EQ(\mathcal{A}) \models s \doteq t$ then we are done. So it remains to investigate the case $EQ(\mathcal{A}) \not\models s \doteq t$. We have to show that $[s], [t] \notin \hat{T}(\mathcal{A})$.

By (IH) $EQ(\mathcal{A}) \models s_i \doteq t_i$ or $s_i, t_i \notin \hat{T}(\mathcal{A})$ for $i = 1, 2$.

(a) $EQ(\mathcal{A}) \models s_1 \doteq t_1$ and $EQ(\mathcal{A}) \models s_2 \doteq t_2$:
Then by definition of $\models$ and by the soundness of paramodulation we get $EQ(\mathcal{A}) \models s \doteq t$, in contradiction to the assumption.

(b) $EQ(\mathcal{A}) \models s_1 \doteq t_1$ and $[s_2], [t_2] \notin \hat{T}(\mathcal{A})$.

(b1) $s_2 \doteq t_2$ is the active clause:
Let us assume that $s \doteq t$ is obtained by paramodulating s_2 into a subterm in s_1 at position λ, i.e. $(s_1)_\lambda = s_2$ and $s \doteq t$ is of the form $s_1[t_2]_\lambda \doteq t_1$. All other cases, where s_2 is paramodulated into a subterm in t_1, or t_2 is paramodulated into a subterm in s_1 or in t_1 are obviously symmetric. Now let us assume that $[s_1[t_2]_\lambda]$ is in $\hat{T}(\mathcal{A})$. By Lemma 3.108 $\hat{T}(\mathcal{A})$ is closed under subterms; i.e. if $[s] \in \hat{T}(\mathcal{A})$ and r is a subterm of s then $[r] \in \hat{T}(\mathcal{A})$. But this implies that $[t_2] \in \hat{T}(\mathcal{A})$, contradicting the assumption. If, on the other hand, $t_1 \in \hat{T}(\mathcal{A})$ then also $s_1 \in \hat{T}(\mathcal{A})$ and for every subterm r of s_1 $[r] \in \hat{T}(\mathcal{A})$. But, as s_2 is a subterm of s_1, this implies $[s_2] \in \hat{T}(\mathcal{A})$, contradicting the choice of s_2. So we have shown that neither s nor t are in $\hat{T}(\mathcal{A})$.

(b2) $s_1 \doteq t_1$ is the active clause:
It is enough to investigate the replacement of s_1 by t_1 in s_2; all other cases are obviously symmetric. So $s \doteq t$ is of the form

$$s_2[t_1]_\lambda \doteq t_2.$$

If paramodulation takes place on the top of s_2, i.e. $\lambda = \epsilon$, then $s_1 = s_2$ and $s \doteq t$ is of the form $t_1 \doteq t_2$. By assumption $t_2 \notin \hat{T}(\mathcal{A})$; we have to show that also $t_1 \notin \hat{T}(\mathcal{A})$. But $t_1 \in \hat{T}(\mathcal{A})$ and $EQ(\mathcal{A}) \models s_1 \doteq t_1$ implies $s_1 \in \hat{T}(\mathcal{A})$; by $s_1 = s_2$ this gives $s_2 \in \hat{T}(\mathcal{A})$, contradicting the choice of s_2. So we have shown $t_1 \notin \hat{T}(\mathcal{A})$.

Now let us assume that s_2 is a function term $f(r_1, \ldots, r_m)$ and $\lambda \neq \epsilon$. We have to show that $s_2[t_1]_\lambda \notin \hat{T}(\mathcal{A})$. Let us assume that $s_2[t_1]_\lambda$ is an element of $\hat{T}(\mathcal{A})$. Then by Lemma 3.108 $[t_1] \in \hat{T}(\mathcal{A})$ and, by $EQ(\mathcal{A}) \models s_1 \doteq t_1$, we get $EQ(\mathcal{A}) \models s_2[t_1]_\lambda \doteq s_2[s_1]_\lambda$. But $s_2[s_1]_\lambda$ is just s_2 and so

$$EQ(\mathcal{A}) \models s_2[t_1]_\lambda \doteq s_2.$$

So, by $s_2[t_1]_\lambda \in \hat{T}(\mathcal{A})$ we also get $s_2 \in \hat{T}(\mathcal{A})$, contradicting the choice of s_2. Therefore, also in this case, $[s], [t] \notin \hat{T}(\mathcal{A})$.

(c) $[s_1], [t_1] \not\in \hat{T}(\mathcal{A})$, $EQ(\mathcal{A}) \models s_2 \doteq t_2$: symmetric to case (b).

(d) $[s_1], [t_1] \not\in \hat{T}(\mathcal{A})$ and $[s_2], [t_2] \not\in \hat{T}(\mathcal{A})$:
It is enough to investigate the case where $s_1 \doteq t_1$ is the active clause and s_1 is replaced by t_1 in s_2. Then $s \doteq t$ is of the form $s_2[t_1]_\lambda \doteq t$. It is sufficient to show that $s_2[t_1]_\lambda \not\in \hat{T}(\mathcal{A})$. Now $s_2[t_1]_\lambda$ contains the term t_1 which, by assumption, is not in $\hat{T}(\mathcal{A})$. Therefore, by Lemma 3.108, $s_2[t_1]_\lambda \not\in \hat{T}(\mathcal{A})$ too. ◇

Remark: Lemma 3.110 shows that, by extending $EQ(\mathcal{A})$ to $EQ^*(\mathcal{A})$, it is impossible to derive a new equation containing a term in $\hat{T}(\mathcal{A})$. ♯

DEFINITION 3.111 Let $\mathcal{A}$ be a GEARM and H be a Herbrand universe containing that of $\mathcal{A}$. Then $\sim_e$ is defined to be the equivalence relation given by

$$s \sim_e t \text{ iff } EQ^*(\mathcal{A}) \models s \doteq t.$$

The equivalence class of a term t is denoted by $[t]_e$. ♯

If $\hat{T}(\mathcal{A}) \neq [H]$ then $[H]$, the set of all equivalence classes under $\sim$, may be infinite. The next proposition shows that $\sim_e$ decomposes the Herbrand universe into finitely many equivalence classes; these equivalence classes will become the domains of finite models.

PROPOSITION 3.1 *Let $\mathcal{A}$ be a GEARM and $EQ(\mathcal{A}), EQ^*(\mathcal{A})$ be as in the Definitions 3.104 and 3.106. Then we have*

(1) If $[s] \in \hat{T}(\mathcal{A})$ then $[s] = [s]_e$.

(2) Either $H|_{\sim_e} = \hat{T}(\mathcal{A})$ or there exists a term t in H with $H|_{\sim_e} = \hat{T}(\mathcal{A}) \cup \{[t]_e\}$.

Proof: By Lemma 3.110 $EQ^*(\mathcal{A}) \models s \doteq t$ and $[s] \in \hat{T}(\mathcal{A})$ implies $EQ(\mathcal{A}) \models s \doteq t$; therefore $r \in [s]_e$ implies $r \in [s]$ and $[s] = [s]_e$. This proves (1).

(2):
If $H|_\sim = \hat{T}(\mathcal{A})$ then, by definition of $EQ^*(\mathcal{A})$, we have $EQ^*(\mathcal{A}) = EQ(\mathcal{A})$ and so $\sim_e = \sim$.

Therefore it remains to investigate the case $\hat{T}(\mathcal{A}) \subset H|_\sim$. According to the definition of $EQ^*(\mathcal{A})$ we have to distinghish two cases:

(a) There exists a $c \in \mathrm{CS}(H)$ with $c \not\in \hat{T}(\mathcal{A})$.
Then, by definition of $EQ(\mathcal{A})$, we have

$$EQ^*(\mathcal{A}) - EQ(\mathcal{A}) = \{c \doteq d \mid d \in \mathrm{CS}(H) - \hat{T}(\mathcal{A})\} \cup$$

$\{f(r_1,\dots,r_k) \doteq c \mid [r_1],\dots,[r_k] \in \hat{T}(\mathcal{A}) \cup \{c\}, [f(r_1,\dots,r_k)] \notin \hat{T}(\mathcal{A})\}$.

We prove by induction on the term depth $\tau(t)$ of a term t that $[t] \notin \hat{T}(\mathcal{A})$ implies $[t]_e = [c]_e$.

$\tau(t) = 0$:
Then t is a constant symbol with $[t] \notin \hat{T}(\mathcal{A})$. By definition of $EQ^*(\mathcal{A})$ we have $t \doteq c \in EQ^*(\mathcal{A})$ and therefore $[t]_e = [c]_e$.

(IH) Assume that for all t with $\tau(t) \leq n$ and $[t] \notin \hat{T}(\mathcal{A})$ we have $[t]_e = [c]_e$.
Now let t be a term with $\tau(t) = n+1$ and $t \notin \hat{T}(\mathcal{A})$. Then t is of the form $f(s_1,\dots,s_m)$ where $f \in \mathrm{FS}_m(H)$ and $s_1,\dots,s_m \in H$. Now $\tau(s_i) \leq n$ for $i = 1,\dots,m$ and, by (IH), $[s_i] \in \hat{T}(\mathcal{A}) \cup \{[c]\}$ for $i = 1,\dots,m$. So we have $[f(s_1,\dots,s_m)] \notin \hat{T}(\mathcal{A})$ and for all s_i $[s_i] \in \hat{T}(\mathcal{A}) \cup \{[c]\}$ and, by definition, $EQ^*(\mathcal{A})$ contains the equation $f(s_1,\dots,s_m) \doteq c$. Therefore $EQ^*(\mathcal{A}) \models f(s_1,\dots,s_m) \doteq c$ and $[t]_e = [c]_e$.

(b) $\mathrm{CS}(H) = \mathrm{CS}(\hat{T}(\mathcal{A}))$:
In this case there exists a term $t \in H$ s.t., for all proper subterms s of t, $[s] \in \hat{T}(\mathcal{A})$, but $[t] \notin \hat{T}(\mathcal{A})$. By definition of $EQ^*(\mathcal{A})$ we have

$$EQ^*(\mathcal{A}) - EQ(\mathcal{A}) = \{f(r_1,\dots,r_m) \doteq t \mid [r_1],\dots,[r_m] \in \hat{T}(\mathcal{A}) \cup \{[t]\}, [f(r_1,\dots,r_m)] \notin \hat{T}(\mathcal{A}), f \in \mathrm{FS}_m(H)\}.$$

We prove that for all terms r with $[r] \notin \hat{T}(\mathcal{A})$ $[r]_e = [t]_e$; like in case (a) we proceed by induction on term depth.

$\tau(r) = 0$:
such terms do not exist and the statement is trivially true.

(IH) Assume that for all terms r with $\tau(r) \leq n$ and $[r] \notin \hat{T}(\mathcal{A})$ we have $[r]_e = [t]_e$;

Now let r be a term with $\tau(r) = n+1$ and $[r] \notin \hat{T}(\mathcal{A})$. Then r is of the form $f(s_1,\dots,s_m)$ where $f \in \mathrm{FS}_m(H)$ and $s_1,\dots,s_m \in H$.

By $\tau(s_i) \leq n$ we may apply (IH) and obtain

$$[s_1] \in \hat{T}(\mathcal{A}) \cup \{[t]\},\dots,[s_m] \in \hat{T}(\mathcal{A}) \cup \{[t]\}.$$

By $[f(s_1,\dots,s_m)] \notin \hat{T}(\mathcal{A})$ $EQ^*(\mathcal{A})$ contains the equation $f(s_1,\dots,s_m) \doteq t$. Therefore

$$EQ^*(\mathcal{A}) \models f(s_1,\dots,s_m) \doteq t$$

and thus $[f(s_1,\ldots,s_m)]_e = [t]_e$. ◇

EXAMPLE 3.112 Let

$$\mathcal{A} = \{P(a),\ P(g(a)),\ f(a) \doteq b, f(b) \doteq a\}$$

like in Example 3.107. We have shown that

$$\begin{array}{rcl} EQ^*(\mathcal{A}) & = & \{f(a) \doteq b,\ f(b) \doteq a,\ f(g(a)) \doteq g(b),\ f(g(b)) \doteq g(b), \\ & & g(g(a)) \doteq g(b), g(g(b)) \doteq g(b)\}. \end{array}$$

Recall that $s \sim_e t$ iff $EQ^*(\mathcal{A}) \models s \doteq t$. Indeed we have

$$H|_{\sim_e} = \hat{T}(\mathcal{A}) \cup \{[g(b)]\} = \{[a],[b],[g(a)],[g(b)]\}.$$

Consider, for example, the term $f(g(f(g(a))))$. Then by using the equations $f(g(a)) \doteq g(b), g(g(b)) \doteq g(b)$ and $f(g(b)) \doteq g(b)$ we derive $f(g(f(g(a)))) \doteq g(b)$, i.e.

$$EQ^*(\mathcal{A}) \models f(g(f(g(a)))) \doteq g(b) \text{ and } f(g(f(g(a)))) \sim_e g(b).\ \sharp$$

The equivalence classes under $\sim_e$ can serve as a finite domain of a model realizing the Herbrand model $\mathcal{M}(\mathcal{A},\Sigma)$.

DEFINITION 3.113 Let $\mathcal{A}$ be a GEARM and Σ be a signature with $\Sigma(\mathcal{A}) \subseteq \Sigma$. We define a finite interpretation $\mathcal{F}(\mathcal{A},\Sigma)$ in the following way:

$$\begin{array}{rcl} \mathcal{F}(\mathcal{A},\Sigma) & = & (D,\Phi), \text{ where} \\ D & = & H(\Sigma)|_{\sim_e}, \\ \Phi(c) & = & [c]_e \text{ for } c \in \mathrm{CS}(\Sigma), \\ \Phi(f([s_1]_e,\ldots,[s_m]_e) & = & [f(s_1,\ldots,s_m)]_e \\ & & \text{for } f \in \mathrm{FS}_m(\Sigma),\ s_1,\ldots,s_m \in H, \\ \Phi(P)([s_1]_e,\ldots[s_m]_e) & = & \textbf{true} \text{ iff } \mathcal{A} \models P(s_1,\ldots,s_m) \\ & & \text{for } P \in \mathrm{PS}_m(\Sigma).\ \sharp \end{array}$$

PROPOSITION 3.2 *Let $\mathcal{A}$ be a GEARM and Σ be a signature with $\Sigma(\mathcal{A}) \subseteq \Sigma$; then $\mathcal{F}(\mathcal{A},\Sigma)$ is well-defined.*

Proof: The only potential problem in Definition 3.113 is the semantics of the predicate symbols, where it is not completely trivial that unique truth values are actually obtained. What we have to show is that the truth value under $\mathcal{A}$ only depends on the equivalence class, and not on specific representants of the class. We distinguish the cases $\mathcal{A} \models A$ and $\mathcal{A} \not\models A$:

(a) $\mathcal{A} \models P(s_1, \ldots, s_m)$ and $[s_i]_e = [t_i]_e$ for $i = 1, \ldots, m$ implies $\mathcal{A} \models P(t_1, \ldots, t_m)$.
By Definition of $\sim_e$ $[s_i]_e = [t_i]_e$ means that

$$EQ^*(\mathcal{A}) \models s_i \doteq t_i \text{ for } i = 1, \ldots, m.$$

By $\mathcal{A} \models P(s_1, \ldots, s_m)$ there exists an atom $P(r_1, \ldots, r_m) \in \mathcal{A}$ with

$$EQ(\mathcal{A}) \models r_1 \doteq s_1, \ldots EQ(\mathcal{A}) \models r_m \doteq s_m.$$

By definition of $\hat{T}(\mathcal{A})$ we have $[r_1], \ldots, [r_m] \in \hat{T}(\mathcal{A})$ and, by definition of $\sim$, also $[s_1], \ldots, [s_m] \in \hat{T}(\mathcal{A})$.

Now $EQ^*(\mathcal{A}) \models s_i \doteq t_i$ and $[s_i] \in \hat{T}(\mathcal{A})$ implies $EQ(\mathcal{A}) \models s_i \doteq t_i$ by Lemma 3.110.

Putting things together we have $\mathcal{A} \models P(s_1, \ldots, s_m)$ and $EQ(\mathcal{A}) \models s_i \doteq t_i$; by definition of $\models$ we eventually obtain $\mathcal{A} \models P(t_1, \ldots, t_m)$.

(b) $\mathcal{A} \not\models P(s_1, \ldots, s_m)$ and $[s_i]_e = [t_i]_e$ for $i = 1, \ldots, m$ implies $\mathcal{A} \not\models P(t_1, \ldots, t_m)$.
If $[s_i] = [t_i]$ for $i = 1, \ldots, m$ then the property holds by definition of $\sim$ and of $\models$.

If, for some $j \in \{1, \ldots, m\}$, $[s_j] \neq [t_j]$ but $[s_j]_e = [t_j]_e$ then, by Proposition 3.1, $[s_j]_e = [t_j]_e = [t]_e$ for a term $t \in H$ with $[t] \notin \hat{T}(\mathcal{A})$. But this means $[t_j] \notin \hat{T}(\mathcal{A})$ and, by definition of the semantics of GEARMs and of $\hat{T}(\mathcal{A})$, $\mathcal{A} \not\models P(t_1, \ldots, t_m)$. ◇

PROPOSITION 3.3 *Let A be a nonequational ground atom over a signature Σ and let $\mathcal{A}$ be a GEARM with $\Sigma(\mathcal{A}) \subseteq \Sigma$ (we assume that Σ always contains the constants of $H(\mathcal{A})$). Let $\mathcal{M} = \mathcal{M}(\mathcal{A}, \Sigma)$ and $\mathcal{F} = \mathcal{F}(\mathcal{A}, \Sigma)$; then $v_{\mathcal{M}}(A) = v_{\mathcal{F}}(A)$.*

Proof: Let $A = P(s_1, \ldots, s_m)$. We distinguish two cases:

(a) $v_{\mathcal{M}}(A) = \mathbf{true}$:
By definition of $\mathcal{M}(\mathcal{A}, H)$ we get $\mathcal{A} \models A$.
Moreover, by definition of $\mathcal{F}(\mathcal{A}, \Sigma)$ and for $\mathcal{F}(\mathcal{A}, \Sigma) = (D, \Phi)$ we obtain

$$\Phi(P)([s_1]_e, \ldots, [s_m]_e) = \mathbf{true}.$$

An easy induction on term depth shows that for all ground terms s we have $t_{\mathcal{F}}(s) = [s]_e$; this, in turn, yields

$$(*) \quad \begin{array}{rcl} \Phi(P)([s_1]_e, \ldots, [s_m]_e) & = & \Phi(P)(t_{\mathcal{F}}(s_1), \ldots, t_{\mathcal{F}}(s_m)) \\ & = & v_{\mathcal{F}}(P(s_1, \ldots, s_m)), \end{array}$$

and so $v_{\mathcal{F}}(P(s_1, \ldots, s_m) = \textbf{true}$.

(b) $v_{\mathcal{M}}(A) = \textbf{false}$:
Then $\mathcal{A} \not\models P(s_1, \ldots, s_m)$ and, by definition of $\mathcal{F}$,

$$\Phi(P)([s_1]_e, \ldots [s_m]_e) = \textbf{false}.$$

But then, using (∗) again, we obtain $v_{\mathcal{F}}(A) = \textbf{false}$. ◇

Remark: Proposition 3.3 does not hold for equational ground atoms: Let $\mathcal{A} = \{P(a), f(a) \doteq a\}$ and $\Sigma = \{a, b, f, P\}$. Then, clearly, $[f^n(a)] = [a]$ for all $n \in \mathbb{N}$, but on the other hand $[f^n(b)] \neq [f^m(b)]$ for $n \neq m$; in particular

$$\mathcal{A} \not\models b \doteq f(b).$$

By definition of $\sim_e$ we have

$$\begin{aligned} EQ^*(\mathcal{A}) &= \{f(a) \doteq a, f(b) \doteq b\} \text{ and} \\ H|_{\sim_e} &= \{[a]_e, [b]_e\}. \end{aligned}$$

Therefore we get $[b]_e = [f(b)]_e$ and

$$t_{\mathcal{F}}(b) = t_{\mathcal{F}}(f(b)) = [b]_e$$

for the term evaluation function $t_{\mathcal{F}}$ corresponding to $\mathcal{F}(\mathcal{A}, \Sigma)$. But, for every equality interpretation $\mathcal{N}$,

$$v_{\mathcal{N}}(s \doteq t) = \textbf{true} \text{ iff } t_{\mathcal{N}}(s) = t_{\mathcal{N}}(t).$$

This gives us $v_{\mathcal{F}}(b \doteq f(b)) = \textbf{true}$ and, at the same time, $v_{\mathcal{M}}(b \doteq f(b)) = \textbf{false}$ and so

$$v_{\mathcal{M}}(b \doteq f(b)) \neq v_{\mathcal{F}}(b \doteq f(b)).$$

The example above shows that, in interpreting *equational* sets of clauses, the truth values of the interpretations $\mathcal{M}(\mathcal{A}, \Sigma)$ and $\mathcal{F}(\mathcal{A}, \Sigma)$ may differ. In particular $\mathcal{F}(\mathcal{A}, \Sigma)$ cannot be a finite realization of all equational Herbrand models $\mathcal{M}(\mathcal{A}, \Sigma)$ of a set of clauses $\mathcal{C}$.

Just take $\mathcal{C} = \{\neg f(b) \doteq b\}$. Clearly $\mathcal{C}$ is true in the equational Herbrand model $\mathcal{M}(\mathcal{A}, \Sigma)$ (for $\mathcal{A}$ defined above), but false in $\mathcal{F}(\mathcal{A}, \Sigma)$. This phenomenon requires a separate analysis of finite model building for equational clause logic. ♯

The following theorem shows that, for nonequational sets of clauses, GEARMS $\mathcal{A}$ can be replaced by finite interpretations via $\mathcal{F}(\mathcal{A}, \Sigma)$.

THEOREM 3.1 *Let $\mathcal{C}$ be a set of nonequational clauses and let $\mathcal{M}$ be a Herbrand interpretation of $\mathcal{C}$ with $\mathcal{M} = \mathcal{M}(\mathcal{A}, \Sigma)$ for a GEARM $\mathcal{A}$. Let $\mathcal{F} = \mathcal{F}(\mathcal{A}, \Sigma)$, then*

(1) $v_{\mathcal{F}}(\mathcal{C}) = v_{\mathcal{M}}(\mathcal{C})$,

(2) If $\mathcal{M}$ is a model of $\mathcal{C}$ then $\mathcal{F}$ is a finite realization of $\mathcal{M}$.

Proof: It suffices to prove

(a) $v_{\mathcal{F}}(\mathcal{A}) = v_{\mathcal{M}}(A)$ for all ground atoms A over Σ,

(b) $v_{\mathcal{F}}(\mathcal{C}) = v_{\mathcal{M}}(\mathcal{C})$.

Clearly (b) coincides with point (1) of the theorem and (2) directly follows from (a).

(a) was shown in Proposition 3.3, so it remains to prove (b). The proof of (b) can be further reduced to

$v_{\mathcal{M}}(C) = v_{\mathcal{F}}(C)$ for all $C \in \mathcal{C}$.

To this aim we follow the path of Lemma 3.68 and of Proposition 1.5 in Section 1.

To every v-instance $\mathcal{M}_I$ of $\mathcal{M}$ we assign the canonic image $\mathcal{F}_{\Phi(I)}$: $(D, \Psi, \Phi(I))$ where

$$\begin{aligned} \Phi(I) &= \{(x_1, [t_1]_e), \ldots, (x_m, [t_m]_e)\} \\ \text{for } I &= \{(x_1, t_1), \ldots, (x_m, t_m)\}. \end{aligned}$$

Looking at the proof of Poposition 1.5 we see that the result is largely independent of the specific equivalence relation [ ] defined on ground terms. So we are going to prove the analogous version of Lemma 3.68 expressed below:

(*) Let $\mathcal{M}_I$ be a v-instance of $\mathcal{M}$ and $\mathcal{F}_J$ be the canonic image of $\mathcal{M}_I$. Then, for all literals L over $\Sigma(\mathcal{C})$ we have $v_{\mathcal{M}_I}(L) = v_{\mathcal{F}_J}(L)$.

Clearly it is enough to prove (*) for atoms; the extension to negative literals is obvious.

So let

$$\begin{aligned} A &= P(s_1, \ldots, s_n) \text{ and} \\ I &= \{(x_1, t_1), \ldots, (x_m, t_m)\}, \end{aligned}$$

and η be the substitution $\{x_1 \leftarrow t_1, \ldots, x_m \leftarrow t_m\}$. As $\mathcal{M}$ is a Herbrand interpretation we obtain

$$\begin{aligned} (I)\ v_{\mathcal{M}_I}(P(s_1, \ldots, s_n)) &= \Phi(P)(t_{\mathcal{M}_I}(s_1), \ldots, t_{\mathcal{M}_I}(s_n)) \\ &= \Phi(P)(s_1\eta, \ldots, s_n\eta). \end{aligned}$$

By Proposition 3.2 $\mathcal{F}$ is well-defined, i.e. $\mathcal{A} \models P(s_1\eta, \ldots, s_n\eta)$ implies $\mathcal{A} \models P(r_1, \ldots, r_n)$ for all r_i $(i = 1, \ldots, n)$ with $[r_i]_e = [s_i\eta]_e$. So

$$(II)\ \Phi(P)(s_1\eta, \ldots, s_n\eta) = \textbf{true} \quad \text{iff} \quad \mathcal{A} \models P(s_1\eta, \ldots, s_n\eta)$$
$$\text{iff} \quad \Psi(P)([s_1\eta]_e, \ldots, [s_n\eta]_e) = \textbf{true}$$

by definition of $\mathcal{F}$ from $\mathcal{M}$.

Let $J = \{(x_1, [t_1]_e), \ldots, (x_m, [t_m]_e)\}$ be the v-instance over D corresponding to I. Our aim is to prove

$$(**) \qquad [s_i\eta]_e = t_{\mathcal{F}_J}(s_i) \text{ for } i = 1, \ldots, n$$

which, in turn, implies

$$(III)\ \Psi(P)([s_1\eta]_e, \ldots, [s_n\eta]_e) = \Psi(P)(t_{\mathcal{F}_J}(s_1), \ldots, t_{\mathcal{F}_J}(s_n))$$
$$= v_{\mathcal{F}_J}(P(s_1, \ldots, s_n)).$$

Putting (I), (II) and (III) together we finally get

$$v_{\mathcal{M}_I}(P(s_1, \ldots, s_n)) = v_{\mathcal{F}_J}(P(s_1, \ldots, s_n)).$$

It remains to prove $(**)$:
More generally we prove that $t_{\mathcal{F}_J}(s) = [s\eta]_e$ for all terms s over $\Sigma(\mathcal{C})$ with variables in $\{x_1, \ldots, x_m\}$. We proceed by induction on the term depth τ of s:

$\tau(s) = 0$:

(a) s is a constant symbol.
Then, by definition of $\mathcal{F}$, $\Psi(s) = [s]_e$ and so

$$t_{\mathcal{F}_J}(s) = \Psi(s) = [s]_e = [s\eta]_e.$$

(b) s is a variable.
Then $s = x_i$ for some $i \in \{1, \ldots, m\}$ and the definitions of J and of η yield

$$t_{\mathcal{F}_J}(x_i) = J(x_i) = [t_i]_e = [x_i\eta]_e.$$

(IH) Assume $t_{\mathcal{F}_J}(s) = [s\eta]_e$ for $\tau(s) \leq k$.
Now let s be a term with $\tau(s) = k + 1$. Then there exists a function symbol f and terms $r_1, \ldots, r_p$ of depth $\leq k$ such that $s = f(r_1, \ldots, r_p)$. So we obtain

$$t_{\mathcal{F}_J}(f(r_1, \ldots, r_p)) = \Psi(f)(t_{\mathcal{F}_J}(r_1), \ldots, t_{\mathcal{F}_J}(r_p))$$
$$=_{IH} \Psi(f)([r_1\eta]_e, \ldots, [r_p\eta]_e).$$

By definition of Ψ we have

$$\Psi(f)([r_1\eta]_e, \ldots, [r_p\eta]_e) = [f(r_1\eta, \ldots, r_p\eta)]_e.$$

But

$$[f(r_1\eta, \ldots, r_p\eta)]_e = [f(r_1, \ldots, r_p)\eta]_e = [s\eta]_e$$

and therefore

$$t_{\mathcal{F}_J}(s) = [s\eta].$$

$\diamond$

COROLLARY 3.1 *Let $\mathcal{C}$ be a set of clauses having a Hebrand model representable by a GEARM $\mathcal{A}$. Then $\mathcal{C}$ has a finite model.*

Proof: The finite model of $\mathcal{C}$ is $\mathcal{F}(\mathcal{A}, \Sigma(\mathcal{C}))$. $\diamond$

EXAMPLE 3.114 Let $\mathcal{C}$ be an arbitrary set of (nonequational) clauses over the signature $\Sigma = \{P, a, b, f, g\}$ having a Hebrand model $\mathcal{M}$ represented by

$$\mathcal{A} = \{P(a),\ P(g(a)),\ f(a) \doteq b,\ f(b) \doteq a\}.$$

We construct $\mathcal{F}:\ (D, \Phi)$ for $\mathcal{F} = \mathcal{F}(\mathcal{A}, \Sigma)$:
Per definition of $\mathcal{F}$ $D = H|_{\sim_e}$ where $\sim_e$ is defined by the set of equations $EQ^*(\mathcal{A})$. Thus, from Example 3.112, we derive

$$D = \{[a],\ [b],\ [g(a)],\ [g(b)]\}.$$

It remains to define Φ:
By Definition 3.113 we obtain

$$\begin{array}{rcl} \Phi(a) & = & [a], \\ \Phi(b) & = & [b], \\ \Phi(f) & = & \{([a],[b]),\ ([b],[a]),\ ([g(a)],[g(b)]),\ ([g(b)],[g(b)])\}, \\ \Phi(g) & = & \{([a],[g(a)]),\ ([b],[g(b)]),\ ([g(a)],[g(b)]),\ ([g(b)],[g(b)])\}, \\ \Phi(P) & = & \{([a],\mathbf{true}),\ ([b],\mathbf{false}),\ ([g(a)],\mathbf{true}),\ ([g(b)],\mathbf{false}). \end{array}$$

$\sharp$

Clause evaluation:

Theorem 3.1 gives us a mean for algorithmic *clause evaluation* over GEARMs. Let $\mathcal{A}$ be a GEARM and $\mathcal{M}$ be the corresponding Herbrand model w.r.t. a signature Σ. Furthermore let $\mathcal{C}$ be a set of clauses over Σ. Then, by Theorem 3.1, $v_{\mathcal{M}}(\mathcal{C}) = v_{\mathcal{F}}(\mathcal{C})$ for the corresponding finite realization $\mathcal{F}$. But, as the domain of $\mathcal{F}$ is finite, the value $v_{\mathcal{F}}(\mathcal{C})$ is computable in finitely many steps. So let $\mathcal{A}$ be a GEARM, $\mathcal{C}$ be a set of (nonequational) clauses and Σ be a signature containing these of $\mathcal{A}$ and $\mathcal{C}$. Then the value $v_{\mathcal{M}(\mathcal{A},\Sigma)}(\mathcal{C})$ can be computed in the following way:

- compute $\mathcal{F} : \mathcal{F}(\mathcal{A}, \Sigma)$;
- compute $v_{\mathcal{F}}(\mathcal{C})$.

In order to compute $\mathcal{F}$ we have to use an algorithm computing the congruence relation $\sim$ and then, via computation of $EQ^*(\mathcal{A})$, the equivalence classes under $\sim_e$; the remaining computations are straightforward.

We have shown that Theorem 3.1 does not hold for equational sets of clauses. Nevertheless the method can used for constructing finite models on the class $\mathcal{PVD}_g^=$; we only need to include the ground equations of the clause set $\mathcal{C} \in \mathcal{PVD}_g^=$ into the set of basic ground terms which define the set $\hat{T}(\mathcal{A})$. We illustrate the problem in the example below:

EXAMPLE 3.115 Let

$$\begin{aligned} \mathcal{A} &= \{P(a), a \doteq f(a)\} \text{ and} \\ \mathcal{C} &= \{c \not\doteq f(c)\}. \end{aligned}$$

as in the remark after Proposition 3.3. We have seen that $\mathcal{C}$ is true in $\mathcal{M}(\mathcal{A}, \Sigma)$ (for $\Sigma = \Sigma(\mathcal{A}) \cup \Sigma(\mathcal{C})$), but false in $\mathcal{F}(\mathcal{A}, \Sigma)$. This comes from the identification of c and $f(c)$ in $\mathcal{F}(\mathcal{A}, \Sigma)$. Now let us extend the set $\mathcal{A}$ to

$$\mathcal{A}^* = \{P(a), a \doteq f(a), c \doteq c, f(c) \doteq f(c)\}.$$

Clearly $\mathcal{A}$ and $\mathcal{A}^*$ define the same EH-models, i.e. $\mathcal{M}(\mathcal{A}, \Sigma) = \mathcal{M}(\mathcal{A}^*, \Sigma)$. But the inclusion of the terms c and $f(c)$ changes the set of terms T; indeed we obtain

$$T(\mathcal{A}^*) = \{a, f(a), c, f(c)\}.$$

Under the equivalence relation $\sim$, defined via $\mathcal{A}^*$ we obtain $[a] = [f(a)]$, but $[c] \neq [f(c)]$. Therefore

$$\hat{T}(\mathcal{A}^*) = \{[a], [c], [f(c)]\}.$$

The first term t with $[t] \notin \hat{T}(\mathcal{A}^*)$ is $f(f(c))$. This term plays the key role in the definition of $EQ^*(\mathcal{A}^*)$ and we obtain

$$\begin{aligned} EQ^*(\mathcal{A}^*) &= \{a \doteq f(a), f(f(f(c))) \doteq f(f(c))\}, \text{ while} \\ EQ^*(\mathcal{A}) &= \{a \doteq f(a), f(c) \doteq c\}. \end{aligned}$$

Let us compare $\mathcal{F} : \mathcal{F}(\mathcal{A}, \Sigma)$ and $\mathcal{F}' : \mathcal{F}(\mathcal{A}^*, \Sigma)$: the domain D of $\mathcal{F}$ is $\{[a]_e, [c]_e\}$, but the domain D' of $\mathcal{F}'$ is $\{[a]_e, [c]_e, [f(c)]_e, [f(f(c))]_e\}$. Moreover

$$\begin{array}{rcl} t_{\mathcal{F}'}(c) & = & [c]_e, \\ t_{\mathcal{F}'}(f(c)) & = & [f(c)]_e \text{ and therefore} \\ t_{\mathcal{F}'}(c) & \neq & t_{\mathcal{F}'}(f(c)). \end{array}$$

We see that $\mathcal{F}'$ is a model of $\mathcal{C}$, while $\mathcal{F}$ is none. ♯

Note that for the evaluation of a nonequational clause set $\mathcal{C}$ over a GEARM $\mathcal{A}$ the syntax of $\mathcal{C}$ has virtually no effect on the finite realization $\mathcal{F}(\mathcal{A}, \Sigma)$; but the example above shows that, in the equational case, the size of the finite domain strongly depends on the terms occurring in $\mathcal{C}$.

DEFINITION 3.116 Let $\mathcal{A}$ be a GEARM and let $\mathcal{C}$ be a set of clauses in $\mathcal{PVD}_g^=$. Moreover let GET($\mathcal{C}$) be the set of all terms occurring in the (ground) equational atoms in $\mathcal{C}$. Then we define an extended GEARM by

$$\mathcal{A}^*(\mathcal{C}) = \mathcal{A} \cup \{s \doteq s | s \in \mathrm{GET}(\mathcal{C})\}.$$

♯

By using $\mathcal{A}^*$ instead of $\mathcal{A}$ on $\mathcal{C} \in \mathcal{PVD}_g^=$ we can preserve the truth values of the equational atoms occurring in $\mathcal{C}$:

LEMMA 3.117 *Let $\mathcal{A}$ be a GEARM, $\mathcal{C} \in \mathcal{PVD}_g^=$, $\Sigma = \Sigma(\mathcal{A}) \cup \Sigma(\mathcal{C})$, and $\mathcal{M} : \mathcal{M}(\mathcal{A}, \Sigma)$ be the EH-interpretation defined by $\mathcal{A}$. Let $\mathcal{A}^*(\mathcal{C})$ be as in Definition 3.116 and $\mathcal{F} = \mathcal{F}(\mathcal{A}^*(\mathcal{C}), \Sigma)$. Then for all equational atoms $s \doteq t$ occurring in $\mathcal{C}$ we have*

$$v_{\mathcal{F}}(s \doteq t) \quad = \quad v_{\mathcal{M}}(s \doteq t).$$

Proof: Let $s \doteq t$ be an arbitrary equational atom occurring in $\mathcal{C}$. By definition of $\mathcal{M}$ we have $v_{\mathcal{M}}(s \doteq t) = \mathbf{true}$ iff $\mathcal{A} \models s \doteq t$. By definition of $\mathcal{A}^*(\mathcal{A})$ we have

$$\mathcal{A} \models s \doteq t \quad \text{iff} \quad \mathcal{A}^*(\mathcal{C}) \models s \doteq t.$$

Moreover, by definition of $\sim$ via $\mathcal{A}^*(\mathcal{A})$,

$$\mathcal{A}^*(\mathcal{C}) \models s \doteq t \quad \text{iff} \quad [s] = [t].$$

As $s \doteq t$ occurs in $\mathcal{C}$ we have $[s], [t] \in \hat{T}(\mathcal{A}^*(\mathcal{C}))$. Therefore, by Proposition 3.1, $[s] = [s]_e$ and $[t] = [t]_e$. But this implies

$$(I)\ \mathcal{A}^*(\mathcal{C}) \models s \doteq t \text{ iff } [s]_e = [t]_e$$

and $v_{\mathcal{M}}(s \doteq t) = \mathbf{true}$ iff $[s]_e = [t]_e$.

Now consider $v_{\mathcal{F}}(s \doteq t)$; by definition of the equational semantics

$$v_{\mathcal{F}}(s \doteq t) = \mathbf{true} \text{ iff } t_{\mathcal{F}}(s) = t_{\mathcal{F}}(t).$$

By definition of $\mathcal{F}$ we also get $t_{\mathcal{F}}(s) = [s]_e$ and $t_{\mathcal{F}}(t) = [t]_e$. Therefore

$$(II)\ v_{\mathcal{F}}(s \doteq t) = \mathbf{true} \text{ iff } [s]_e = [t]_e.$$

Finally (I) and (II) together yield

$$v_{\mathcal{M}}(s \doteq t) = v_{\mathcal{F}}(s \doteq t).$$

◇

THEOREM 3.2 *Let $\mathcal{A}$ be a GEARM, $\mathcal{C} \in \mathcal{PVD}_g^=$, $\Sigma = \Sigma(\mathcal{A}) \cup \Sigma(\mathcal{C})$, and $\mathcal{M} : \mathcal{M}(\mathcal{A}, \Sigma)$ be the EH-interpretation defined by $\mathcal{A}$. Let $\mathcal{A}^*(\mathcal{C})$ be as in Definition 3.116 and $\mathcal{F} = \mathcal{F}(\mathcal{A}^*(\mathcal{C}), \Sigma)$. Then $v_{\mathcal{F}}(\mathcal{C}) = v_{\mathcal{M}}(\mathcal{C})$.*

Proof: It is sufficient to prove $v_{\mathcal{F}}(\{C\}) = v_{\mathcal{M}}(\{C\})$ for all $C \in \mathcal{C}$. Now let C be a clause in $\mathcal{C}$. By definition of $\mathcal{PVD}_g^=$ all equational atoms in C are ground and so $\{C\}$ is logically equivalent to a disjunction $D \vee E$, where D is a universal closure of disjunctions of nonequational atoms and E a disjunction of ground literals only. But then

$$v_{\mathcal{M}}(\{C\}) = or(v_{\mathcal{M}}(D), v_{\mathcal{M}}(E)).$$

By definition of $\mathcal{A}^*(\mathcal{C})$ we have $\mathcal{M}(\mathcal{A}, \Sigma) = \mathcal{M}(\mathcal{A}^*(\mathcal{C}), \Sigma)$ and so, by Theorem 3.1, $v_{\mathcal{M}}(D) = v_{\mathcal{F}}(D)$.

By Lemma 3.117 we have $v_{\mathcal{M}}(s \doteq t) = v_{\mathcal{F}}(s \doteq t)$ for all atoms in E and therefore $v_{\mathcal{M}}(E) = v_{\mathcal{F}}(E)$. This eventually yields

$$v_{\mathcal{M}}(\{C\}) \quad = \quad v_{\mathcal{F}}(\{C\}).$$

◇

COROLLARY 3.2 *$\mathcal{PVD}_g^=$ has the finite model property.*

Proof: Assume that $\mathcal{C} \in \mathcal{PVD}_g^=$ and that $\mathcal{C}$ is satisfiable. By Theorem 2.2 MBEQ terminates on $\mathcal{C}$ and produces a GEARM $\mathcal{A}$ of a Herbrand model $\mathcal{M} : \mathcal{M}(\mathcal{A}, \Sigma)$ for $\Sigma = \Sigma(\mathcal{A}) \cup \Sigma(\mathcal{C})$. Let $\mathcal{F} = \mathcal{F}(\mathcal{A}^*(\mathcal{C}), \Sigma)$; then, by Theorem 3.2, $v_{\mathcal{M}}(\mathcal{C}) = v_{\mathcal{F}}(\mathcal{C})$. But $\mathcal{F}$ is a finite interpretation and thus a finite model of $\mathcal{C}$. ◇

Remark: MBEQ + the construction of $\mathcal{F}(\mathcal{A}^*(\mathcal{C}), \Sigma)$ from $\mathcal{A}$ and $\mathcal{C}$ defines a method of finite model building on $\mathcal{PVD}_g^=$ *without backtracking*. In the first step MBEQ constructs a GEARM $\mathcal{A}$ of $\mathcal{C}$. Afterwards

$EQ^*(\mathcal{A}^*(\mathcal{C}))$ and the classes $[s]_e$ are computed, which form the domain of the finite model. Finally the tables for the functions and predicates in $\mathcal{F} : \mathcal{F}(\mathcal{A}^*(\mathcal{C}), \Sigma)$ are computed via $\sim_e$ and the definition of $\mathcal{F}$. We give an example below. ♯

EXAMPLE 3.118 Let

$$\begin{aligned}\mathcal{C} &= \{P(a) \vee P(f(a)),\ f(a) \doteq b,\ f(b) \doteq a,\\ &\quad \neg P(f(f(x))) \vee P(x),\ \neg P(f(x)) \vee \neg P(x)\}\end{aligned}$$

as in Example 3.94. We have shown that $T^i(X_s^*(\mathcal{C}))$ converges to the set

$$\begin{aligned}&\{f(a) \doteq b,\ f(b) \doteq a,\ x \doteq x,\ P(a),\ \neg P(b),\ \neg P(f(a)),\\ &\neg P(f(f(x))) \vee P(x),\ \neg P(f(x)) \vee \neg P(x)\}.\end{aligned}$$

¿From this fixed point we extract the GEARM $\mathcal{A}$ for

$$\mathcal{A} = \{f(a) \doteq b,\ f(b) \doteq a,\ P(a)\}.$$

¿From $\mathcal{A}$, using the construction in Definition 3.113, we obtain the finite model $\mathcal{F} : (D, \Phi)$, where

$$\begin{aligned}\Phi(a) &= [a],\\ \Phi(b) &= [b],\\ \Phi(f) &= \{([a],[b]),\ ([b],[a])\},\\ \Phi(P) &= \{([a],\mathbf{true}),([b],\mathbf{false})\}.\end{aligned}$$

♯

Chapter 4

CONSTRAINT-BASED METHODS

The methods presented in this chapter are based on the use of *constrained formulae*, and are devoted to the principle of *simultaneous search for refutations and models* (respectively proofs and counter-models). They are extensions of proof procedures (resolution and tableaux) obtained by adding new rules, so-called disinference rules, to the standard inference rules. The purpose of these new rules is to discard redundant instances of formulae or of clauses. The expression of these redundant instances (by constraints) prevents some further applications of inference rules and thus leads to a reduction of the search space. In some cases, the method may produce Herbrand models, represented by sets of constrained literals. By the principle of lazy unification, constraint-based inference is efficient and useful also in ordinary refutational theorem proving without features of model building.

Constraint languages are very useful modelling tools, also because they encode conditions arising in a lot of problem specifications in a natural way. They can be considered as a reasonable compromise between expressive power and efficiency. They have been popularized by constraint programming languages where constraints are imposed on variables over specific domains (for example arithmetic, tree- or finite domain constraints see [Marriott and Stuckey, 1998]).

Despite the obvious advantages in using constraints, their use in automated deduction did not become popular before the nineties.

To the best of the authors' knowledge the first (naive) explicit use of constraints in automated deduction is Chang's V-resolution [Chang, 1972]. V-resolution stands for "variable-constrained resolution". Chang has introduced V-resolution in order to implement by resolution, what he called "modified Herbrand's procedures" (i.e. searching through sets M_i

(i=1,2,...) of variants of clauses until possibly detecting a substitution σ such that $M_j\sigma$ is propositionally unsatisfiable).

In V-resolution renaming is forbidden and copies of clauses must therefore be introduced explicitely (see [Chang, 1972] and also [Leitsch, 1997]).

In Chang's terminology a *clause with constrained variables* is a couple $C \parallel \sigma$ where C is a disjunction of literals and σ is a substitution (having the set of constrained variables as domain); the purpose of σ is to keep track of the values already assigned to variables.

As an example, consider the two clauses $C_1 : P(a)$ and $C_2 : \neg P(x) \vee P(f(x))$. As the variable x in C_2 is "rigid" only the clause $P(f(a))$ can be generated by V-resolution instead of infinitely many ones in ordinary resolution. By admitting two copies of C_2, e.g. $\neg P(x) \vee P(f(x))$ and $\neg P(y) \vee P(f(y))$ we can also derive $P(f(f(a)))$, etc...

Our use of constraints is Herbrand-based and thus, in some sense, related to Chang's. Indeed we consider clauses as schemas for (typically infinite) sets of ground instances over a Herbrand universe. Properties of inference (and disinference) among these sets of ground clauses can be conveniently expressed by constraints on the variables of the original clauses (i.e. on the admitted substitutions). The inference system RAMC (**R**esolution **A**nd **M**odel **C**onstruction) to be presented in this chapter is based on a clever use of these "Herbrand" constraints.

Before presenting RAMC itself, we have to introduce the constraint language. This language is defined by *equational formulae* as originally defined and analyzed in [Comon, 1988, Comon and Lescanne, 1989]. Informally speaking, equational formulae are standard first-order formulae containing equality as the only predicate symbol.

1. Equational formulae

1.1 Definition and semantics

DEFINITION 4.1 The set of *equational formulae* is the least subset of first-order formulae inductively defined as follows:

- $\top$ (true) and $\bot$ (false) are equational formulae.
- If s and t are two terms, then $t = s$ is an equational formulae.
- If $\mathcal{F}$ is an equational formula, then $\neg\mathcal{F}$ is an equational formula.
- If $\mathcal{F}$ is an equational formula, and x is a variable, then $\exists x.\mathcal{F}$ and $\forall x.\mathcal{F}$ are equational formulae.
- If $\mathcal{F}$ and $\mathcal{G}$ are equational formulae, then $(\mathcal{F} \vee \mathcal{G})$ and $(\mathcal{F} \wedge \mathcal{G})$ are equational formulae. ♯

As usual, the formula $\neg(t = s)$ is called a *disequation* and will be denoted by $t \neq s$.

Remark: ¿From now on, all equational formulae will be considered to be *rectified*, i.e. distinct quantifiers bind different variables. ♯

It is well known that the validity problem (or equivalently the satisfiability problem) is undecidable for equational formulae. However, by restricting interpretations over the structure of finite trees, the equational theory becomes decidable. This has been shown in the early 70's by [Mal'cev, 1971]. In this interpretation, the equality predicate is simply interpreted as syntactic equality on the Herbrand universe (over the given signature), i.e. the ground equation $t = s$ is true iff t and s are *syntactically* equal. The variables range over the set of *ground* terms defined by the underlying signature. More precisely, the notion of *solution* of an equational formula is defined as follows:

DEFINITION 4.2 Let $\mathcal{F}$ be an equational formula and Σ be a signature. A ground substitution σ of the free variables of $\mathcal{F}$ over Σ is said to be a *solution* of $\mathcal{F}$ iff one of the following conditions holds:

- $\mathcal{F}$ is $\top$.
- $\mathcal{F}$ is of the form $s = t$ and $s\sigma$ is syntactically equal to $t\sigma$.
- $\mathcal{F}$ is of the form $\neg\mathcal{G}$ and σ is not a solution of $\mathcal{G}$.
- $\mathcal{F}$ is of the form $(\mathcal{G} \vee \mathcal{H})$ and σ is a solution of $\mathcal{G}$ or a solution of $\mathcal{H}$.
- $\mathcal{F}$ is of the form $(\mathcal{G} \wedge \mathcal{H})$ and σ is a solution of $\mathcal{G}$ and a solution of $\mathcal{H}$.
- $\mathcal{F}$ is of the form $\exists x.\mathcal{G}$ and there exists a ground term t such that the substitution $\sigma \cup \{x \leftarrow t\}$ is a solution of $\mathcal{G}$.
- $\mathcal{F}$ is of the form $\forall x.\mathcal{G}$ and for all ground terms t, the substitution $\sigma \cup \{x \leftarrow t\}$ is a solution of $\mathcal{G}$.

If $\mathcal{F}$ is an equational formula, we denote by $\mathcal{S}_\Sigma(\mathcal{F})$ the set of solutions of $\mathcal{F}$. Two equational formulae $\mathcal{F}$ and $\mathcal{G}$ are said to be *equivalent* (denoted by $\mathcal{F} \equiv \mathcal{G}$) iff $\mathcal{S}_\Sigma(\mathcal{F}) = \mathcal{S}_\Sigma(\mathcal{G})$. ♯

Remark: The set of solutions of an equational formula does not only depend on the formula itself but also on the signature. Therefore the set of solutions of $\mathcal{F}$ over Σ is denoted by $\mathcal{S}_\Sigma(\mathcal{F})$. Frequently the underlying signature is fixed and evident, and we write $\mathcal{S}(\mathcal{F})$ instead of $\mathcal{S}_\Sigma(\mathcal{F})$. ♯

EXAMPLE 4.3 The formulae:

$$\mathcal{F}_1 : f(x) = x$$

and

$$\mathcal{F}_2 : x \neq a \wedge \forall y.x \neq f(y)$$

are both equational formulae.

$\mathcal{F}_1$ has no solution, no matter which signature we choose, because the terms x and $f(x)$ are not unifiable (due to a positive occur check); i.e. $\mathcal{S}_\Sigma(\mathcal{F}_1) = \emptyset$ for all signatures Σ.

The set of solutions of $\mathcal{F}_2$ depends on the signature Σ:

- If $\Sigma = \{a, f\}$, then $\mathcal{S}_\Sigma(\mathcal{F}_2) = \emptyset$. Indeed, every ground term over Σ is either a or of the form $f(t)$, for $t \in H(\Sigma)$.
- If Σ' contains function symbols or constants not in Σ then $\mathcal{S}_{\Sigma'}(\mathcal{F}_2) \neq \emptyset$. For example, if $\Sigma' = \{a, f, g\}$, where g is a unary function symbol then

$$\mathcal{S}_{\Sigma'}(\mathcal{F}_2) \quad = \quad \{\{x \leftarrow g(s)\} | s \in H(\Sigma')\}.$$

♯

1.2 Transformation into solved form

If an equational formula contains no universally quantified variable, no disjunction and no negation, then the problem of finding the set of solutions of this formula can be reduced to the well known *unification problem* i.e. to decide whether two (possibly non-ground) terms are unifiable or not. Efficient algorithms have been designed for solving this problem (see [Martelli and Montanari, 1982] and Chapter 3).

In particular, every unification problem admits a most general solution, called a *most general unifier* (m.g.u. for short). Therefore, a very natural question is whether there exists a notion analogous to an m.g.u. for equational formulae, i.e. is it possible to compute a solution which is more general than the other ones?

Obviously, this is *not possible* due to the presence of disjunctions; consider for example $F : x = a \vee x = b$; F has the solutions $\{x \leftarrow a\}$ and $\{x \leftarrow b\}$, none of them being more general than the other one. Moreover, it is clear, that, due to the presence of negations, it is even impossible – in general – to describe the set of solutions of an equational formula by a finite disjunction of m.g.u.'s. For example, the solutions of the formula $x = f(u, v) \wedge u \neq v$ (over $\Sigma = \{a, f\}$) *cannot* be described by a (finite) disjunction of unification problems. We see that *disequations are needed* for a finite representation of the set of all solutions.

Fortunately, a notion similar to that of most general unifier can be defined for equational formulae: the notion of a *definition with constraints* [Comon and Lescanne, 1989].

Roughly speaking, a definition with constraints consists in a substitution *plus* a set of negative constraints (i.e. disequations) on the parameters of the substitution.

DEFINITION 4.4 A formula $\mathcal{X}$ is said to be a *definition with constraints* iff it is either $\top$, or $\bot$, or is of the form

$$\exists w_1, \ldots, w_n.[\bigwedge_{j=1}^{m} x_j = s_j] \wedge [\bigwedge_{i=1}^{k} x'_i \neq s'_i]$$

where:

- the x_j's ($1 \leq j \leq m$) are variables appearing only once in $\mathcal{X}$;
- each x'_i's ($1 \leq i \leq k$) are variables syntactically different from the term s'_i;
- if $H(\Sigma)$ is finite (i.e. if Σ does not contain function symbols) then $k = 0$.

The first part of the formula (i.e. $\bigwedge_{j=1}^{m} x_j = s_j$) corresponds, roughly speaking, to a substitution of the variables x_j, whereas the second part imposes some further constraints on the *parameters* of this substitution (i.e. on the variables occurring in s_j for some $j \in \{1, \ldots, m\}$).

EXAMPLE 4.5 We consider the following formulae:

$$\mathcal{F}_1 : x = f(u, v) \wedge y = f(v, u) \wedge v \neq u \wedge v \neq succ(u) \wedge u \neq succ(v),$$

$$\mathcal{F}_2 : x = f(u, v) \wedge u \neq x,$$

$$\mathcal{F}_3 : x = f(u, v) \wedge f(u, v) \neq f(v, u).$$

$\mathcal{F}_1$ is a definition with constraints. Its solutions are substitutions σ such that:

$x\sigma = f(t, s)$,
$y\sigma = f(s, t)$,
t is neither s nor $succ(s)$ and s is not $succ(t)$,
$u\sigma = t$,
$v\sigma = s$.

$\mathcal{F}_2$ and $\mathcal{F}_3$ are *not* definitions with constraints. Indeed the second part of $\mathcal{F}_2$ contains the variable x whereas $\mathcal{F}_3$ contains a disequations on the nonvariable terms $f(u, v)$ and $f(v, u)$.

Note that, for $\mathcal{F}_2$, $x = f(u,v) \wedge u \neq x \equiv x = f(u,v) \wedge u \neq f(u,v)$. Moreover $u = f(u,v)$ has no solution and thus $u \neq f(u,v)$ is equivalent to $\top$. Hence $\mathcal{F}_2$ is equivalent to $x = f(u,v)$ which is a definition with constraints.

By using a unification algorithm we see that $f(u,v) = f(v,u)$ is equivalent to $u = v$, and so $\mathcal{F}_3$ is equivalent to $x = f(u,v) \wedge u \neq v$, which is a definition with constraints. ♯

An interesting feature of definitions with constraints is that *any definition with constraints distinct from $\bot$ has at least one solution.* This phenomenon is essentially based on the following lemma.

LEMMA 4.6 *Assume that $H(\Sigma)$ is an infinite Herbrand universe and $\mathcal{F} : \forall \bar{y}. \bigvee_{i=1}^{n} z_i = u_i$ a formula fulfilling the following conditions:*

(a) $\mathcal{F}$ does not contain free variables.

(b) For all i, z_i is a variable distinct from the term u_i.

Then $\mathcal{F} \equiv \bot$.

Proof: We proceed by induction on $\rho(\mathcal{F})$, the number of variables in the quantifier prefix of $\mathcal{F}$.

If $\rho(\mathcal{F}) = 0$ then $\mathcal{F}$ is identical to $\bot$ and the lemma is trivially true.

(IH) Assume that for all $\mathcal{F}$ fulfilling (a) and (b) with $\rho(\mathcal{F}) \leq m$, $\mathcal{F}$ is equivalent to $\bot$.

Let $\rho(\mathcal{F}) = m + 1$. Then $\mathcal{F}$ is of the form

$$\forall \bar{y}. \forall x. (z_1 = u_1 \vee \cdots \vee z_n = u_n),$$

where $\bar{y}$ is a variable vector of length m.

We distinguish two cases:

1. $x \notin \{z_1, \ldots, z_n\}$:
   Let t be an arbitrary term in $H(\Sigma)$ and let

   $$\mathcal{F}' \quad = \quad \forall \bar{y}. (z_1 = u_1\{x \leftarrow t\} \vee \cdots \vee z_n = u_n\{x \leftarrow t\}).$$

   Then $\mathcal{F}'$ fulfils (a) and (b) and $\mathcal{F} \rightarrow \mathcal{F}' \equiv \top$. But $\rho(\mathcal{F}') \leq m$ and thus, by (IH), $\mathcal{F}'$ is equivalent to $\bot$; so $\mathcal{F} \equiv \bot$.

2. $x \in \{z_1, \ldots, z_n\}$:
   In this case $\mathcal{F}$ is equivalent to $\mathcal{F}'$ for

   $$\mathcal{F}' = \forall \bar{y}. (\forall x. (x = t_1 \vee \cdots x = t_k \vee y_1 = w_1 \vee \cdots \vee y_r = w_r) \vee R)$$

where:

- R is the disjunction of the $z_i = u_i$ not containing x,
- $\{y_1, \ldots, y_r\} \subseteq V(\bar{y})$,
- $x \in w_i$ for $i = 1, \ldots, r$
- and $\{t_1, \ldots, t_k, w_1, \ldots, w_r\} \subseteq \{u_1, \ldots, u_n\}$.

Note that R may be $\bot$ and/or r may be 0.

Let $\mathcal{G}$ be

$$\forall \bar{y}.\forall x.(x = t_1 \vee \ldots x = t_k \vee y_1 = w_1 \vee \cdots \vee y_r = w_r)$$

We prove that $\mathcal{G} \equiv \bot$:

If $x \in V(t_i)$ then, by $x \neq t_i$, we can delete the disjunct $x = t_i$ because $x\sigma = t\sigma \equiv \bot$ for all ground substitutions σ.

After elimination of these equations we obtain a formula $\mathcal{G}'$ which is equivalent to $\mathcal{G}$ for

$$\mathcal{G}' \quad = \quad \forall \bar{y}.\forall x.(x = s_1 \vee \cdots \vee x = s_p \vee y_1 = w_1 \vee \cdots \vee y_r = w_r)$$

where $x \notin V(\{s_1, \ldots, s_p\})$ and $\{s_1, \ldots, s_p\} \subseteq \{t_1, \ldots, t_k\}$.

Now let σ be a ground substitution with $dom(\sigma) = V(\bar{y})$. Then

$$\mathcal{G}'_\sigma : \ \forall x.(x = s_1\sigma \vee \ldots \vee x = s_p\sigma \vee y_1\sigma = w_1\sigma \vee \cdots \vee y_r\sigma = w_r\sigma)$$

is equivalent to $\bot$:

Indeed, as $H(\Sigma)$ is infinite, we simply choose a term t in $H(\Sigma)$ s.t.

$$\tau(t) > \tau(\{s_1\sigma, \ldots, s_p\sigma, y_1\sigma, \ldots, y_r\sigma\}).$$

Then the substitution θ, for $\theta = \{x \leftarrow t\}$, falsifies the matrix of $\mathcal{G}'_\sigma$: Clearly, by definition of t, $t \notin \{s_1\sigma, \ldots, s_p\sigma\}$ and as $x \in V(w_i\sigma)$ $\tau(w_i\sigma\theta) \geq \tau(t) > \tau(y_i\sigma)$.

But this property holds for all ground substitutions σ over $H(\Sigma)$ with domain $V(\bar{y})$ and so $\mathcal{G}' \equiv \bot$.

So in both cases $\mathcal{G} \equiv \bot$. But this, in turn, implies that $\mathcal{F}'$ is equivalent to $\forall \bar{y}.R$, which fulfils (a) and (b). By $\rho(\forall \bar{y}.R) = m$ we may apply (IH) and obtain $\forall \bar{y}.R \equiv \bot$; so also $\mathcal{F}'$ and $\mathcal{F}$ are equivalent to $\bot$. $\diamond$

THEOREM 1.1 *Any formula $\mathcal{F}$ which is a definition with constraints and is syntactically distinct from $\bot$ has at least one solution.*

Proof: If $\mathcal{F}$ is $\top$, then the proof is trivial.

Otherwise it is of the form

$$\exists w_1, \ldots, w_n.[\bigwedge_{j=1}^{m} x_j = s_j] \wedge [\bigwedge_{i=1}^{k} x'_i \neq s'_i].$$

We distinguish two cases, according to the value of the integer k.

1 $k = 0$:

Let $y_1, \ldots, y_l$ be the variables of $\mathcal{F}$ not in $\{x_1, \ldots, x_m\}$ and let σ be $\{y_i \leftarrow t_i | i = 1, \ldots, l\}$ where the t_i are arbitrary ground terms.

Now let $j \in \{1, \ldots, m\}$. Since $V(s_j) \cap \{x_1, \ldots, x_m\} = \emptyset$, we have

$$V(s_j) \subseteq \{y_1, \ldots, y_l\}$$

and thus $s_j\sigma$ is ground.

Let θ be $\{x_j \leftarrow s_j\sigma | j = 1, \ldots, m\}$. Then $\theta \cup \sigma$ is a solution of $\mathcal{F}$.

2 $k > 0$:

In this case, by definition of a definition with constraints, $H(\Sigma)$ must be infinite.

Again let $y_1, \ldots, y_l$ be the variables of $\mathcal{F}$ not in $\{x_1, \ldots, x_m\}$.

Let us consider the formula $\mathcal{G}$ defined as follows:

$$\forall y_1, \ldots, y_l. \bigvee_{i=1}^{k} x'_i = s'_i$$

Since $\mathcal{F}$ is a definition with constraints, none of the variables x_i appear in $\mathcal{G}$, and so $\mathcal{G}$ does not contain free variables.

Moreover x'_i is a variable distinct from s'_i. So $\mathcal{G}$ satisfies the conditions of Lemma 4.6; hence $\mathcal{G} \equiv \bot$.

Therefore, the formula $\bigwedge_{i=1}^{k} x'_i \neq s'_i$ must have a solution σ.

But then $\mathcal{F}\sigma$ is equivalent to

$$\exists w_1, \ldots, w_n.[\bigwedge_{j=1}^{m} x_j = s_j]\sigma.$$

Moreover

$$\mathcal{F}' : \exists w_1, \ldots, w_n.[\bigwedge_{j=1}^{m} x_j = s_j]\sigma$$

is a definition with constraints. By case 1 above $\mathcal{F}'$ has a solution and so $\mathcal{F} \not\equiv \bot$. $\diamond$

Remark: The proof of Theorem 1.1 also provides a way of effectively *enumerating* the (generally infinite) set of solutions of a given definition with constraints. ♯

DEFINITION 4.7 An equational formula $\mathcal{F}$ is said to be in *solved form* iff it is a definition with constraints or a disjunction of definitions with constraints. ♯

In [Comon and Lescanne, 1989], an algorithm is defined transforming an arbitrary equational formula into an equivalent solved form. We will not restate the details of the algorithm, but rather want to give a taste of the techniques used to transform equational formulae (the interested reader may consult [Comon and Lescanne, 1989, Comon and Delor, 1994, Comon, 1988, Comon, 1991]). In particular, we will present a version of the constraint solving algorithm, less efficient and less flexible than that of [Comon and Lescanne, 1989], but simpler and easier to understand, fitting the explanatory goals of this chapter. Some possible improvements and extensions will also be discussed in Section 1.3.

1.2.1 Equational problems

We first restrict our attention to a particular class of equational formulae called *equational problems*.

DEFINITION 4.8 An equational formula is an *equational problem* if it is of the form:

$$\exists \overline{x}.\forall \overline{y}.\mathcal{M}$$

where $\mathcal{M}$ is a quantifier-free formula. The variables in $\overline{y}$ are the called the *parameters* of the problem. ♯

Note that equational problems are more general than solved forms since they may contain universal quantifiers and since $\mathcal{M}$ may be any quantifier free formula. First we prove that the restriction to equational problems does not entail any loss of generality. More precisely, we will show that, any algorithm *SEP* transforming equational problems into equivalent formulae in solved forms can be extended to an algorithm transforming *any* equational formula into solved forms. The corresponding algorithm *SEF* is given below:

- Let $\mathcal{F}$ be an equational formula in prenex form. Then we define $SEF(\mathcal{F})$ as follows:

$$\begin{array}{l} SEF(\mathcal{F}) = SEP(\mathcal{F}) \\ \qquad \text{if } \mathcal{F} \text{ is an equational problem} \\ SEF(\exists x.\mathcal{F}) = \exists x.SEF(\mathcal{F}) \\ SEF(\forall x.\mathcal{F}) = SEP(\forall x.Pr(\neg SEF(Pr(\neg \mathcal{F})))) \end{array}$$

- If $\mathcal{F}$ is not in prenex form we define

$$SEF(\mathcal{F}) \quad = \quad SEF(Pr(\mathcal{F}))$$

where Pr is an operator transforming equational formulae into equivalent prenex normal forms.

Before proving the soundness of this algorithm we give an example illustrating the nature of *SEF*.

EXAMPLE 4.9 Let $\mathcal{F}$ be $\forall x.\exists y.x = y$. We apply *SEF* to $\mathcal{F}$:

$$\begin{array}{rcl}
SEF(\mathcal{F}) & = & SEF(\forall x.\exists y.x = y) \\
& = & SEP(\forall x.Pr(\neg SEF(Pr(\neg \exists y.x = y)))) \\
& = & SEP(\forall x.Pr(\neg SEF(\forall y.x \neq y))) \\
& = & SEP(\forall x.Pr(\neg SEP(\forall y.Pr(\neg SEF(Pr(\neg x \neq y)))))) \\
& = & SEP(\forall x.Pr(\neg SEP(\forall y.Pr(\neg SEF(x = y))))) \\
& = & SEP(\forall x.Pr(\neg SEP(\forall y.Pr(\neg SEP(x = y))))) \\
& = & SEP(\forall x.Pr(\neg SEP(\forall y.x \neq y))
\end{array}$$

As we shall see later $SEP(\forall y.x \neq y) = \bot$. Hence:

$$\begin{array}{rcl}
SEF(\mathcal{F}) & = & SEP(\forall x.Pr(\neg \bot)) \\
& = & SEP(\forall x.\top) \\
& = & \top.
\end{array}$$

♯

DEFINITION 4.10 We say that *SEF* is *sound* if for all equational formulae $\mathcal{F}$ and $\mathcal{G}$ with $SEF(\mathcal{F}) = \mathcal{G}$ we have $\mathcal{F} \equiv \mathcal{G}$ and $\mathcal{G}$ is in solved form. *SEF* is called *terminating*, if for all equational formulae $\mathcal{F}$ $SEF(\mathcal{F})$ is defined.

Similarly *SEP* is called *sound* if for all equational problems $\mathcal{F}$ and all equational formulae $\mathcal{G}$ with $SEP(\mathcal{F}) = \mathcal{G}$ we have $\mathcal{F} \equiv \mathcal{G}$ and $\mathcal{G}$ is in solved form. *SEP* is called *terminating*, if for all equational problems $\mathcal{F}$ $SEP(\mathcal{F})$ is defined. ♯

LEMMA 4.11 *SEF is sound and terminating relative to SEP, i.e.: if SEP is sound and terminating, then SEF is sound and terminating as well.*

Remark: soundness and termination of *SEP* will be shown afterwards.

♯

Proof: If $\mathcal{F}$ is not in prenex form then, by definition, $SEF(\mathcal{F}) = SEF(Pr(\mathcal{F}))$. Furthermore it is easy to see that, once *SEF* is applied to a prenex form, all further recursive calls apply to prenex forms only.

Clearly also $\mathcal{F} \equiv Pr(\mathcal{F})$. Thus in order to prove the lemma we may safely assume that $\mathcal{F}$ is already in prenex form.

Let $\psi(\mathcal{F})$ be the total number of quantifiers in $\mathcal{F}$. We proceed by induction on $\psi(\mathcal{F})$.

- If $\psi(\mathcal{F}) = 0$, then $\mathcal{F}$ is an equational problem, hence $SEF(\mathcal{F}) = SEP(\mathcal{F})$. By assumption *SEP* is sound and terminating.
- If $\psi(\mathcal{F}) > 0$, then $\mathcal{F}$ is either of the form $\exists x.\mathcal{G}$ or of the form $\forall x.\mathcal{G}$, where $\mathcal{G}$ is in prenex normal form.

  $\mathcal{F}$ is of the form $\exists x.\mathcal{G}$:
  Then, by definition we have $SEF(\mathcal{F}) = \exists x.SEF(\mathcal{G})$. Moreover $\psi(\mathcal{G}) < \psi(\mathcal{F})$, hence by induction hypothesis, $SEF(\mathcal{G})$ terminates and returns a solved form equivalent to $\mathcal{G}$. Therefore also $SEF(\mathcal{F})$ terminates and returns a solved form equivalent to $\exists x.\mathcal{G}$, which is equivalent to $\mathcal{F}$.

  $\mathcal{F}$ is of the form $\forall x.\mathcal{G}$:
  Then
  $$SEF(\mathcal{F}) = SEF(\forall x.Pr(\neg SEF(Pr(\neg \mathcal{G})))).$$
  But $\psi(Pr(\neg\mathcal{G})) = \psi(\mathcal{G}) < \psi(\mathcal{F})$, thus, by induction hypothesis, $SEF(Pr(\neg\mathcal{G}))$ terminates and yields a formula $\mathcal{G}'$ in solved form with $\mathcal{G}' \equiv \neg\mathcal{G}$.
  Because $\mathcal{G}'$ is in solved form it does not contain universal quantifiers. Therefore $\forall x.Pr(\neg\mathcal{G}')$ does not contain existential quantifiers and thus is an equational problem. But this implies
  $$SEF(\forall x.Pr(\neg\mathcal{G}')) \quad = \quad SEP(\forall x.Pr(\neg\mathcal{G}')).$$
  By assumption *SEP* is sound and terminating. ◇

Lemma 4.11 tells us that the problem of solving equational formulae can be reduced to solving equational problems. Note that the complexity of *SEF* is non-elementary! [Vorobyov, 1996]. But the main source of complexity is not the function *SEP* presented below, but lies essentially in the full syntax of equational formulae.

1.2.2 Solution of equational problems

Let $\mathcal{P}$ be an equational problem of the form

$$\exists x_1, \ldots, x_n.\forall y_1, \ldots, y_m.\mathcal{M}$$

The algorithm we are going to present consists of two main steps:

1 **Elimination of parameters**: the parameters occurring in the problem are eliminated, leading to formulae of the form $\exists\overline{x}.\mathcal{M}$ where $\mathcal{M}$ is quantifier-free.

2 **Computation of solved forms**: the remaining formulae are transformed into disjunctions (sets) of definition with constraints.

Remark: In the following presentation, we restrict ourselves to the case in which the signature contains at least one function symbol. Indeed, if the signature contains only constant symbols, the universe is finite and the variables can be completely instantiated, leading to a purely propositional formula; however, for the sake of efficiency, instantiations should be avoided if possible. ♯

Reduction rules The solution algorithm will be specified using *reduction rules.* These rules will be used to transform formulae into sets of definition with constraints. They operate on sets of formulae. A reduction on a single formula $\mathcal{F}$ can be understood as a reduction on $\{\mathcal{F}\}$. If ρ is a system of reduction rules we write $\rho(\mathcal{F})$ for any normal form of $\mathcal{F}$ under ρ. If ρ is not confluent (which is the case for most of our transformations below) $\rho(\mathcal{F})$ is considered the result of a fixed but arbitrary deterministic interpretation of ρ.

In order to avoid the use of too many rules making the matter complex and intransparent, we apply the reduction rules *modulo* some basic properties of equational formulae; these properties are the commutativity of the logical symbols $\vee, \wedge, \exists, \forall$ (i.e. $\exists x \exists y$ is the same as $\exists y \exists x$) and the symmetry of $=$. Consider, for instance, the standard replacement rule (as defined for example in [Jouannaud and Kirchner, 1991]):

$$x = t \wedge R \ \rightarrow x = t \ \wedge R\{x \leftarrow t\}$$

This rule can be applied to the formula : $f(x) = g(y) \wedge a = x$, yielding $x = a \wedge f(a) = g(y)$, though it is clear that the replacement rule does *not* syntactically apply to the initial formula. An application of this rule is possible *only* by taking into account the commutativity of the symbols $\wedge$ and $=$, i.e. $f(x) = g(y) \wedge a = x$ is transformed to $x = a \wedge f(x) = g(y)$.

The complete (syntactic) version of the replacement rule would be:

$$R_1 \wedge Q \wedge R_2 \rightarrow R_1\{x \leftarrow t\} \wedge x = t \wedge R_2\{x \leftarrow t\}$$

where Q is either $x = t$ or $t = x$.

However, writing down such complete rules leads to complex and long formulations and thus should be avoided. Therefore, we choose to apply all reduction rules modulo the following properties:

$$\begin{array}{lcl}
\mathcal{F}_1 \vee \mathcal{F}_2 & \equiv & \mathcal{F}_2 \vee \mathcal{F}_1 \\
\mathcal{F}_1 \wedge \mathcal{F}_2 & \equiv & \mathcal{F}_2 \wedge \mathcal{F}_1 \\
t = s & \equiv & s = t \\
\exists x \exists y \mathcal{F} & \equiv & \exists y \exists x \mathcal{F} \\
\forall x \forall y \mathcal{F} & \equiv & \forall y \forall x \mathcal{F} \\
\bot \vee \mathcal{F} & \equiv & \mathcal{F} \\
\bot \wedge \mathcal{F} & \equiv & \bot \\
\top \vee \mathcal{F} & \equiv & \top \\
\top \wedge \mathcal{F} & \equiv & \mathcal{F} \\
\forall x.\mathcal{F} & \equiv & \mathcal{F} \text{ if } x \text{ does not occur in } \mathcal{F}. \\
\exists x.\mathcal{F} & \equiv & \mathcal{F} \text{ if } x \text{ does not occur in } \mathcal{F}. \\
\neg\bot & \equiv & \top \\
\neg\top & \equiv & \bot
\end{array}$$

Remark: Clearly it is possible to "compile" these rules into sequences of standard rules without equivalences. ¿From now on the equivalences above will be considered as holding implicitly and will never be applied explicitly. This assumption is particularly important in proving termination and completeness, and also for implementations. ♯

DEFINITION 4.12 We say that a rule system ρ operating on sets of equational formula is *sound* iff for any pair of sets of equational formulae (S, S') such that

$$S \rightarrow_\rho S'$$

we have:

$$\bigcup_{\mathcal{F} \in S} \mathcal{S}(\mathcal{F}) = \bigcup_{\mathcal{F} \in S'} \mathcal{S}(\mathcal{F})$$

Remark: If the set of formulae S consists only of the element $\mathcal{F}$ and $S' = \{\mathcal{F}'\}$ then soundness means just $\mathcal{S}(\mathcal{F}') = \mathcal{S}(\mathcal{F})$. ♯

At this point we start to present the details of the solution algorithm for equational problems.

1.2.3 Elimination of parameters

Transformation into c.n.f. The first step consists in transforming the matrix of the formula into conjunctive normal form (cnf) via the standard cnf-transformation [Baaz et al., 2001]. Note that the standard transformation may entail an exponential blow-up of the size of the formula. But the structural cnf-transformation, which is polynomial, cannot be applied as this would lead to an extension of the signature and thus to formulae outside equational logic. Therefore, only the standard transformation via distributivity remains.

Simplification of equations and disequations The second step consists in simplifying the disequations occurring in the problem. We use the standard Decomposition, Clash and Occur-check rules (see for example [Jouannaud and Kirchner, 1991]), or more exactly the dual version of these rules, dealing with disjunctions of disequations instead of conjunctions of equations.

DEFINITION 4.13 The *dis-unification* (or *DisUnif*) rules are defined as follows:

($\neq$-Decomposition)	$f(t_1,\ldots,t_n) \neq f(s_1,\ldots,s_n)$	$\rightarrow$	$\bigvee_{i=1}^n t_i \neq s_i$
($\neq$-Clash)	$f(t_1,\ldots,t_n) \neq g(s_1,\ldots,s_m)$ if f is different from g	$\rightarrow$	$\top$
($\neq$-Occur-Check)	$x \neq t$ if x occurs in t	$\rightarrow$	$\top$
$\forall$-Shift	$\forall x.(\mathcal{F} \wedge \mathcal{G})$	$\rightarrow$	$\forall x.\mathcal{F} \wedge \forall x.\mathcal{G}$

The set of these rules is denoted by *DisUnif*. ♯

LEMMA 4.14 *DisUnif is sound.*

Proof: Trivial.

LEMMA 4.15 *DisUnif is terminating (on any formula $\mathcal{F}$).*

Proof: The $\neq$-Decomposition, $\neq$-Clash and $\neq$-Occur Check rules strictly reduce the size of atomic subformulae;

the $\forall$-Shift rule does not change the atomic formulae, but strictly reduces the number of conjunction symbols in the scope of a universal quantifier. ◇

EXAMPLE 4.16 We illustrate how the dis-unification rules work on

$$\mathcal{F} \;:\; \forall y,z,y',z'.(c \neq d \;\wedge\; [g(u,a) \neq g(b,u) \;\vee\; g(x,y) \neq g(f(y),z) \;\vee\; g(x',f(z') = g(y',f(b))]),$$

where a, b, c, d denote constant symbols and f, g function symbols of arity $1, 2$, respectively. The application of the decomposition rule yields:

$$\mathcal{F}' \;:\; \forall y,z,y',z'.(c \neq d \;\wedge\; [u \neq b \;\vee\; a \neq u \;\vee\; x \neq f(y) \;\vee\; y \neq z \;\vee\; g(x',f(z')) = g(y',f(b))]).$$

Then, after application of the clash rule on $c \neq d$, we obtain

$$\mathcal{F}_2 \;:\; \forall y,z,y',z'.(\top \;\wedge\; [u \neq b \;\vee\; a \neq u \;\vee\; x \neq f(y) \;\vee\; y \neq z \vee g(x',f(z')) = g(y',f(b))]),$$

which, by $\top \wedge \mathcal{G} \equiv \mathcal{G}$ reduces to :

$$\begin{aligned} \mathcal{F}_3 \ : \ & \forall y, z, y', z'.(u \neq b \ \vee \ a \neq u \ \vee \ x \neq f(y) \ \vee \ y \neq z \ \vee \\ & g(x', f(z')) = g(y', f(b))). \ \sharp \end{aligned}$$

Before we specify the rules for eliminating parameters, we have to establish the following lemma, which characterizes the form of formulae which are normal forms under *DisUnif*.

LEMMA 4.17 *Let $\mathcal{F}$ be an equational problem, then DisUnif($\mathcal{F}$) is of the form:*

$$\exists x_1, \ldots, x_n. \bigwedge_{i=1}^{n} \forall y_{i1}, \ldots, y_{im_i}.C_i$$

where C_i is a disjunction of literals (i.e. of equations or disequations). Moreover, any disequation occurring in C_i is of the form $x \neq t$, where x is a variable.

Proof: Any disequation that is not of this form is reducible by one of the following rules:

$\neq$-Decomposition, $\neq$-Clash, or $\neq$-Occur-check.

Moreover any universal formula over a conjunction is reduced by $\forall$-Shift. ◇

The *PElim* rule

The following set of rules *PElim* is called "elimination of parameters".

$$\textit{PElim:} \quad \forall y.(y \neq t \vee R) \quad \rightarrow \quad \textit{DisUnif}(R\{y \leftarrow t\})$$
If y does not occur in t.

LEMMA 4.18 *PElim is sound and terminating.*

Proof: Let σ be a solution of $\forall y.(y \neq t \vee R)$. By definition, σ must be a solution of $s \neq t \vee R\{y \leftarrow s\}$ for any ground term s (note that y does not occur in t); therefore σ is also a solution of $t\sigma \neq t \vee R\{y \leftarrow t\sigma\}$, and thus of $R\{y \leftarrow t\sigma\}$.

Now, assume that σ is a solution of $R\{y \leftarrow t\}$. Let s be a ground term. If s is equal to $t\sigma$ then $\sigma \in \mathcal{S}(R\{y \leftarrow s\})$. Otherwise $\sigma \in \mathcal{S}(s \neq t)$.

PElim is also terminating since the number of parameters is strictly reduced. ◇

For any formula $\mathcal{F}$, we denote by $R_{\text{presolve}}(\mathcal{F})$ the following procedure:

$$R_{\text{presolve}}(\mathcal{F}) = \textit{PElim}(\textit{DisUnif}(\mathcal{F}))$$

EXAMPLE 4.19 We apply the *PElim* rule on the formula of Example 4.16.

Let $\mathcal{F}$ be

$$\forall y, z, y', z'.(u \neq b \vee a \neq u \vee x \neq f(y) \vee y \neq z \vee g(x', f(z')) = g(y', f(b)))$$

Then $R_{\text{presolve}}(\mathcal{F}) = PElim(\mathcal{F})$ and $PElim(\mathcal{F})$ is

$$\forall z, y', z'.u \neq b \vee a \neq u \vee x \neq f(z) \vee g(x', f(z')) = g(y', f(b)). \ \sharp$$

The Explosion rule The previous rule allowed us to eliminate parameters occurring at "root" positions in disequations, but how can we eliminate parameters occurring at deeper positions? To solve this problem we will use the so-called *explosion rule*, which is based on the domain closure axiom. Informally speaking, the domain closure axiom states that there are no other terms except those in $H(\Sigma)$. It means in particular, that any term t must be of the form $f(t_1, \ldots, t_n)$, with $f \in \Sigma$ (this corresponds to the extremal clause in inductive definitions). For example, if $\Sigma = \{0, succ, +\}$, any term t must be either 0, $succ(s)$ or $t_1 + t_2$ (for terms s, t_1, t_2 in $H(\Sigma)$).

Before we give a formal description of explosion, we present an illustrating example.

Let $\mathcal{F} : \forall y.x \neq f(y, g(y))$, be an equational problem. It is clear that we cannot eliminate the parameter y from the problem without using additional information on the underlying domain, i.e. on the considered signature. Indeed, the solutions of this formula essentially depend on the set of function symbols.

Assume that $\Sigma = \{a, g, f\}$ where g is of arity 1 and f of arity 2. Let s be a ground term (to be substituted for x); then s is either a or of the form $g(t_1)$, or of the form $f(t_1, t_2)$ for some terms $t_1, t_2 \in H(\Sigma)$. This is expressed by the formula

$$(x = a) \vee (\exists y_1.x = g(y_1)) \vee (\exists y_2, y_3.x = f(y_2, y_3)).$$

which is valid over Σ.

Therefore we can add this formula to $\mathcal{F}$ without changing the set of solutions; the result is:

$$\mathcal{G} : \forall y.x \neq f(y, g(y)) \wedge (x = a \vee \exists y_1.x = g(y_1) \vee \exists y_2, y_3.x = f(y_2, y_3))$$

Afterwards the problem can be splitted into three subproblems by the $\wedge$-distributivity rule. So we obtain the following three problems:

- $\mathcal{G}_1 : \ \forall y.x \neq f(y, g(y)) \wedge x = a.$

- $\mathcal{G}_2 : \forall y.x \neq f(y, g(y)) \wedge \exists y_1.x = g(y_1)$.
- $\mathcal{G}_3 : \forall y.x \neq f(y, g(y)) \wedge \exists y_2, y_3.x = f(y_2, y_3)$.

By replacing x by its value, $\mathcal{G}_1$ can be reduced to:

$$\forall y.a \neq f(y, g(y)) \wedge x = a$$

i.e. to $x = a$ (by the Clash rule and by simplification).

Similarly, the second formula is reduced to $\exists y_1.x = g(y_1)$

Finally, the third formula is reduced to

$$\mathcal{G}_3' : \exists y_2, y_3.(\forall y.f(y_2, y_3) \neq f(y, g(y)) \wedge x = f(y_2, y_3)).$$

Then $\neq$-decomposition on $\mathcal{G}_3'$ yields

$$\mathcal{G}_3'' : \exists y_2, y_3.(\forall y.(y_2 \neq y \vee y_3 \neq g(y)) \wedge x = f(y_2, y_3)).$$

But in $\mathcal{G}_3''$ we can eliminate y by *PElim* and obtain:

$$\exists y_2, y_3.(y_3 \neq g(y_2) \wedge x = f(y_2, y_3)).$$

Eventually we obtain the following formula, which is a disjunction of definitions with constraints:

$$x = a \vee (\exists y_1)x = g(y_1) \vee (\exists y_2, y_3)(y_3 \neq g(y_2) \wedge x = f(y_2, y_3))$$

We see that the initial parameter has been eliminated, at the cost of additional existential quantifiers and of an increase of the formula.

This simple example illustrates the general principle of the elimination of quantifiers: first use the domain closure axiom to instantiate the variables, then use decomposition to shift the quantifiers towards root positions in terms, and finally, eliminate the parameters by applying *PElim*. In general, of course, several instantiations of variables are needed (depending on the *minimal* depth of an occurrence of the considered parameter). This example also shows that we have to consider several different solutions according to the list of symbols in the signature. Thus this rule operates on *sets* of formulae rather than on single formulae in an essential way. These sets of formulae have to be interpreted as disjunctions, i.e. the solutions of the initial formula is the *union* of the sets of solutions of the obtained set of formulae.

The formal definition of the *explosion* rule is given below:

DEFINITION 4.20 (EXPLOSION) Let $\mathcal{P}$ be an equational problem containing a disequation $x \neq t$ where t contains a parameter. Then we define the reduction

$$\begin{aligned}\mathcal{P} \quad \rightarrow \quad & \{R_{\text{presolve}}(\exists x_{f1}, \dots, x_{fn}.(\mathcal{P}\{x \leftarrow f(x_{f1}, \dots, x_{fn})\} \wedge \\ & x = f(x_{f1}, \dots, x_{fn}))|f \in \Sigma, n \in \mathbb{N}, arity(f) = n\}.\end{aligned}$$

where for any $f \in \Sigma$ of arity n, $x_{f1}, \ldots, x_{fn}$ are new free variables. By *Explosion* we denote the system of all explosion rules. ♯

LEMMA 4.21 *Explosion is sound.*

Proof: The soundness of the explosion rule is an immediate consequence of the domain closure axiom, i.e. for any ground term t over a signature Σ the following formula is valid:

$$\bigvee_{f \in \Sigma, n \in \mathbb{N}, arity(f)=n} \exists x_1, \ldots, x_n . t = f(x_1, \ldots, x_n).$$

◇

Using explosion in solving equational problems can be very expensive. More precisely, explosion may significantly increase the size of formulae and introduce some redundancy into the solution process. We will see later how to (partially) overcome these problems.

EXAMPLE 4.22 Let us consider the formula $\mathcal{F}$ obtained in Example 4.19:

$$\mathcal{F} : \forall z, y', z'.(u \neq b \vee a \neq u \vee x \neq f(z) \vee g(x', f(z')) = g(y', f(b))).$$

In order to eliminate the remaining parameters occurring in disequations (namely z), we need to apply the explosion rule on the variable x.

Assume that the signature consists just of the symbols : a, b, f, g occurring in $\mathcal{F}$.

Then, we get the following set of problems $\{\mathcal{F}_1, \mathcal{F}_2, \mathcal{F}_3, \mathcal{F}_4\}$.

$$\begin{aligned}
\mathcal{F}_1 &: \forall z, y', z'.(u \neq b \vee a \neq u \vee a \neq f(z) \vee g(x', f(z')) = g(y', f(b))) \\
&\quad \wedge x = a, \\
\mathcal{F}_2 &: \forall z, y', z'.(u \neq b \vee a \neq u \vee b \neq f(z) \vee g(x', f(z')) = g(y', f(b))) \\
&\quad \wedge x = b, \\
\mathcal{F}_3 &: \exists x_1, x_2.[\forall z, y', z'.(u \neq b \vee a \neq u \vee g(x_1, x_2) \neq f(z) \vee \\
&\quad g(x', f(z')) = g(y', f(b))) \wedge x = g(x_1, x_2)], \\
\mathcal{F}_4 &: \exists x_1.[\forall z, y', z'.(u \neq b \vee a \neq u \vee f(x_1) \neq f(z) \vee \\
&\quad g(x', f(z')) = g(y', f(b))) \wedge x = f(x_1)].
\end{aligned}$$

The original problem is equivalent to the disjunction of the four new problems. Now, it is clear that the three first problems can be simplified by the clash rule and the basic equivalences; the result is

$$\begin{aligned}
\mathcal{F}'_1 &: x = a, \\
\mathcal{F}'_2 &: x = b, \\
\mathcal{F}'_3 &: (\exists x_1, x_2) x = g(x_1, x_2).
\end{aligned}$$

The last problem is more difficult to solve.

We have to apply the decomposition rule, which gives us the following formula:

$$\mathcal{F}'_4 : \exists x_1.[\forall z, y', z'.\mathcal{Q}''_4(u \neq b \vee a \neq u \vee x_1 \neq z \vee g(x', f(z')) = g(y', f(b)))$$
$$\wedge x = f(x_1)].$$

Now z occurs at the root position in a disequation, which allows us to apply *PElim*, leading to:

$$\mathcal{F}''_4 : \exists x_1.[\forall z, y', z'.(u \neq b \vee a \neq u \vee g(x', f(z')) = g(y', f(b))) \wedge x = f(x_1)].$$

z has been eliminated from the formula – at the cost of a significant increase of the size of the formula, which has been replaced by a set of formulae $\{\mathcal{F}'_1, \mathcal{F}'_2, \mathcal{F}'_3, \mathcal{F}''_4\}$ - containing new existential variables. ♯

Termination of *Explosion* is slightly more difficult to prove than for the other rules.

LEMMA 4.23 *Explosion is terminating (on equational problems).*

Proof: We will define a weight function on equational problems and a well-founded ordering on these weights such that, after each application of the explosion rule, the value of a formula strictly decreases.

Let $\mathcal{P}$ be an equational problem in cnf. Then $\mathcal{P}$ is of the form

$$\exists \overline{x}.\forall \overline{y}. \bigwedge_{i=1}^{n} .C_i.$$

We introduce the following measure $\mathcal{I}$:

$$\mathcal{I}(\mathcal{P}) = \{\mathcal{I}'(C_i) | 1 \leq i \leq n\}$$

where $\mathcal{I}'$ is defined as follows:

- For any term t, *param*(t) is the number of positions in t containing a parameter.
- For any disequation $t \neq s$, $\phi(t \neq s) = \max(\mathit{param}(t), \mathit{param}(s))$.
- For any formula $\mathcal{F}$, $\psi(\mathcal{F})$ is the number of distinct parameters occurring in $\mathcal{F}$.

- Let $\mathcal{F}$ be a disjunction of equations and disequations. Let E be the set of disequations in $\mathcal{F}$ containing at least one parameter. Then, we define:
$$\mathcal{I}'(\mathcal{F}) = (\psi(\mathcal{F}), \{\phi(d) | d \in E\}).$$

$\mathcal{I}(\mathcal{P})$ is ordered using the lexicographic and multiset extension of the standard ordering on natural numbers.

Let $\mathcal{P}, \mathcal{P}'$ be two equational problems such that $\mathcal{P} \rightarrow_{Explosion} \mathcal{P}'$. We prove that

$$\mathcal{I}(\mathcal{P}) > \mathcal{I}(R_{\text{presolve}}(\mathcal{P}')).$$

First, we show that

$$\mathcal{I}(\mathcal{P}) \quad > \quad \mathcal{I}(DisUnif(\mathcal{P}')).$$

Each disequation occurring in $\mathcal{P}'$ is of the form $(t \neq s)\{x \leftarrow f(x_1, \ldots, x_n)\}$, where $t \neq s$ occurs in $\mathcal{P}$. Since $f(x_1, \ldots, x_n)$ does not contain any parameter, we have

$$param(t \neq s\{x \leftarrow f(x_1, \ldots, x_n)\}) \quad = \quad param(t \neq s).$$

Hence $\mathcal{I}(\mathcal{P}) = \mathcal{I}(\mathcal{P}')$.

It is clear that the simplification rules do not increase the number of parameters, hence ψ does not increase. By definition of ϕ, the application of the decomposition, clash, or occur-check rules on *parameterless* disequations does not modify ϕ. Moreover, any application of the decomposition, clash, or occur-check rule on a disequation containing a parameter strictly decreases *param*, and thus strictly decreases ϕ.

Moreover, since $\mathcal{P}$ contains a disequation $x \neq t$, where t is not a variable, $\mathcal{P}'$ must contain a disequation $f(x_1, \ldots, x_n) \neq t$, hence either the decomposition or the clash rule applies.

Therefore, $\mathcal{I}$ decreases strictly and so

$$\mathcal{I}(\mathcal{P}') > \mathcal{I}(DisUnif(\mathcal{P}')).$$

It only remains to prove that

$$\mathcal{I}(DisUnif(\mathcal{P}')) \quad \geq \quad \mathcal{I}(R_{\text{presolve}}(\mathcal{P}')),$$

which is straightforward: if *PElim* is applied, then ψ is strictly decreased, else we have $R_{\text{presolve}}(\mathcal{P}') = DisUnif(\mathcal{P}')$, thus

$$\mathcal{I}(DisUnif(\mathcal{P}')) \geq \mathcal{I}(R_{\text{presolve}}(\mathcal{P}')).$$

Therefore we have proven that

$$\mathcal{I}(\mathcal{P}) > \mathcal{I}(R_{\text{presolve}}(\mathcal{P}')).$$

Since $>$ is well-founded *Explosion* terminates. ◇

The Elimination of remaining parameters

If $\mathcal{P}$ is irreducible under *Explosion* and *PElim* then any parameter y occurring in $\mathcal{P}$ must occur in an equation. Indeed, otherwise, either $\mathcal{P}$ contains a disequation $y \neq t$ and *PElim* applies, or it contains a disequation $x \neq t$, where y occurs in t and *Explosion* can be applied. The elimination of the remaining parameters is based on Lemma 4.6. This lemma will allow us the remove all non trivial equations containing parameters, hence to obtain parameter-free problems.

First we can assume that for any equation $s = t$ occurring in the problem, there is no variable x occurring at the same position p in s and t. Indeed, if this is the case, we simply replace these two occurrences by any ground term thus obtaining an equivalent equation; that this is justified follows from the lemma below.

DEFINITION 4.24 Let $t = s$ be an equation. Let P be the set of positions p in t and s such that $t_{|p}$ is syntactically equivalent to $s_{|p}$. Let t' and s' the terms obtained from t and s by replacing each term at a position $p \in P$ by the same ground term r (where r may be arbitrary). Then we define the function *Clean* as follows :

$$\mathit{Clean}(t = s) = (t' = s')$$

Clean can be easily extended to any kind of formulae: it maps any formula $\mathcal{F}$ to the formula obtained from $\mathcal{F}$ by replacing each equation $s = t$ by $\mathit{Clean}(s = t)$. ♯

LEMMA 4.25 *For any equational formula* $\mathcal{F}$ *we have:*

$$\mathcal{F} \equiv \mathit{Clean}(\mathcal{F})$$

Proof: Let $s = t$ be an equation and $s' = t' = \mathit{Clean}(s' = t')$. By decomposition and elimination of trivial equations, $s = t$ and $s' = t'$ can be reduced to the same formulae, hence $s = t \equiv s' = t'$. ◇

DEFINITION 4.26 Let $\mathcal{F}$ be a formula.

Let $\mathit{PClean}(\mathcal{F})$ be the formula obtained by replacing each equation containing a parameter by $\perp$.

After application of *Clean* we can finally get rid of the parameters altogether:

LEMMA 4.27 *Let* $\mathcal{F}$ *be an equational formula,* $\mathcal{F}' \in R_{presolve}(\mathcal{F})$, *and* $\mathcal{P} = \mathit{Clean}(\mathcal{F}')$. *Then* $\mathcal{S}(\mathcal{P}) = \mathcal{S}(\mathit{PClean}(\mathcal{P}))$.

Proof: Each conjunct of $\mathcal{P}$ is of the following form: $\mathcal{F} : \forall \bar{y}. \bigvee_{i=1}^{n} t_i = s_i \vee R$ where R does not contain parameters, and each equation $t_i = s_i$ contains at least one parameter.

Assume that there exists a solution σ of $\forall \bar{y}. \bigvee_{i=1}^{n} t_i = s_i \vee R$.

Then by iterated application of the decomposition rule for equations and of symmetry the formula

$$\mathcal{G}_\sigma : \forall \bar{y}. \bigvee_{i=1}^{n} t_i\sigma = s_i\sigma$$

can be transformed to the equivalent formula

$$\mathcal{G}'_\sigma : \forall \bar{y}. \bigvee_{j=1}^{k} z_j = r_j,$$

where, for each j z_j is a variable and r_j is a term different from z_j and $\{z_1, \ldots, z_k\} \subseteq V(\bar{y})\}$. But then $\mathcal{G}'_\sigma$ satisfies the conditions of Lemma 4.6. Indeed, $t_i\sigma$ is syntactically distinct from $s_i\sigma$ (since t_i and s_i contains at least one occurrence of a parameter, occurring at *distinct positions* in t_i and s_i) and it contains at least one parameter.

Hence $\mathcal{G}'_\sigma$ is equivalent to $\perp$ and also $\mathcal{G}_\sigma \equiv \top$.

Therefore

$$\mathcal{S}(\forall \bar{y}. \bigvee_{i=1}^{n} t_i = s_i \vee R) = \mathcal{S}(R),$$

i.e. $\mathcal{S}(\mathcal{P}) = \mathcal{S}(PClean(\mathcal{P}))$. $\diamond$

DEFINITION 4.28 Let $\mathcal{F}$ be a formula. We denote by $PFree(\mathcal{F})$ the following set of formulae:

$$PFree(\mathcal{F}) \quad = \quad \{PClean(Clean(\mathcal{F}')) | \mathcal{F}' \in R_{\text{presolve}}(\mathcal{F})\}.$$

EXAMPLE 4.29 Let the formula $\mathcal{F}''_4$ from Example 4.22, i.e.

$$\mathcal{F}''_4 = \exists x_1.[\forall z, y', z'.(u \neq b \vee a \neq u \vee g(x', f(z')) = g(y', f(b))) \wedge x = f(x_1)].$$

Then $PFree(\mathcal{F}''_4) = \mathcal{F}_4^{(3)}$ for

$$\mathcal{F}(3)_4 = \exists x_1.((u \neq b \vee a \neq u) \wedge x = f(x_1)).$$

$\sharp$

Computation of solved forms So far we have obtained problems without any parameters. It only remains to transform those problems into solved forms.

To this aim we use the usual simplification rules (decomposition, clash and occur check) and their dual version for negation, together with a distributivity rule (the so-called Disjunction Explosion rule, shorthand (DE)).

DEFINITION 4.30 The set of the rules below is called *Unify*:

(DE)	$\{\exists \bar{x}.((\mathcal{P} \vee \mathcal{Q}) \wedge \mathcal{R})\} \rightarrow \{\exists \bar{x}.(\mathcal{P} \wedge \mathcal{R}), \exists \bar{x}.(\mathcal{Q} \wedge \mathcal{R})\}$.
(R)	$x = t \wedge \mathcal{P} \rightarrow x = t \wedge \mathcal{P}\{x \leftarrow t\}$.
	if x occurs in $\mathcal{P}$, but not in t.
(D)	$f(t_1, \ldots, t_n) = f(s_1, \ldots, s_n) \rightarrow \bigwedge_{i=1}^{n} t_i = s_i$.
(Clash)	$f(t_1, \ldots, t_n) = g(s_1, \ldots, s_m) \rightarrow \bot$
	if f is different from g.
(OC)	$x = t \rightarrow \bot$ if x occurs in t.
(NS)	$\neg(\bigwedge_{i=1}^{n} \mathcal{F}_n) \rightarrow \bigvee_{i=1}^{n} \neg \mathcal{F}_n$.

Remark: As *Explosion*, (DE) operates on sets of formulae instead on single formulae. Note that R may be $\top$ in (DE). (R) (for Replacement), Clash, Occur Check and (D) (for Decomposition) are the standard replacement rules, as defined in [Jouannaud and Kirchner, 1991]. (NS) stands for (Negation-Shift) and (OC) for (Occur-Check). ♯

DEFINITION 4.31 Let $\mathcal{P}$ be an equational problem of the form

$$\exists \bar{x}. \bigwedge_{i=1}^{n} \mathcal{P}_i$$

where the $\mathcal{P}_i$ are disjunctions, negations or atomic formulae (possibly with $n = 1$).

A variable x in $\mathcal{P}$ is called *solved* if x occurs only once in $\mathcal{P}$ and there exists a $\mathcal{P}_i$ with $\mathcal{P}_i$ syntactically equal to $x = t$, where x does not occur in t. ♯

LEMMA 4.32 *Unify terminates on any set of parameter-free equational problems.*

Proof: We define the following measure $\mathcal{I}'$ on equational problems $\mathcal{P}$:

$$\mathcal{I}'(\mathcal{P}) = (\mathit{unsolved}(\mathcal{P}), \mathit{size}(\mathcal{P}), \mathit{negation}(\mathcal{P})),$$

where:

- *unsolved*($\mathcal{P}$) is the number of unsolved variables in $\mathcal{P}$,

- *size*($\mathcal{P}$) is the number of occurrences of non logical symbols in $\mathcal{P}$,
- *negation*($\mathcal{P}$) is the number of occurrences of $\wedge$ in the scope of a negation.

$\mathcal{I}'$ is ordered using the lexicographic extension of the standard integer ordering.

If (DE), (D), (Clash) or (OC) is applied, then the number of unsolved variables does not increase. Moreover, the size of the formula decreases strictly.

If (R) is applied on a conjunct $x = t \wedge \mathcal{F}$ then the number of unsolved variables is strictly reduced: indeed the rule is applicable only if x occurs in $\mathcal{F}$, but not in t. Thus x is unsolved in $x = t \wedge \mathcal{F}$. But (R) yields $x = t \wedge \mathcal{F}\{x \leftarrow t\}$ where x is solved.

If (NS) is applied, then the number of unsolved variables and the size of the formula do not increase. Moreover, *negation*($\mathcal{F}$) strictly decreases. ◇

The following theorem ensures the completeness of our algorithm:

THEOREM 1.2 *(Completeness) Any parameterless equational problem $\mathcal{P}$ which is irreducible under Unify is a definition with constraints.*

Proof: By irreducibility under (DE), $\mathcal{P}$ is disjunction-free, and thus is of the form

$$\exists \overline{y}.(\bigwedge_{i=1}^{n} t_i = s_i \wedge \bigwedge_{i=1}^{m} t'_i \neq s'_i).$$

By irreducibility under (D) and (Clash) t_i and t'_i must be variables.

Moreover, by irreducibility under (OC) and (R), t_i must be solved, i.e. it occurs only once.

Finally, by irreducibility under (OC) t'_i does not occur in s'_i. ◇

EXAMPLE 4.33 The formula

$$\mathcal{F} : \exists x_1.((u \neq b \vee a \neq u) \wedge x = f(x_1))$$

is not a definition with constraints, but is transformed by (DE) into $\{\mathcal{F}_1, \mathcal{F}_2\}$ for:

$$\begin{aligned} \mathcal{F}_1 &= \exists x_1.(u \neq b \wedge x = f(x_1)), \\ \mathcal{F}_2 &= \exists x_1.(u \neq a \wedge x = f(x_1)). \end{aligned}$$

Obviously $\mathcal{F}_1$ and $\mathcal{F}_2$ are definitions with constraints. ♯

Remark: Example 4.33 also illustrates the drawback of the our solution method. Indeed, since $a \neq b$, $u \neq a \vee u \neq b$ is a valid formula, $\mathcal{F}$ is equivalent to

$$\exists x_1.x = f(x_1).$$

When we go back to a former stage of transformation we find the disjunct $\mathcal{G} : g(u, a) \neq g(b, u)$ in the initial formula. But $\mathcal{G}$ is valid and so *the whole formula* is equivalent to $\top$! Therefore, the algorithm, though preserving the set of solutions, does not give the "simplest" way of expressing this set, resulting in considerable redundancy in the solution process. ♯

1.3 Improvements of the solution method

Looking at the last example we realize that the algorithm presented in Subsection 1.2 is rather naive and needs some improvements. In particular, the original algorithm described in [Comon and Lescanne, 1989], though based on similar ideas, is actually much more flexible. By imposing far less constraints on the application of the rules, it provides a general framework to describe several distinct constraint solving algorithms. But the corresponding termination proofs are more involved and less comprehensible. The basic difference is that the elimination of parameters and the transformation into normal form are performed *simultaneously*, which allows to prune the search space, in particular if an incompatibility is detected during the second step. For example $a = b \wedge \mathcal{P}$ can be directly transformed to $\perp$, without considering the formula $\mathcal{P}$ at all.

The solution method in Subsection 1.2, a version chosen for the sake of simplicity, and even the more complicated algorithm in [Comon and Lescanne, 1989] are both highly redundant. Indeed, transformation into cnf, as well as blind application of *Explosion* and (DE) entails the duplication of significant parts of the formula, thus leading to an increase of the size of the formula and to high redundancy. For this reason several improvements of the original algorithm presented in [Comon and Lescanne, 1989] have been proposed. These methods aim at reducing the complexity of the algorithm. We will not give a detailed description of these new algorithms here (since the proofs of soundness and termination of these algorithms are in general much more involved than for the original one); instead we give a sketch of the main underlying ideas and some examples.

The key points in the improvements of the algorithm consists in reducing the number of applications of costly rules (like distributivity and explosion), that are the main source of complexity.

1.3.1 The binary explosion rule

In [Peltier, 1997b], a new definition of "solved form" (admitting the occurrence of *some* universal quantifiers) and a new explosion rule is proposed. The aim of this approach is to reduce the high branching degree of the explosion rule defined in [Comon and Lescanne, 1989]. Before we specify this rule formally, we explain the principle of **Binary Explosion** via an example.

EXAMPLE 4.34 Let $\Sigma = \{a, b, f, g\}$ and let

$$\mathcal{F} :\equiv \neg\exists x.(y = f(x) \wedge x \neq z)$$

be an equational formula. In order to find the solutions of $\mathcal{F}$ we have to eliminate the variable x. For doing so, we distinguish several cases, according to the value of the variable y. More precisely, we transform $\mathcal{F}$ into a disjunction of 4 formulae by a case analysis on the possible values for y.

$$\begin{array}{lcl} \mathcal{F}_1 & : & \neg\exists x(y = f(x) \wedge x \neq z) \wedge y = a, \\ \mathcal{F}_2 & : & \neg\exists x(y = f(x) \wedge x \neq z) \wedge y = b, \\ \mathcal{F}_3 & : & \exists u, v.[\neg\exists x.(y = f(x) \wedge x \neq z) \wedge y = g(u, v)], \\ \mathcal{F}_4 & : & \exists u.[\neg\exists x.(y = f(x) \wedge x \neq z) \wedge y = f(u)]. \end{array}$$

The first three formulae are reduced to (respectively) $y = a, y = b$ and $\exists u, v.y = g(u, v)$ by applying (R), (Clash) and standard simplifications .

The fourth formula is reduced by (R) to

$$\exists u.[\neg\exists x.(f(u) = f(x) \wedge x \neq z) \wedge y = f(u)]$$

which, by (D), reduces to

$$\exists u.[\neg\exists x.(u = x \wedge x \neq z) \wedge y = f(u)].$$

By *PElim* we get eventually obtain

$$\exists u.(u \neq z \wedge y = f(u)).$$

By inspecting $\mathcal{F}$ more carefully, we see that in transforming $\mathcal{F}$ into $\mathcal{F}_1 \vee \mathcal{F}_2 \vee \mathcal{F}_3 \vee \mathcal{F}_4$, we need not distinguish the cases $y = a, y = b$ and $y = g(u, v)$.

Indeed, the three corresponding problems will be treated *exactly in the same way*. In order to avoid triplicating the corresponding problem, we merge these 3 cases into a single one. This is achieved by adding the formula $\mathcal{G} :\ (\forall u.y \neq f(u))$.

Note that $\mathcal{G}$ is equivalent to $y = a \vee y = b \vee \exists u, v.y = g(u, v)$. Hence we *do not* eliminate all parameters from the formula. We even introduced new universal quantifiers which need *not* be removed explicitly. ♯

DEFINITION 4.35 (BINARY EXPLOSION) Let $\mathcal{P}$ be an arbitrary equational formula and let us assume that the equation $x = t$ occurs at position p in $\mathcal{P}$. Then we define the rule

$$P[x = t]_p \quad \rightarrow \quad (\forall \overline{u}.x \neq \rho(t) \wedge P[\bot]_p) \vee \exists \overline{u}.(x = \rho(t) \wedge P[\rho(t) = t]_p)$$

where $\rho(t)$ is obtained by replacing each variable in t by a new one, and $\overline{u}$ is a vector of variables occurring in $\rho(t)$. The system of all rules of this form is called *BinExp*. ♯

Remark: An important property of *BinExp* is that, in contrast to *Explosion*, the increase of the size of the formula does *not* depend on $\tau(t)$.

For example, the formula

$$\forall y.x \neq f(u, g(v, f(y, y)))$$

is transformed (by several simplifications) into

$$\begin{array}{c} \forall y_1, y_2.x \neq f(u, g(v, f(y_1, y_2))) \\ \vee \exists y_1, y_2.(x = f(u, g(v, f(y_1, y_2))) \wedge y_1 \neq y_2). \end{array}$$

♯

A new definition of solved form In contrast to *Explosion*, *BinExp* does not eliminate all universal quantifiers from the formula. Indeed, the universal quantifiers introduced by the rule itself will clearly remain in the formula and should not be eliminated. Therefore the method leads to a new kind of solved form, replacing the definition with constraints of [Comon and Lescanne, 1989]. These solved forms are more expressive than the definition with constraints (in the sense that they allow to specify sets of solutions in a more compact way) and admit the occurrence of *some universal quantifiers*. However, the use of universal quantifiers is very restricted in order to preserve the key properties of the definition with constraints: i.e., any formula distinct from $\bot$ must have a solution, and it must be possible to enumerate these solutions in an *efficient* way. The new definition of solved form is based on that of a linear complement problem.

DEFINITION 4.36 A *linear complement problem* (or *lcp*) on a variable x is a formula of the form

$$\mathcal{F} : \bigwedge_{i=1}^{n} \forall \bar{u}_i.x \neq t_i$$

where $\bar{u}_i$ is a vector of all variables in the term t_i and each variable in t_i occurs only once (i.e. t_i is linear). If $\mathcal{F}$ is an lcp, we denote by $E(\mathcal{F})$ the set $\{t_1, \ldots, t_n\}$.

Remark: Every lcp $\forall \bar{u}.x \neq t$ can be replaced by an equivalent disjunction of unification problems. That means all inequalities of this form can be represented by a finite set of equations. We made use of this property in constructing finite models from linear atomic representations in Subsection 1.4 of Section 1. ♯

EXAMPLE 4.37 Let $\Sigma = \{a, f, g\}$.

The formulae

$$\mathcal{F}_1 : \forall x.y \neq f(x) \wedge y \neq a$$

and

$$\mathcal{F}_2 : \forall x_1.y \neq f(x_1) \wedge \forall x_1, x_2.y \neq g(x_1, x_2) \wedge y \neq a$$

are lcps on y.

$$\mathcal{S}(\mathcal{F}_1) \quad = \quad \{y \leftarrow g(s_1, s_2) | s_1, s_2 \in H(\Sigma)\}.$$

$\mathcal{F}_2$ has *no solutions*, as all terms in $H(\Sigma)$ are either f-terms or g-terms, or a. ♯

The following definition generalizes solved forms.

DEFINITION 4.38 A formula $\exists \bar{x}.\mathcal{F}$ is said to be in *solved form* iff it is $\top, \bot$ or if the disjunctive normal form of $\mathcal{F}$ is a disjunction of formulae of the form

$$\bigwedge_{i=1}^{n} x_i = t_i \wedge \bigwedge_{i=1}^{k} \bigwedge_{j=1}^{m_i} (y_i \neq s_{ij} \wedge C_i) \ (\star).$$

where

- C_i $(1 \leq i \leq k)$ is a lcp on y_i with an infinite number of solutions;
- For $1 \leq i \leq n$, x_i occurs only once in the conjunction;
- For $1 \leq i \leq k$, $1 \leq j \leq m_i$, y_i does not occur in s_{ij};
- y_i is different from y_j for $i \neq j$.

In order to check that a formula is in solved form, we have to decide whether the lcp has an infinite number of solutions. This problem, though still NP-complete, is obviously much simpler than solving equational formulae. In practice the considered lcps are relatively simple and the test quite fast.

In [Peltier, 1997b], a complete algorithm for transforming equational formulae into solved form is given.

1.3.2 The algorithm of Comon and Delor

In [Comon and Delor, 1994], a more general algorithm is presented, dealing directly with equational formulae instead of equational problems. This algorithm is more flexible and, in particular, avoids the systematic transformation into cnf or dnf which is required in [Comon and Lescanne, 1989]. Dual rules are presented for eliminating *existential* quantifiers (instead of universal ones as in the previous section) when the existential quantifiers are "universal" by their polarity.

1.3.3 Pichler's Method

In [Pichler, 1999], an algorithm is presented which is significantly more efficient than this in [Comon and Lescanne, 1989]. The algorithm mainly focuses on the elimination of parameters. The ideas are the following ones:

1 Decomposition rules on equations occurring in disjunctions are not performed explicitely; thus applications of distributivity are avoided and the number of disjuncts is substantially reduced. It is proved in [Pichler, 1999] that decomposition of equations is not always necessary for eliminating parameters, based on a distinction of useful and useless variables.

2 The problem of eliminating parameters from equational problems is reduced to the so-called *co-TTC problem* (the *Term Tuple Cover* problem). The co-TTC problem is the following: given a set M of k-tuples of terms, does there exist a tuple of ground terms that is not an instance of an element of M ? Furthermore disjuncts of the form

$$\forall \bar{y}.\mathcal{P} \vee \bar{z} \neq \bar{t}$$

are replaced by

$$[\exists \bar{y}(\bar{z} = \bar{t} \wedge \mathcal{P})] \vee [\forall \bar{z}.\bar{z} \neq \bar{t}].$$

Note that the second disjunct

$$\forall \bar{z}.\bar{z} \neq \bar{t}$$

corresponds exactly to a co-TTC problem, whereas the first one contains no parameters.

3 An efficient algorithm for solving co-TTC problem is defined in [Gottlob and Pichler, 2001]. This algorithm, like this defined in Subsection 1.2 uses the domain closure axioms, thus behaving similarly as the explosion rule, but is carefully designed to avoid unnecessary instantiations of the variables and to reduce redundancy.

Pichler's algorithm yields some interesting theoretic result as well, namely that deciding the solvability of equational problems with a matrix in cnf (let us denote it by EQP-cnf) can be reduced to co-TTC in polynomial time.

Since co-TTC was known to be NP-complete, the reduction gave the NP-completeness of EQP-cnf.

2. Constrained clause logic

First-order clauses C can be interpreted as schemas for the set of all ground instances $\mathcal{C}$ over some Herbrand universe. Equational formulae, used as constraints for clauses, can restrict the sets $\mathcal{C}$ to subsets $\mathcal{C}'$ which are nonredundant in a sense to be described below. These restrictions can be turned into a powerful tool to build and to express Herbrand models which are not representable just as ARMs (see Chapter 3).

For the remaining part of this chapter we consider ground clauses as **multisets of literals**. Thus $P(a) \vee P(a)$ and $P(a)$ are considered as different, but $P(a) \vee P(b)$ and $P(b) \vee P(a)$ are considered as equal. For comfort we keep the notation of ground clauses as disjunctions of literals, provided no confusion may arise.

DEFINITION 4.39 Let C be a disjunction of literals and $\mathcal{X}$ be an equational formula; then the pair $D: \ (C, \mathcal{X})$, denoted by

$$[\![C : \mathcal{X}]\!]$$

is called a *constrained clause.* C is called the *clause part* and $\mathcal{X}$ the *constraint part* of D.

If C is unit then D is called a constrained literal (or *c-literal*). If C is the empty clause and $\mathcal{X}$ has a solution then $[\![C : \mathcal{X}]\!]$ is denoted by $\Box$ (*the empty c-clause*); if, on the other hand, $\mathcal{X} \equiv \top$ then D can be denoted by C itself. $\sharp$

NOTATION 2.1 *Let S be a set of c-clauses. We denote by* **Unit**(S) *the set of unit c-clauses in S.* $\sharp$

The semantics of a c-clause $[\![C : \mathcal{X}]\!]$ is defined as the set of ground instances $C\sigma$ where σ is a solution of $\mathcal{X}$. Note that $[\![C : \top]\!]$ (corresponding to C itself) represents the set of all ground instances of C. More formally:

DEFINITION 4.40 Let $D : [\![C : \mathcal{X}]\!]$ be a constrained clause and H be a Herbrand universe. By $\mathcal{S}_H(D)$ we denote the set

$$\{C\sigma | \sigma \in \mathcal{S}_H(\mathcal{X})\}.$$

If S is a set of c-clauses then $\mathcal{S}_H(S)$ denotes the set $\bigcup_{D \in S} \mathcal{S}_H(D)$. $\sharp$

Remark: The set $\mathcal{S}_H([\![C : \mathcal{X}]\!])$ depends not only on C and $\mathcal{X}$ but also on the Herbrand universe H. This is the reason that we have to write $\mathcal{S}_H$ instead of $\mathcal{S}$. But, in most cases, the Herbrand universe is given in advance or clear from the context; thus, if it is "safe", we will frequently write $\mathcal{S}([\![C : \mathcal{X}]\!])$ instead of $\mathcal{S}_H([\![C : X]\!])$. $\sharp$

The following definition allows us to extend the usual notion of validity (hence of satisfiability) to c-clauses and sets of c-clauses.

DEFINITION 4.41 Let $\mathcal{I}$ be an interpretation and C be a c-clause. $\mathcal{I}$ *validates* C (denoted by $\mathcal{I} \models C$) iff for all ground clauses $D \in \mathcal{S}(C)$, $\mathcal{I} \models D$. $\sharp$

EXAMPLE 4.42 Let C be the c-clause

$$[\![P(x, y) : x = a \wedge y = f(x, z)]\!]$$

and $H = H(\{f, a\})$. Then

$$\mathcal{S}_H(C) \quad = \quad \{P(a, f(a, t)) | t \in H\}.$$

In this case C is equivalent to the "ordinary" clause $P(a, f(a, z))$.

Let D be

$$[\![P(x, y) \vee Q(y, z) : x \neq y \wedge x \neq z]\!]$$

and H as above. Then

$$\mathcal{S}_H(D) \quad = \quad \{P(t, t') \vee Q(t', t'') | t, t', t'' \in H, t \neq t', t \neq t''\}.$$

In contrast to C, $\mathcal{S}_H(D)$ cannot be represented as as set of ground instances of an ordinary clause. Clearly for $H' = \{a, b\}$ we have

$$\mathcal{S}_{H'}(D) = \{P(a, b) \vee Q(b, b),\ P(b, a) \vee Q(a, a)\}.$$

and thus $\mathcal{S}_{H'}(D)$ itself is even finite. For $H'' = \{a\}$ we have $\mathcal{S}_{H''}(D) = \emptyset$. This shows the importance of the universe to the semantics.

If $\mathcal{I}$ is $\{P(t, t) | t \in H\}$ then C and D are false in $\mathcal{I}$. For

$\mathcal{J} : \{P(a, f(a, t)) | t \in H\}$

we have $\mathcal{J} \models C$, but $\mathcal{J} \not\models D$. $\sharp$

The following properties are immediate consequences of the definitions; For all c-clauses $D : [\![C : \mathcal{X}]\!]$, we have:

- If $\mathcal{X} \equiv \top$, then D is equivalent to C: standard clauses are particular cases of c-clauses.

- If $\mathcal{X} \equiv \bot$ then $\mathcal{S}(D) = \emptyset$ and thus $[\![C : \mathcal{X}]\!]$ is valid, i.e. any c-clause with an unsatisfiable constraint is a tautology.

Remark: Constraints are interpreted in the empty theory, as defined in 1. In particular, since $a = b$ has no solution, the c-clause $[\![P(x) : a = b]\!]$ is equivalent to $[\![P(x) : \bot]\!]$ and thus represents the empty set of ground clauses.

DEFINITION 4.43 Two c-clauses C and D are called *equivalent* iff they represent the same set of ground clauses i.e. iff

$$\mathcal{S}(C) = \mathcal{S}(D)$$

This relation will be denoted by $C \cong D$. Similarly, if S_1 and S_2 are two sets of c-clauses, we will write $S_1 \cong S_2$ if

- For all $C \in S_1$ there is a $D \in S_2$ with $C \cong D$ and

- For all $C \in S_2$ there is a $D \in S_1$ with $C \cong D$. ♯

Remark: The relation $\cong$ is much weaker than that of H-subsumption presented in Chapter 5. For example, the sets $\{P(a), P(f(x))\}$ and $\{P(x)\}$ are equivalent w.r.t. H-subsumption (on the signature $\Sigma = \{a, f\}$) but are not in $\cong$.

Consider also the clauses $C : [\![P(f(x)) : \top]\!]$ and $D : [\![P(y) \vee P(f(x)) : \top]\!]$. C and D have the same condensation normal form and thus are logically equivalent, but $\mathcal{S}_H(C) \cap \mathcal{S}_H(D) = \emptyset$ for every Herbrand universe H. Even if we put ground clauses into condensation normal forms we get $\mathcal{S}(C) \subset \mathcal{S}(D)$, but not $\mathcal{S}(C) = \mathcal{S}(D)$. ♯

PROPOSITION 2.1 *Let $\mathcal{X}$ and $\mathcal{Y}$ be two equational formulae with the same set of free variables and let C be a standard clause.*

If $\mathcal{S}(\mathcal{X}) \subseteq \mathcal{S}(\mathcal{Y})$ then $\mathcal{S}([\![C : \mathcal{X}]\!]) \subseteq \mathcal{S}([\![C : \mathcal{Y}]\!])$, and $[\![C : \mathcal{Y}]\!] \models [\![C : \mathcal{X}]\!]$.

Proof: Trivial. ◇

DEFINITION 4.44 The result of the application of a substitution σ to a c-clause $[\![C : \mathcal{X}]\!]$ is the c-clause

$$[\![C\sigma : \mathcal{X}\sigma]\!]$$

and is denoted by $[\![C : \mathcal{X}]\!]\sigma$.

Remark: Let σ be a ground substitution, and $[\![C : \mathcal{X}]\!]$ a c-clause.

If $\sigma \in \mathcal{S}(\mathcal{X})$ then $[\![C : \mathcal{X}]\!]\sigma \cong C\sigma$, else $[\![C : \mathcal{X}]\!]\sigma$ is the empty set of ground clauses (representing a tautology). ♯

¿From now on we do not consider c-clauses themselves, but rather their equivalence classes under $\cong$. For example the c-clauses $[\![P(x) : x = a]\!]$ and $[\![P(a) : \top]\!]$ will be considered as equal, because they define the same set of ground instances.

Moreover, we will frequently identify a c-clause C (or a set of c-clauses S) with the corresponding set of ground instances $\mathcal{S}(C)$ (or $\mathcal{S}(S)$). We will also write $C \subseteq D$, for $\mathcal{S}(C) \subseteq \mathcal{S}(D)$ or $C \cap D$ for $\mathcal{S}(C) \cap \mathcal{S}(D)$.

This convention is justified by the following theorem.

THEOREM 2.1

1 *The emptiness problem for sets of c-clauses S is decidable, i.e. there exists an algorithm that decides whether $S \cong \emptyset$.*

2 *For all sets of c-clauses S_1, S_2 there are sets of c-clauses S_3, S_4, S_5 (which can be constructed effectively) such that*

(a) $\mathcal{S}(S_3) = \mathcal{S}(S_1) \cup \mathcal{S}(S_2)$;

(b) $\mathcal{S}(S_4) = \mathcal{S}(S_1) \cap \mathcal{S}(S_2)$;

(c) $\mathcal{S}(S_5) = \mathcal{S}(S_1) \setminus \mathcal{S}(S_2)$.

Proof:

1 It suffices to decide whether for all c-clauses $[\![C : \mathcal{X}]\!]$ in S, $\mathcal{S}(\mathcal{X}) = \emptyset$, which is decidable by Theorem 1.1.

2 (a) It suffices to define $S_3 = S_1 \cup S_2$.

(b) We may assume w.l.o.g. that the sets S_1 and S_2 are singletons; the general case can be reduced to the singleton one using distributivity of the operators $\cup, \cap$.

Let $S_1 = \{[\![C_1 : \mathcal{X}_1]\!]\}$ and $S_2 = \{[\![C_2 : \mathcal{X}_2]\!]\}$ s.t. S_1 and S_2 are variable disjoint and

$$C'_1 = \bigvee_{i=1}^{l_1} P_{1,i}(\bar{t}_{1,i}),$$

$$C'_2 = \bigvee_{i=1}^{l_2} P_{2,i}(\bar{t}_{2,i}).$$

For disjunctions C_1', C_2' representing C_1, C_2. If $l_1 \neq l_2$ then, by definition of the semantics, $\mathcal{S}(S_1) \cap \mathcal{S}(S_2) = \emptyset$. In this case we may define $S_4 = [\![C : \bot]\!]$ for an arbitrary clause C.

So let us assume that $l_1 = l_2 = n$ for some $n \in \mathbb{N}$.

Let Π be the set of all permutations π on $\{1, \ldots, n\}$ such that $P_{1,i} = P_{2,\pi(i)}$ and let $\mathcal{F}$ be the formula

$$\bigvee_{\sigma \in \Pi} \bigwedge_{i=1}^{n} \bar{t}_{1,i} = \bar{t}_{2,\sigma(i)} \wedge \mathcal{X}_2$$

We define $S_4 = \{[\![C_1 : \mathcal{X}_1 \wedge \mathcal{F}]\!]\}$. Then

$$\mathcal{S}(S_4) \quad = \quad \mathcal{S}(S_1) \cap \mathcal{S}(S_2).$$

We show

(i) $\mathcal{S}(S_4) \subseteq \mathcal{S}(S_1) \cap \mathcal{S}(S_2)$ and

(ii) $\mathcal{S}(S_1) \cap \mathcal{S}(S_2) \subseteq \mathcal{S}(S_4)$.

For (i) let $\theta \in \mathcal{S}(\mathcal{X}_1 \wedge \mathcal{F})$.

Then $C_1\theta = \bigvee_{i=1}^{n} P_{1,i}(\bar{t}_{1,i})\theta$ where $\theta \in \mathcal{S}(\mathcal{X}_1)$ and $\theta \in \mathcal{S}(\mathcal{F})$, i.e.

- $C_1\theta \in \mathcal{S}(S_1)$;
- there exists a $\pi \in \Pi$ with $P_{1,i} = P_{2,\sigma(i)}$, and $\theta \in \mathcal{S}(\bar{t}_{1,i} = \bar{t}_{2,\sigma(i)})$;
- $\theta \in \mathcal{S}(\mathcal{X}_2)$.

Thus, as clauses are multisets of literals, we have $C_1\theta = C_2\theta$ and $\theta \in \mathcal{S}(S_1) \cap \mathcal{S}(S_2)$; this proves (i).

For (ii) let us assume that $\sigma \in \mathcal{S}(\mathcal{X}_1)$ and $\sigma' \in \mathcal{S}(\mathcal{X}_2)$ and $C_1\sigma = C_2\sigma'$. Then, as S_1 and S_2 are variable disjoint we may define $\theta = \sigma \cup \sigma'$. Then

$$C_1\theta = C_2\theta \text{ and } C_1\theta \in \mathcal{S}(S_1) \cap \mathcal{S}(S_2).$$

By definition of $\mathcal{F}$ we have $\theta \in \mathcal{S}(\mathcal{X}_1 \wedge \mathcal{F})$ and thus $C_1\theta \in \mathcal{S}(S_4)$.

(c) We may assume as in the point above that $S_1 = \{[\![C_1 : \mathcal{X}_1]\!]\}$ and $S_2 = \{[\![C_2 : \mathcal{X}_2]\!]\}$ for clauses C_1, C_2 and equational formulae $\mathcal{X}_1, \mathcal{X}_2$. If the number of literals in C_1, C_2 are different then we define $S_5 = S_1$. Otherwise we define $S_5 = \{[\![C_1 : \mathcal{X}_1 \wedge \mathcal{G}]\!]\}$ for

$$\mathcal{G} = \neg \bigvee_{\sigma \in \Pi} \exists \bar{x}_2 \bigwedge_{i=1}^{n} \bar{t}_{1,i} = \bar{t}_{2,\sigma(i)} \wedge \mathcal{X}_2$$

where n is the number of literals in C_1 and C_2 and $\bar{x}_2 = V(S_2)$.

◇

Remark: Theorem 2.1 jusifies the identification of sets of c-clauses with the equivalence classes under $\cong$. Note that, e.g. $C_1 : [\![P(f(x)) : \top]\!]$ and $C_2 : [\![P(x) : x \neq a]\!]$ are different as constrained clauses; C_1 is equivalent to C_2 over H for $H = H(\{a, f\})$, but not over H' for $H' = H(\{a, b, f\})$ for some constant symbol b. Thus the equivalence of constrained clauses does *not* only depend on their syntactic form but also on the universe.

$\sharp$

$\cap, \setminus, \cup$ will often be applied to unit c-clauses. The following table defines the meaning of these operators in this special case, which is simpler than the general one (see also [Comon, 1988]).

Notation	Definition
$[\![P(\bar{t}) : \mathcal{X}]\!] \cup [\![P(\bar{s}) : \mathcal{Y}]\!]$	$[\![P(\bar{x}) : \exists \bar{y}.\bar{x} = \bar{t} \wedge \mathcal{X} \vee \exists \bar{z}.\bar{x} = \bar{s} \wedge \mathcal{Y}]\!]$ where $\bar{y} = \mathcal{V}ar([\![P(\bar{t}) : \mathcal{X}]\!])$ and $\bar{z} = \mathcal{V}ar([\![P(\bar{s}) : \mathcal{Y}]\!])$
$[\![P(\bar{t}) : \mathcal{X}]\!] \cap [\![P(\bar{s}) : \mathcal{Y}]\!]$	$[\![P(\bar{x}) : \exists \bar{y}.\bar{x} = \bar{t} \wedge \mathcal{X} \wedge \exists \bar{z}.\bar{x} = \bar{s} \wedge \mathcal{Y}]\!]$ where $\bar{y} = \mathcal{V}ar([\![P(\bar{t}) : \mathcal{X}]\!])$ and $\bar{z} = \mathcal{V}ar([\![P(\bar{s}) : \mathcal{Y}]\!])$
$[\![P(\bar{t}) : \mathcal{X}]\!] \setminus [\![P(\bar{s}) : \mathcal{Y}]\!]$	$[\![P(\bar{x}) : \exists \bar{y}.\bar{x} = \bar{t} \wedge \mathcal{X} \wedge \neg(\exists \bar{z}.\bar{x} = \bar{s} \wedge \mathcal{Y})]\!]$ where $\bar{y} = \mathcal{V}ar([\![P(\bar{t}) : \mathcal{X}]\!])$ and $\bar{z} = \mathcal{V}ar([\![P(\bar{s}) : \mathcal{Y}]\!])$
$[\![P(\bar{t}) : \mathcal{X}]\!] \setminus [\![P'(\bar{s}) : \mathcal{Y}]\!]$	$[\![P(\bar{t}) : \mathcal{X}]\!]$ where $P \neq P'$

We introduce the following notations.

DEFINITION 4.45 A c-literal L *belongs* to a c-clause $C : [\![P_1 \vee \ldots \vee P_n : \mathcal{X}]\!]$ iff there exists an i with $L = [\![P_i : \mathcal{X}]\!]$ (we write $L \in C$).

DEFINITION 4.46 For all c-literals L with $L = [\![P(\bar{t}) : \mathcal{X}]\!]$ we write $\neg L$ for $[\![\neg P(\bar{t}) : \mathcal{X}]\!]$.

DEFINITION 4.47 two constrained literals L and L' are said to be complementary iff $L \cap \neg L' \neq \emptyset$ i.e. if there are complementary literals in the corresponding sets of ground instances.

3. The RAMC method

We now describe the inference method RAMC (Resolution And Model Construction), as originally presented in [Caferra and Zabel, 1992].

3.1 Informal presentation

RAMC is an extension of the resolution principle, based on the use of constrained clauses. In contrast to the method proposed in Chapter 3, the aim of RAMC is not to extract models from saturated sets of clauses, but rather to *avoid* the generation of new clauses. The basic idea is to intervene in the core of the inference rule.

In standard binary resolution it is essential to check conditions making two atoms *equal*, i.e. to find out whether they have an m.g.u. RAMC maintains this approach, but *also* considers conditions *preventing* atoms from becoming identical. Clearly both principles have to be combined under preservation of completeness. The corresponding conditions on clauses can be expressed, in a very natural way, in the constraint part. This allows us to discard some instance of the c-clauses, and therefore to reduce the search space.

The following simple example might serve as an illustration:

EXAMPLE 4.48 Let us consider the following set of c-clauses

$$S = \{C_1, \ldots, C_5\}$$

for

$$\begin{aligned}
C_1 &= [\![P(a) : \top]\!], \\
C_2 &= [\![\neg P(x) \vee R(x, f(y)) : \top]\!], \\
C_3 &= [\![\neg R(a, x) \vee \neg Q(y) \vee Q(g(y)) \vee Q_1(x) : \top]\!], \\
C_4 &= [\![Q_1(f(x)) : \top]\!] \\
C_5 &= [\![Q(a) : \top]\!].
\end{aligned}$$

Looking at C_1 and C_2 we see that these clauses can be resolved; the corresponding resolvent is:

$$C_6 = [\![R(a, f(y)) : \top]\!].$$

Now, considering clause C_2, we see that by substituting a for x we obtain $\neg P(a) \vee R(a, f(y))$, a clause subsumed by C_6. That means, all ground instances $C_2\theta$ with $x \leftarrow a \in \theta$ are subsumed by C_6 (or even by ground instances of C_6). Thus it is justified to discard all these ground instances of C_2. This can be achieved by adding the negative constraint $x \neq a$ to the constraint part of C_2. Therefore, we simply *delete* clause C_2 and replace it by a new clause C_7 for

$$C_7 = [\![\neg P(x) \vee R(x, f(y)) : x \neq a]\!].$$

Discarding instances of a clause that are subsumed by existing c-clauses suggests a new rule, which we call the *dissubsumption* rule. The formal description of the rule will be given in the next subsection.

Next consider the c-literal $[\![\neg P(x) : x \neq a]\!]$ in C_7. It is easy to see that the only positive occurrence of P in S, namely $[\![P(a) : \top]\!]$ cannot be resolved with $[\![\neg P(x) : x \neq a]\!]$. In fact there exist no ground instances which are complementary. Therefore, in any set of ground instances S' of S, the instances of $[\![\neg P(x) : x \neq a]\!]$ are *pure* in S' and thus can be added to S' without destroying sat-equivalence. But note that adding $[\![\neg P(x) : x \neq a]\!]$ merely preserves sat-equivalence, but not the logical one.

Now let $S_1 = \{C_1, \ldots, C_7\}$ and

$$C_8 \quad = \quad [\![\neg P(x) : x \neq a]\!].$$

Then $S_2 : S_1 \cup \{C_8\}$ is sat-equivalent to S_1 and even *implies* S_1. Thus if S_1 is satisfiable then S_2 is satisfiable too, and *every* model of S_2 is also a model of S_1. The rule generating C_8 will be called GPL (generate-pure-literal).

Looking upon S_2 we see that any instance of C_7 is subsumed by C_8, hence C_7 can be deleted altogether (this is another application of the so-called *dissubsumption* rule).

Now, consider the literal $\neg R(a, x)$ in C_3. The only literal occuring in S_2 which is complementary to $\neg R(a, x)$ is the literal $R(a, f(y))$ in clause C_6. If we could discard any instance of the literal $\neg R(a, x)$ that is of the form $R(a, f(y))$, then the literal became pure. But, thanks to the expressive power of equational problems, this can be achieved simply by adding the condition

$$\forall y.x \neq f(y)$$

to C_3. Thus the instances of C_3 can be split into those where x is replaced by a term $f(t)$ and the remaining ones. But consider

$$C_3' : \; [\![\neg R(a, x) \vee \neg Q(y) \vee Q(g(y)) \vee Q_1(x) : x = f(y)]\!].$$

Then all ground instances of C_3' are subsumed by C_4 and so C_3' is redundant. Thus by dissumption (C_4 on C_3) we obtain

$$C_9 \quad = \quad [\![\neg R(a, x) \vee \neg Q(y) \vee Q(g(y)) \vee Q_1(x) : \forall u.x \neq f(u)]\!].$$

and C_3 itself may be deleted. But now all ground instances of $\neg R(a, x)$ in C_9 are pure and, like before, we may generate a new unit clause C_{10} for

$$C_{10} \quad = \quad [\![\neg R(a, x) : \forall .ux \neq f(u)]\!].$$

By the explosion rule we can transform $\forall y.x \neq f(y)$ into a definition with constraints of the form:

$$x = a \vee \exists u.x = g(u)$$

Therefore, we can delete C_{10} and replace it by:

$$C_{11} = [\![\neg R(a,x) : x = a \vee \exists u.x = g(u)]\!].$$

Now, using C_{11}, we can apply the dissubsumption rule to clause C_9. The result is

$$C_{12} = [\![\neg R(a,x) \vee \neg Q(y) \vee Q(g(y)) \vee Q_1(x) : \mathcal{F}]\!]$$

where

$$\mathcal{F} = \neg(x = a \vee \exists u.x = g(u)) \wedge \forall v.x \neq f(v))$$

$\mathcal{F}$ is equivalent to:

$$\mathcal{F} \equiv x \neq a \wedge \forall u.x \neq g(u) \wedge \forall v.x \neq f(v).$$

It is easy to check that, over the Herbrand universe of S, $\mathcal{F}$ is unsolvable. Hence C_{12} is a tautology and may be deleted.

At this point, we have the following set of clauses

$$S' = \{C_1, C_4, C_5, C_6, C_8, C_{11}\}$$

for

$$\begin{aligned} C_1 &= [\![P(a) : \top]\!] \\ C_4 &= [\![Q_1(f(x)) : \top]\!] \\ C_5 &= [\![Q(a) : \top]\!] \\ C_6 &= [\![R(a, f(y)) : \top]\!] \\ C_8 &= [\![\neg P(x) : x \neq a]\!] \\ C_{11} &= [\![\neg R(a,x) : x = a \vee \exists u.x = f(u)]\!]. \end{aligned}$$

Obviously S' is a satisfiable set of unit c-clauses. Like in the case of ARMs, S' can be interpreted a representation of a Herbrand model of S. In fact the conditions below define a set of models M.

$$\begin{array}{rll} P(s) \text{ is true} & \text{if} & s = a, \\ P(s) \text{ is false} & \text{if} & s \neq a, \\ Q(s) \text{ is true} & \text{if} & s = a, \\ Q_1(s) \text{ is true} & \text{if} & s = f(t), \text{ for } t \in H, \\ R(s,r) \text{ is true} & \text{if} & s = a, r = f(t) \text{ for } t \in H, \\ R(s,r) \text{ is false} & \text{if} & s = a, r = a, \\ R(s,r) \text{ is false} & \text{if} & s = a, r = f(t), \text{ for } t \in H. \end{array}$$

M can be seen as a *partial* model, in the sense that it does not provide unique truth values to any atomic formula. But we may easily select a single $\mathcal{M} \in M$ for defining the instances of all positive unit clauses as **true** and all other instances to **false**. This way we can consider

$$\{[\![P(a) : \top]\!], [\![Q_1(f(x)) : \top]\!], [\![Q(a) : \top]\!], [\![R(a, f(y)) : \top]\!]\}$$

as a (unique) representation of a Herbrand model of S, in the same sense as ordinary ARMs represent Herbrand models. ♯

3.2 From Inference to Disinference

In this subsection we give a formal description of RAMC.

Remark: All rules presented in this section should be applied modulo the standard properties of commutativity and associativity of $\vee$. ♯

3.2.1 Inference rules

Like in ordinary clause logic the *inference rules* serve the purpose of constructing refutations of clause sets. We only have to adapt unification, resolution and factoring to c-clauses. Unification is not performed directly, but is replaced by adding constraints stating conditions for atoms to become equal. A typical advantage of representing unification in the syntax is that unification need not be performed immediately, i.e. the constraints need not to be solved (lazy unification). In fact it is possible to wait until the empty clause is derived and then to check whether the constraints are satisfiable or not.

More formally, the *binary constrained resolution rule* (or *c-resolution* for short) is defined below.

DEFINITION 4.49 (C-RESOLUTION) Let C_1, C_2 be two c-clauses with $V(C_1) \cap V(C_2) = \emptyset$ and consider the rule

$$\frac{C_1 :\ [\![P(\bar{t}_1) \vee D : \mathcal{X}]\!] \quad C_2 :\ [\![\neg P(\bar{t}_2) \vee E : \mathcal{Y}]\!]}{C :\ [\![D \vee E : \mathcal{X} \wedge \mathcal{Y} \wedge \bar{t}_1 = \bar{t}_2]\!].}$$

C is called the *c-resolvent* of C_1 and C_2 upon $\neg P(\overline{t_1})$ and $P(\overline{t_2})$ and is denoted *c-res*$(C_1, \neg P(\bar{t}_1), C_2, \neg P(\bar{t}_2))$. ♯

EXAMPLE 4.50 Let C_1, C_2 be two c-clauses with

$$\begin{aligned} C_1 &= [\![P(x) \vee R(x) : x \neq a]\!], \\ C_2 &= [\![\neg P(z) \vee Z(y) : z \neq f(y)]\!]. \end{aligned}$$

The c-clause

$$C_3 : [\![R(x) \vee Z(y) : x \neq a \wedge z \neq f(y) \wedge z = x]\!]$$

is the *c-resolvent of* C_1 *and* C_2.

C_3 can be reduced by the simplification rules to

$$[\![R(x) \vee Z(y) : x \neq a \wedge x \neq f(y)]\!]. \ \sharp$$

Constrained factorization (or *c-factorization*) can be defined in a similar way:

DEFINITION 4.51 (C-FACTORIZATION) Consider the rule:

$$\frac{C: \ [\![P(\bar{t}_1) \vee P(\bar{t}_2) \vee E : \mathcal{X}]\!]}{C': \ [\![P(\bar{t}_1) \vee E : \mathcal{X} \wedge \bar{t}_1 = \bar{t}_2]\!]}$$

C' is called a *c-factor* of C and is denoted *c-fact*$(C, P(\bar{t}_1), P(\bar{t}_2))$. $\sharp$

Remark: In contrast to resolution, in its definition used in Chapter 3, we separate factoring from the resolution rule. This makes the presentation simpler, but admits only a weaker form of subsumption. In principle the whole inference system RAMC can be reformulated for resolution with a built-in factoring rule, but for the sake of simplicity we decide for the original version of RAMC. $\sharp$

The following lemma is quite easy to prove and shows that our formalism is "conservative" over ground clause logic, i.e. in the ground case, c-resolution and c-factoring coincide with with standard resolution and factoring (contraction) for ground clauses. Keep in mind that every standard clause C can be represented as $C : \top$.

LEMMA 4.52 *Let* C_1 *and* C_2 *be two ground clauses.*

- *If* C *is a resolvent of* C_1 *and* C_2 *upon the literals* $P(\bar{t}_1)$ *and* $\neg P(\bar{t}_2)$, *then* $C =$ *c-res*$(C_1, P(\bar{t}_1), C_2, \neg P(\bar{t}_2))$.
- *Similarly, if* C_2 *is a factor of* C_1 *upon* $P(\bar{t})$ *and* $P(\bar{t}_2)$ *then* $C_2 =$ *c-fact*$(C_1, P(\bar{t}_1), P(\bar{t}_2))$.

Proof: Let

$$\begin{array}{rcl} D_1 & : & P(\bar{t}_1) \vee E \text{ and} \\ D_2 & : & \neg P(\bar{t}_2) \vee F \end{array}$$

be two ground clauses with $\bar{t}_1$ (syntactically) equal to $\bar{t}_2$. Then there exists a resolvent $D : E \vee F$ of D_1 and D_2.

D_1, D_2 can be presented as c-clauses C_1, C_2 for

$$\begin{array}{rcl} C_1 & = & [\![P(\bar{t}_1) \vee E : \top]\!] \text{ and} \\ C_2 & = & [\![\neg P(\bar{t}_2) \vee F : \top]\!]. \end{array}$$

By Definition 4.49

$$c\text{-}res(C_1, \neg P(\bar{t}_1), C_2, \neg P(\bar{t}_2)) = [\![E \vee F : \top \wedge \top \wedge \bar{t}_1 = \bar{t}_2]\!].$$

But $\top \wedge \top \wedge \bar{t}_1 = \bar{t}_2 \equiv \top$ and thus the c-resolvent is $E \vee F : \top$, which – presented as standard clause – is D.

The proof for c-factorization is similar. ◇

Soundness and refutational completeness of the inference system consisting of c-resolution and c-factoring then follows from Lemma 4.52 and from the *lifting lemma* below.

LEMMA 4.53

(a) Let $C_1 : [\![P(\bar{t}_1) \vee D : \mathcal{X}]\!]$ *and* $C_2 : [\![\neg P(\bar{t}_2) \vee E : \mathcal{Y}]\!]$ *be two variable-disjoint c-clauses and let* σ *be a substitution with* $dom(\sigma) = V(\{C_1, C_2\})$ *with* $V(C_1\sigma) \cap V(C_2\sigma) = \emptyset$. *Then*

$$c\text{-}res(c_1, P(\bar{t}_1), c_2, \neg P(\bar{t}_2))\sigma = c\text{-}res(c_1\sigma, P(\bar{t}_1)\sigma, c_2\sigma, \neg P(\bar{t}_2)\sigma).$$

(b) Let $C : [\![P(\bar{t}_1) \vee P(\bar{t}_2) \vee E : \mathcal{X}]\!]$ *be a c-clause and* σ *a substitution with* $dom(\sigma) = V(C_1)$. *Then*

$$c\text{-}fact(C, P(\bar{t}_1), P(\bar{t}_2))\sigma = c\text{-}fact(C\sigma, P(\bar{t}_1)\sigma, P(\bar{t}_2)\sigma).$$

Proof: (a):
By definition of c-resolution, we have

$$c\text{-}res(C_1, P(\bar{t}_1), C_2, \neg P(\bar{t}_2)) = [\![D \vee E : \mathcal{X} \wedge \mathcal{Y} \wedge \bar{t}_1 = \bar{t}_2]\!].$$

By definition of substitutions on contrained clauses we get

$$\begin{aligned} c\text{-}res(C_1, P(\bar{t}_1), C_2, \neg P(\bar{t}_2))\sigma &= [\![D \vee E : \mathcal{X} \wedge \mathcal{Y} \wedge \bar{t}_1 = \bar{t}_2]\!]\sigma \\ &= [\![D\sigma \vee E\sigma : \mathcal{X}\sigma \wedge \mathcal{Y}\sigma \wedge \bar{t}_1\sigma = \bar{t}_2\sigma]\!]. \end{aligned}$$

By assumption $C_1\sigma$ and $C_2\sigma$ are variable disjoint, and thus, by definition of c-resolution, the last c-clause is indeed

$$c\text{-}res(C_1\sigma, P(\bar{t}_1)\sigma, C_2\sigma, \neg P(\bar{t}_2)\sigma).$$

(b): directly by the definition of substitutions on c-clauses. ◇

LEMMA 4.54 *c-resolution and c-factorization are sound.*

Proof: Let $C_1 : [\![D_1 : \mathcal{X}_1]\!]$ and $C_2 : [\![D_2 : \mathcal{X}_2]\!]$ be two variable disjoint c-clauses and let $C : [\![D : \mathcal{X}]\!]$ be a c-resolvent of C_1 and C_2. Let $\mathcal{I}$ be a Herbrand interpretation with $\mathcal{I} \models \mathcal{S}(C_1)$ and $\mathcal{I} \models \mathcal{S}(C_2)$. We prove that $\mathcal{I} \models \mathcal{S}(C)$. By definition of c-resolution C is of the form

$$C = \textit{c-res}(C_1, P(\bar{t}_1), C_2, \neg P(\bar{t}_2)).$$

Let σ be in $\mathcal{S}(\mathcal{X})$. Then Lemma 4.53 yields

$$C\sigma = \textit{c-res}(C_1\sigma, P(\bar{t}_1)\sigma, C_2\sigma, \neg P(\bar{t}_2)\sigma).$$

We may extend σ to a substitution θ s.t. $\sigma = \theta$ on V(C) and $dom(\theta) = V(\{C_1, C_2\})$. Then clearly

$$C\sigma = \textit{c-res}(C_1\theta, P(\bar{t}_1)\theta, C_2\theta, \neg P(\bar{t}_2)\theta).$$

As $\sigma \in \mathcal{S}(\mathcal{X})$ $C\sigma$ is the ground clause $D\sigma$. Thus $\theta \in \mathcal{S}(\mathcal{X}_1) \cap \mathcal{S}(\mathcal{X}_2)$ and there are ground clauses $D_1\theta$, $D_2\theta$ corresponding to $C_1\theta$ and $C_2\theta$, respectively. By Lemma 4.52 c-resolution correponds to standard resolution on ground clauses; thus $D\sigma$ is just a resolvent of $D_1\theta$ and $D_2\theta$. By assumption $\mathcal{I} \models D_1\theta$ and $\mathcal{I} \models D_2\theta$ and, by the correctness of standard ground resolution, $\mathcal{I} \models D\sigma$.

The proof of soundness of c-factorization is straightforward. ◇

DEFINITION 4.55 (CR-DEDUCTION) Let S be a set of c-clauses. Then a sequence

$$\gamma : C_1, \ldots, C_n$$

is called a *cR-deduction* of C from S if the following conditions are fulfilled:

- $C = C_n$ and
- For all $i \in \{1, \ldots, \}$ either
  1 $C_i \in S$ or
  2 C_i is a c-factor of C_j for some $j < i$ or
  3 C_i is a c-resolvent of variants of C_j and C_k for $j, k < i$.

The *length* of γ (notation $l(\gamma)$) is defined as n (i.e. the number of clauses occurring in γ).

A cR-deduction of the empty clause is called a *cR-refutation* of S. ♯

DEFINITION 4.56 Let $\gamma : C_1, \ldots, C_n$ and $\delta : D_1, \ldots, D_n$ be two cR-deductions having the same length.

Then γ is *more general* than δ (notation $\gamma \leq_s \delta$) if $C_i \leq_s D_i$ for $i = 1, \ldots, n$. ♯

After these preparatory definitions we are going to show the completeness of *cR−deduction*. Like in standard resolution the key to this result is the lifting theorem.

THEOREM 3.1 *Let S be a set of c-clauses and S' be a set of instances of clauses in S. Then for every cR-deduction δ from S' there exists a cR-deduction γ from S with $\gamma \leq_s \delta$.*

Proof: By induction on $l(\delta)$:

- If $l = 1$ then $\delta = C'$ for a C' which is an instance of a clause C in S. Then $\gamma : C$ is a cR-deduction from S and $\gamma \leq_s \delta$.

(IH) Assume that the theorem holds for all S, S' and for all deductions δ with $l(\delta) \leq n$.
Now let δ be cR-deduction from S' with

$$\delta = D_1, \ldots, D_n, D_{n+1}.$$

By (IH) there exists a cR-deduction $C_1, \ldots, C_n$ from S with $C_i \leq_s D_i$ for $i = 1, \ldots, n$. We distinguish the following cases:

(a) $D_{n+1} \in S'$: Then D_{n+1} is an instance of a clause C_{n+1} in S and we define
$\gamma = C_1, \ldots, C_n, C_{n+1}$.

(b) D_{n+1} is a c-factor of D_j for $j \leq n$. But by $C_j \leq_s D_j$ and by Lemma 4.53 D_{n+1} is an instance of a c-factor C_{n+1} of C_j. Then $\gamma : C_1, \ldots, C_n, C_{n+1}$ is a cR-deduction from S with $\gamma \leq_s \delta$.

(c) D_{n+1} is a c-resolvent of D'_i, D'_j where $i, j \leq n$ and D'_i, D'_j are variable-disjoint variants of clauses $D_i, D_j \in S'$. Then, by (IH), there are variable-disjoint variants C'_i, C'_j of clauses $C_i, C_j \in S$ and a substitution σ with $C'_i\sigma = D'_i$ and $C'_j\sigma = D'_j$. But then the conditions of Lemma 4.53 are fulfilled and there exists a c-resolvent C_{n+1} of C'_i, C'_j s.t. $C_{n+1}\sigma = D_{n+1}$. Again
$\gamma : C_1, \ldots, C_n, C_{n+1}$ is a cR-derivation from S with $\gamma \leq_s \delta$. ◇

THEOREM 3.2 *cR-deduction is complete, i.e. for any unsatisfiable set of c-clauses S there exists a cR-refutation of S.*

Proof: Let S be an unsatisfiable set of c-clauses. By definition of the semantics this implies that $\mathcal{S}(S)$ is unsatisfiable. By the compactness theorem there exists a finite unsatisfiable subset S' of $\mathcal{S}(S)$. By the completeness of standard ground resolution there exists a ground refutation δ of S'. But S' can also be considered as a finite set of instances of c-clauses in S (with a constraint part equivalent to $\top$). Thus δ is also a cR-refutation of S'. By Theorem 3.1 there exists a cR-deduction γ from S with $\gamma \leq_s \delta$. γ is also a cR-refutation as $C \leq_s \Box$ implies $C = \Box$. $\diamond$

3.2.2 Disinference rules

In contrast to the inference rules (c-resolution and c-factoring) the *disinference rules* "take care" of satisfiable sets of clauses; by generating conditions preventing the application of inference rules the new rules are aimed at finding models of sets of c-clauses. The main idea is to discard redundant instances of c-clauses by strenghtening the constraints.

We start with the principle of *c-disresolution.*

DEFINITION 4.57 (C-DISRESOLUTION) The rule

$$\frac{C_1 \ : [\![L(\bar{t}_1) \vee E : \mathcal{X}]\!] \quad C_2 \ : [\![L^d(\bar{t}_2) : \mathcal{Y}]\!]}{C \ : [\![L(\bar{t}_1) \vee E : \mathcal{X} \wedge \forall \bar{x}.(\neg\mathcal{Y} \vee \bar{t}_1 \neq \bar{t}_2)]\!]}$$

is called (unit) *c-disresolution* (where $\bar{x} = V(\bar{t}_2) \cup var(\mathcal{Y})$).

The conclusion C is called a *bc-disresolvent* of C_2 with C_1 upon $L(\bar{t}_1)$ and is denoted by *c-Dres*$(C_2, L(\bar{t}_1), C_1)$. ♯

Remark: The constraint in the conclusion C of the disresolution rule is equivalent to $\neg\exists\bar{x}.(\mathcal{Y} \wedge \bar{t}_1 = \bar{t}_2)$. ♯

EXAMPLE 4.58 Consider the following two c-clauses:

$$\begin{aligned} C_1 &: [\![\neg P(y) \vee Z(y) : \top]\!] \text{ and} \\ C_2 &: [\![P(f(x)) : \top]\!]. \end{aligned}$$

The c-clause $C : [\![\neg P(y) \vee Z(y) : \forall x.f(x) \neq y]\!]$ is the *c-dis-resolvent of* C_2 with C_1 upon $\neg P(y)$.

C can be reduced by the simplification rules (on the signature $\Sigma = \{b, f\}$) to

$$[\![\neg P(y) \vee Z(y) : y = b]\!]$$

and then transformed into the normal form

$$[\![\neg P(b) \vee Z(b) : \top]\!].$$

Note that C_1 is equivalent to the conjunction of C and the c-clause $Z(f(x))$ which is the c-resolvent of C_1 and C_2. ♯

The soundness of disresolution rule is easy to prove; indeed the set $\mathcal{S}(C)$ instances of C in Definition 4.57 is strictly contained in the set $\mathcal{S}(C_1)$, since the constraint of C is *stricter* than this of C_1 and the clausal parts of C and C_1 are equal; thus $C_1 \models C$.

But we can do more: C_1 can actually be *deleted* from the set of c-clause and *replaced* by C as will be shown below. This shows that disresolution (as most other disinference rules), in contrast to most inference rules, *does not just add* new clauses, but rather *modifies* existing ones by *strengthening* their constraints, which leads to a deletion of some of the ground instances.

LEMMA 4.59 *Let $\mathcal{I}$ be a Herbrand interpretation and let C_1, C_2 be two c-clauses of the form*

$$\begin{aligned} C_1 &= [\![P(\bar{t}_1) \vee E : \mathcal{X}]\!], \\ C_2 &= [\![\neg P(\bar{t}_2) : \mathcal{Y}]\!]. \end{aligned}$$

Then $\mathcal{I} \models \{C_1, C_2\}$ iff

- $\mathcal{I} \models C_2$,
- $\mathcal{I} \models$ *c-Dres*$(C_2, P(\bar{t}_1), C_1)$ *and*
- $\mathcal{I} \models$ *c-res*$(C_1, P(\bar{t}_1), C_2, \neg P(\bar{t}_2))$.

Proof: Let C = *c-Dres*$(C_2, P(\bar{t}_1), C_1)$ and D = *c-res*$(C_1, P(\bar{t}_1), C_2, \neg P(\bar{t}_2))$. Then

$$\begin{aligned} C &= [\![P(\bar{t}_1) \vee E : \mathcal{X} \wedge \forall \bar{x}.(\neg \mathcal{Y} \vee \bar{t}_1 \neq \bar{t}_2)]\!], \\ D &= [\![E : \mathcal{X} \wedge \mathcal{Y} \wedge \bar{t}_1 = \bar{t}_2]\!]. \end{aligned}$$

The $\Rightarrow$-part of the equivalence is the easy one:
$\mathcal{I} \models C_2$ is trivial. $\mathcal{I} \models C$ because $\mathcal{S}(C) \subseteq \mathcal{S}(C_1)$ and $\mathcal{I} \models D$ as c-resolution is sound.

$\Leftarrow$:
Our assumption is $\mathcal{I} \models \{C_2, C, D\}$. It remains to show that $\mathcal{I} \models C_1$. Let σ be a ground substitution with $dom(\sigma) = V(C) \cup V(D)$; then $V(C_1) \subseteq dom(\sigma)$. Let us assume, for contradiction, that $\sigma \in \mathcal{S}(\mathcal{X})$ and $\mathcal{I} \not\models C_1\sigma$.

Then, $\sigma \notin \mathcal{S}(\forall \bar{x}.(\neg \mathcal{Y} \vee \bar{t}_1 \neq \bar{t}_2))$; as otherwise σ is a solution of the constraint of C and thus $\mathcal{I} \models C\sigma$; but note that $C_1\sigma = C\sigma$ and we obtain a contradiction.

Therefore $\sigma \in \mathcal{S}(\exists \bar{x}.(\mathcal{Y} \wedge \bar{t}_1 = \bar{t}_2)$. But, in this case (by $\sigma \in \mathcal{S}(\mathcal{X})$) $E\sigma \in \mathcal{S}(D)$. By assumption $\mathcal{I} \models D$ and so $\mathcal{I} \models E\sigma$. But $E\sigma$ is a subclause of $C_1\sigma$ and so $\mathcal{I} \models C_1\sigma$, too. But this leads to a contradiction again. ◇

Therefore, c-disresolution is rather a deletion rule (for ground instances) than an inference rule. In c-clause logic this kind of deletion can be achieved by strenghtening the constraints.

The *c-disfactorization rule* is of a similar flavour.

DEFINITION 4.60 (C-DISFACTORIZATION) The rule

$$\frac{C : [\![L(\bar{t}_1) \vee L(\bar{t}_2) \vee E : \mathcal{X}]\!]}{C' : [\![L(\bar{t}_1) \vee L(\bar{t}_2) \vee E : \mathcal{X} \wedge \bar{t}_1 \neq \bar{t}_2]\!]}$$

is called *c-disfactorization* (upon $L(\bar{t}_1)$ and $L(\bar{t}_2)$); C' is called a *c-disfactor* of C and is denoted *c-Dfact*$(C, L(\bar{t}_1), L(\bar{t}_2))$. ♯

EXAMPLE 4.61 When c-factorization is applied to

$$C : [\![P(g(x, f(y))) \vee P(g(f(z), x)) \vee Q(x, y) : \top]\!]$$

upon $P(g(x, f(y)))$ and $P(g(f(z), x))$, we obtain the bc-factor:

$$C_1 : [\![P(g(f(z), f(z))) \vee Q(f(z), z) : \top]\!].$$

When, on the other hand, c-disfactorization is applied to C upon $P(g(x, f(y)))$ and $P(g(f(z), x))$, we obtain (with additional constraint simplification) the c-disfactor

$$C' : \ [\![P(g(x, f(y))) \vee P(g(f(z), x)) \vee Q(x, y) : x \neq f(z) \vee y \neq z]\!]. \ \sharp$$

LEMMA 4.62 *Let $\mathcal{I}$ be a Herbrand interpretation and*

$$C \quad = \quad [\![P(\bar{t}_1) \vee P(\bar{t}_2) \vee E : \mathcal{X}]\!].$$

Then $\mathcal{I} \models C$ iff

- $\mathcal{I} \models$ *c-fact*$(C, P(\bar{t}_1), P(\bar{t}_2)$ *and*
- $\mathcal{I} \models$ *c-Dfact*$(c, P(\bar{t}_1), P(\bar{t}_2))$.

Proof: The proof is analogous to this of Lemma 4.59. ◇

In standard resolution provers the use of subsumption (see Chapter 3) is of central importance because it provides a mean to strongly reduce the search space by deleting redundant clauses. The same idea can be extended to constrained clauses, by taking advantage of the expressive power of equational formulae. Instead of deleting subsumed clauses, we will discard *instances* of the c-clauses that are subsumed by another c-clause; this, in turn, leads to the deletion in the corresponding set of ground clauses.

The *c-dissubsumption rule* computes constraints *preventing* a c-clause C_1 from being subsumed by a c-clause C_2. It is defined formally as follows:

DEFINITION 4.63 (C-DISSUBSUMPTION) The rule

$$\frac{C_1 \;: [\![\bigvee_{i=1}^{n} L_i(\bar{s}_i) : \mathcal{X}]\!] \quad C_2 \;: [\![\bigvee_{i=1}^{n} L_i(\bar{t}_i) \vee E : \mathcal{Y}]\!]}{C_3 \;: [\![L_1(\bar{t}_1) \vee \ldots \vee L_n(\bar{t}_n) \vee E : \mathcal{Y} \wedge \forall \bar{x}.[\neg \mathcal{X} \vee \bar{s}_1 \neq \bar{t}_1 \vee \ldots \vee \bar{s}_n \neq \bar{t}_n]]\!]}$$

is called *dissubsumption* of C_2 by C_1 (where $\bar{x} = V(C_1)$). The conclusion C_3 is denoted by *c-Dsub*(C_2, C_1). ♯

LEMMA 4.64 *Let C_1, C_2 be two c-clauses over a Herbrand universe H and let $\mathcal{I}$ be a Herbrand interpretation of $\{C_1, C_2\}$ over H. Then*

$$\mathcal{I} \models \{C_1, C_2\} \textbf{ iff } \mathcal{I} \models \{C_1, \textit{c-Dsub}(C_2, C_1)\}.$$

Proof: As for Lemma 4.59. ◇

In the same way as we can modify subsumption to dissubsumption, we can modify tautology elimination to a distautology rule.

DEFINITION 4.65 (DISAUTOLOGY-RULE) The rule

$$\frac{C_1 \;: [\![P(\bar{t}_1) \vee \neg P(\bar{t}_2) \vee E : \mathcal{X}]\!]}{C_2 \;: [\![P(\bar{t}_1) \vee \neg P(\bar{t}_2) \vee E : \mathcal{X} \wedge \bar{t}_1 \neq \bar{t}_2]\!]}$$

is called the c-distautology rule; C_2 is denoted by *c-Dtaut*(C_1). ♯

LEMMA 4.66 *Let $\mathcal{I}$ be a Herbrand interpretation over a Herbrand universe H and let C_1 be a c-clause over H with $C_2 = \textit{c-Dtaut}(C_1)$. Then $\mathcal{I} \models C_1$ iff $\mathcal{I} \models C_2$.*

Proof: That $\mathcal{I} \models C_1$ implies $\mathcal{I} \models C_2$ is trivial. For the other direction consider the set $\mathcal{D} : \mathcal{S}(C_1) \setminus \mathcal{S}(C_2)$. By definition of the c-distautology rule $\mathcal{D}$ consists of tautologies only. Therefore $\mathcal{S}(C_2) \cup \mathcal{D}$ is true in $\mathcal{I}$ iff $\mathcal{S}(C_2)$ is true in $\mathcal{I}$. ◇

The GPL rule is the most important rule in the process of model building. It generates “pure literals”. Essentially we adapt the notion of pure literals (see for example [Loveland, 1978]) to sets of constrained clauses.

DEFINITION 4.67 Let S be a set of c-clauses. A literal $[\![L(\bar{t}) : \mathcal{X}]\!]$ is said to be *pure* in S if for all clauses $C \in S$ and for all c-literals of the form $[\![L^d(\bar{s}) : \mathcal{Y}]\!]$ in C, we have:

$$\mathcal{X} \wedge \mathcal{Y} \wedge \bar{s} = \bar{t} \ \equiv \ \bot. \ \sharp$$

Intuitively, pure literals can be safely evaluated to $\top$, since S does not contain complementary instances.

LEMMA 4.68 *Let L be a literal which is pure in S. Then S is sat-equivalent to $S \cup \{L\}$.*

Proof: Clearly the satisfiability of $S \cup \{L\}$ implies the satisfiability of S, we even have $S \cup \{L\}$ implies S.

For the other direction let us assume that L is of the form $[\![P(\bar{t}) : \mathcal{X}]\!]$ (if L is negative the argument is analogous) and that S is satisfiable. Let

$$S' \cup \{P(\bar{t}_1), \ldots, P(\bar{t}_n)\}$$

be a finite subset of $\mathcal{S}(S \cup \{L\})$. Then S' is satisfiable as S is satisfiable. But $[\![P(\bar{t}) : \mathcal{X}]\!]$ is pure in S and therefore no literal $\neg P(\bar{t}_i)$ for $i = 1, \ldots, n$ appears in S'; note that, otherwise, there would be a literal $[\![\neg P(\bar{s}) : \mathcal{Y}]\!]$ in S with

$$\mathcal{X} \wedge \mathcal{Y} \wedge \bar{s} = \bar{t} \ \not\equiv \ \bot.$$

Thus there exists a model $\mathcal{I}$ of S' with $v_{\mathcal{I}}(P(\bar{t}_i)) = \textbf{true}$ for $i = 1, \ldots, n$. By definition $\mathcal{I}$ is also a model of $S' \cup \{P(\bar{t}_1), \ldots, P(\bar{t}_n)\}$.

As this holds for all finite subsets of $S \cup \{L\}$, $S \cup \{L\}$ is satisfiable by the compactness theorem. $\diamond$

Lemma 4.68 suggests a natural way of building a model on a set of clauses S: for each clause C and for each literal L in C we try to transform L into a pure literal. The way we realizes this idea is by setting conditions (coded in the constraints), restricting the instances of literals in order to avoid unification with arguments of dual literals in S. The newly generated literal is then added to S. If its constraint is different from $\bot$ then this may lead to further applications of the dissubsumption rule, thus leading to a further reduction of redundancy.

DEFINITION 4.69 (GPL-RULE) Let S be a set of c-clauses and let

$$\begin{array}{rcl} C & : & [\![L(\bar{t}) \vee E : \mathcal{X}]\!] \text{ and} \\ D & : & [\![L^d(\bar{s}) \vee F : \mathcal{Y}]\!] \end{array}$$

be clauses in S. We define

$$\mathcal{X}(L(\bar{t}), D, L^d(\bar{s})) \quad : \quad \forall \bar{x}.(\neg \mathcal{Y} \vee \bar{s} \neq \bar{t})$$

where $\bar{x}$ is a vector of $V(D)$. Let $S[L]$ be the subset of all clauses in S containing a literal of the form $L^d(\bar{s})$. Then we define the constraint $\mathcal{X}_{pure}$ by

$$\mathcal{X}_{pure} \quad = \quad \bigwedge \{\mathcal{X}(L(\bar{t}), D, L^d(\bar{s})) | D \in S[L], L^d(\bar{s}) \text{ in } D\}.$$

Then the GPL-rule (generate-pure-literal rule) is defined by

$$\frac{[\![L(\bar{t}) \vee E : \mathcal{X}]\!] \quad S}{[\![C' : L(\bar{t}) : \mathcal{X} \wedge \mathcal{X}_{pure}]\!]}$$

C' is called the *result of GPL* on $(C, L(\bar{t}), S)$. ♯

Intuitively $\mathcal{X}_{pure}$ expresses the conditions sufficient for $L(\bar{t})$ to be pure in S: it simply prevents unification between $L(\bar{t})$ and any complementary literal in S.

Remark: In using GPL we have to take care that the literal is pure also in the clause it occurs. For illustration let S be the clause set

$$S \quad = \quad \{C_1 : [\![P(a) : \top]\!],\ C_2 : [\![\neg P(x) \vee P(f(x)) : \top]\!]\}.$$

Clearly S is satisfiable. We apply GPL on $\neg P(x)$ in C_2. Then

$$\begin{aligned} \mathcal{X}(\neg P(x), C_2\{x \leftarrow y\}, P(f(y))) \quad &= \quad \forall y.(\neg \top \vee x \neq f(y)), \\ \mathcal{X}(\neg P(x), C_1, P(a)) \quad &= \quad \neg \top \vee x \neq a. \end{aligned}$$

Therefore $\mathcal{X}_{pure} \equiv \forall y.x \neq f(y) \wedge x \neq a$ and so $\mathcal{X}_{pure} \equiv \bot$.

Indeed there is no literal of the form $\neg P(t)$ which is pure in the set of all ground instances S' of S.

Now let us assume we "forget" to apply C_2 to itself in using the GPL-rule. Then we obtain $\mathcal{X}_{pure} \equiv x \neq a$ and the new unit clause

$$D : [\![\neg P(x) : x \neq a]\!].$$

But adding D to S we see that

$$c\text{-}Dsub(D, C_2) \quad = \quad [\![\neg P(a) \vee P(f(a)) : \top]\!].$$

Now, if we throw away C_2 and replace it by c-$Dsub(D, C_2)$ we obtain the new set S' for

$$S' = \{[\![P(a) : \top]\!],\ [\![\neg P(a) \vee P(f(a)) : \top]\!],\ [\![\neg P(x) : x \neq a]\!]\}$$

But S' is unsatisfiable! This proves that a careless use of GPL may result in incorrect inferences. $\sharp$

Since, by GPL, a new unit c-clause has been added, the simplification rules introduced above may now be invoked. So some of the c-clauses may be shortened, and the equational problems of others may be strengthened. If the equational problem of a c-clause is reduced to $\bot$, this c-clause is simply deleted from the set of c-clauses by our structural rules.

LEMMA 4.70 *Let S be a set of c-clauses and let C : $[\![L(\bar{t}) : \mathcal{X} \wedge \mathcal{X}_{pure}]\!]$ be a c-clause generated by the GPL rule. Then S is sat-equivalent to $S \cup \{C\}$.*

Proof: By definition of $\mathcal{X}_{pure}$, the constrained literal $[\![L(\bar{t}) : \mathcal{X} \wedge \mathcal{X}_{pure}]\!]$ is pure in S. Therefore the result follows from Lemma 4.68. $\diamond$

Remark: Note that the GPL rule *is not an inference rule* in the usual sense. Indeed, the conclusion does *not* follow from the premises; instead the conclusion is merely *consistent* with the set of c-clauses at hand. $\sharp$

Disinference rules and constraint handling

As shown above, the disinference rules aim at discarding instances of c-clauses, resulting in a reduction of search space. The prize to pay is, of course, the additional computation needed for solving the constraints. Notice that the disinference rules are *not needed* for refutational completeness, hence it is possible to restrict the application of these rules to equational formulae of some particular form admitting fast constraint solving algorithms. Possible applications of the disinference rules leading to constraints that are not of this particular form can simply be ignored.

3.3 Simplification rules

These rules aim at simplifying the set of c-clauses by transforming c-clauses into more convenient forms. The general form is the following one:

$$\frac{[\![C : \mathcal{P}]\!]}{[\![C : \mathcal{P}']\!]}$$

if $\mathcal{P}' \equiv \mathcal{P}$ and, according to some ordering, if $\mathcal{P}'$ is simpler than $\mathcal{P}$.

$$\frac{[\![C : \exists x.\mathcal{P}]\!]}{[\![C : \mathcal{P}]\!]}$$

if x is a free variable in $\mathcal{P}$ not occurring in C

$$\frac{[\![C : \bot]\!]}{\mathbf{true}}$$

for all clauses C.

3.4 Examples

In this section, we give examples and applications of the method.

EXAMPLE 4.71 Building a model by several applications of the GPL rule and unit c-dissubsumption:

Let S be the set of c-clauses $\{C_1, C_2\}$ below:

$C_1 : [\![P(f(g(a, x))) \vee Q(x, x) : \forall y.x \neq f(y)]\!]$.

$C_2 : [\![\neg P(f(g(y, b))) \vee \neg Q(z, y) : \top]\!]$.

$C_3 : [\![P(f(g(a, x))) : x = a \vee \exists y, z.x = g(y, z)]\!]$ by GPL upon C_1.

$C_4 : [\![P(f(g(a, x))) \vee Q(x, x) : x = b]\!]$ replaces C_1 by dissubsumption with C_3.

$C_5 : [\![\neg P(f(g(y, b))) : y \neq a]\!]$ by GPL upon C_2.

$C_6 : [\![\neg P(f(g(y, b))) \vee \neg Q(z, y) : y = a]\!]$ replaces C_2 by dissubsumption with C_5.

Now let us see how $\neg Q(z, y)$ in C_6 and $Q(x, x)$ in C_4 will be recognized as pure:

$C_7 : [\![\neg Q(z, y) : y = a]\!]$ by GPL upon C_6.

$C_8 : [\![Q(x, x) : x = b]\!]$ by GPL upon C_4.

$C_9 : [\![P(f(g(a, x))) \vee Q(x, x) : \bot]\!]$ replaces C_4 by dissubsumption with C_8, and C_9 is deleted.

$C_{10} : [\![\neg P(f(g(y, b))) \vee \neg Q(z, y) : \bot]\!]$ replaces C_6 by dissubsumption with C_7, and is deleted as well.

Thus all nonunit c-clauses have been deleted and we are left with the unit c-clauses:

$$\begin{array}{lcl} C_5 & : & [\![\neg P(f(g(y, b))) : y \neq a]\!], \\ C_7 & : & [\![\neg Q(z, y) : y = a]\!], \\ C_8 & : & [\![Q(x, x) : x = b]\!]. \end{array}$$

$\{C_5, C_7, C_8\}$ define in fact a *set* of models of $\{C_1, C_2\}$; but any completion of the missing truth values leads to a model. For instance we may assign all undefined truth values to true. Note that by the nature of constraints we can represent a Herbrand model by the following set of *positive* c-unit clauses as well (in analogy to ARMS):

$$\begin{aligned} C'_5 &: [\![P(f(g(y,b))) : y = a]\!], \\ C'_7 &: [\![Q(z,y) : y \neq a]\!], \\ C_8 &: [\![Q(x,x) : x = b]\!]. \ \sharp \end{aligned}$$

EXAMPLE 4.72 (Aunt Agatha's Killer).

Someone who lives in Dreadsbury Mansion killed Aunt Agatha. Agatha, the butler, and Charles live in Dreadsbury Mansion, and are the only people who live therein. A killer always hates his victim, and is never richer than his victim. Charles hates no one that Aunt Agatha hates. Agatha hates everyone except the butler. The butler hates everyone not richer than Aunt Agatha. The butler hates everyone Agatha hates. No one hates everyone. Agatha is not the butler.

Who killed Aunt Agatha ?

In formalizing the example we need three constant symbols, a, b, c representing Agatha, the Butler and Charles, thus $\Sigma = \{a, b, c\}$. The predicate symbols are self-explaining; K for *is the killer of*, H *hates* and R *is richer than*. The problem in form of constrained clauses is given below:

$$\begin{aligned} &1^\circ\ [\![K(a,a) \vee K(b,a) \vee K(c,a) : \top]\!] \\ &2^\circ\ [\![\neg K(x,y) \vee H(x,y) : \top]\!] \\ &3^\circ\ [\![\neg K(x,y) \vee \neg R(x,y) : \top]\!] \\ &4^\circ\ [\![\neg H(a,x) \vee \neg H(c,x) : \top]\!] \\ &5^\circ\ [\![H(a,x) : x \neq b]\!] \\ &6^\circ\ [\![R(x,a) \vee H(b,x) : \top]\!] \\ &7^\circ\ [\![\neg H(a,x) \vee H(b,x) : \top]\!] \\ &8^\circ\ [\![\neg H(x,a) \vee \neg H(x,b) \vee \neg H(x,c) : \top]\!] \end{aligned}$$

We start the derivation:

9° $[\![\neg H(c,x) : x \neq b]\!]$ from *c-res*(4.1,5)
10° $[\![\neg H(a,x) \vee \neg H(c,x) : x = b]\!]$ replaces 4 by *c-Dsub*(4,9)
11° $[\![H(b,x) : x \neq b]\!]$ from *c-res*(7.1,5)
12° $[\![\neg H(a,x) \vee H(b,x) : x = b]\!]$ replaces 7 by *c-Dsub*(7,11)
13° $[\![\neg H(x,a) \vee \neg H(x,b) \vee \neg H(x,c) : x \neq c]\!]$ replaces 8 by *c-Dsub*(8,9)
14° $[\![\neg H(x,b) \vee \neg H(x,c) : x = b]\!]$ from *c-res*(13.1,11)
15° $[\![\neg H(x,b) : x = b]\!]$ from *c-res*(14.2,11)
16° $[\![\neg H(x,b) \vee \neg H(x,c) : \bot]\!]$ replaces 14 by *c-Dsub*(14,15)
17° $[\![\neg H(x,a) \vee \neg H(x,b) \vee \neg H(x,c) : x \neq c \wedge x \neq b]\!]$
replaces 13 by *c-Dsub*(13,15)
18° $[\![\neg H(x,b) \vee \neg H(x,c) : x = a]\!]$ from *c-res*(17.1,5)
19° $[\![\neg H(x,b) : x = a]\!]$ from *c-res*(18.2,5)
20° $[\![\neg H(x,b) \vee \neg H(x,c) : \bot]\!]$ replaces 18 by *c-Dsub*(18,19)
21° $[\![\neg H(x,a) \vee \neg H(x,b) \vee \neg H(x,c) : \bot]\!]$ replaces 17 by *c-Dsub*(17,19)

At this stage, we have deduced four new unit clauses; currently the unit clauses are 5, 9, 11, 15 and 19. To get a more compact derivation we merge 5 and 11 (the H-unit clauses) and 9,15 and 19 (the $\neg H$-unit clauses). This merging (a kind of anti-normalization of c-clauses) is not necessary from a computational point of view, but makes the whole derivation shorter and more intelligible. So we obtain

22° $[\![H(x,y) : x \neq c \wedge y \neq b]\!]$ (from 5 and 11)
23° $[\![\neg H(x,y) : (x = c \wedge y \neq b) \vee (x \neq c \wedge y = b)]\!]$
(from 9, 15 and 19)

The other c-clauses at hand are 1, 2, 3, and 6 (the c-clauses 12 and 10 are both subsumed by 23).

24° $[\![R(x,a) \vee H(b,x) : x = b]\!]$ replaces 6 by *c-Dsub*(6,22)
25° $[\![\neg K(x,y) \vee H(x,y) : x = c \vee y = b]\!]$ replaces 2 by *c-Dsub*(2,22)
26° $[\![\neg K(x,y) : (x = c \wedge y \neq b) \vee (x \neq c \wedge y = b)]\!]$ from *c-res*(25.2,23)
27° $[\![\neg K(x,y) \vee H(x,y) : x = c \wedge y = b]\!]$ replaces 25 by *c-Dsub*(25,26)
28° $[\![\neg K(x,y) \vee \neg R(x,y) : (x = c \wedge y = b) \vee (x \neq c \wedge y \neq b)]\!]$
replaces 3 by *c-Dsub*(3,26)
29° $[\![R(x,y) : x = b \wedge y = a]\!]$ from *c-res*(23,24.2)
30° $[\![\neg K(x,y) : (x = b \wedge y = a)]\!]$ from *c-res*(29,28.2)
31° $[\![K(a,a) \vee K(b,a) : \top]\!]$ from *c-res*(1.3, 26)
32° $[\![K(a,a) : \top]\!]$ from *c-res*(31.2, 30)

The answer to the question: "who killed Aunt Agatha?" is found!
Obviously Aunt Agatha committed suicide.

33° $[\![K(a,a) \vee K(b,a) : \bot]\!]$ replaces 31 by *c-Dsub*(31, 32)
34° $[\![K(a,a) \vee K(b,a) \vee K(c,a) : \bot]\!]$ replaces 1 by *c-Dsub*(1, 32)
35° $[\![\neg R(x,y) : x = a \wedge y = a]\!]$ from *c-res*(28.1, 32)

The c-clause 28 is replaced by:

$$36^\circ\ [\![\neg K(x,y) \vee \neg R(x,y) : \mathcal{F}]\!]$$

where:

$$\mathcal{F} = [(x = c \wedge y = b) \vee (x \neq c \wedge y \neq b)] \wedge (x \neq a \vee y \neq a).$$

At this stage, we have the unit c-clauses 22, 23, 26, 29, 32, 35, and three non unit c-clauses 24, 27 and 36. We show that the GPL rule can be applied and that we can get rid of the remaining nonunit clauses.

37° $[\![\neg K(x,y) : \mathcal{F}]\!]$ from GPL, 36
38° $[\![\neg K(x,y) \vee \neg R(x,y) : \perp]\!]$ replaces 36 by *c-Dsub*(36,37).
39° $[\![\neg K(x,y) \vee H(x,y) : \perp]\!]$ replaces 27 by *c-Dsub*(27,37).
40° $[\![R(b,a) : \top]\!]$ from GPL, 24.1
41° $[\![R(x,a) \vee H(b,x) : \perp]\!]$ replaces 24 by *c-Dsub*(24,40).

Now only the unit c-clauses $22, 23, 26, 29, 32, 35, 37, 40$ are left. They define a (partial) model of the Dreadsbury Mansion crime. Every ground literal which does not occur in $\mathcal{S}(\{22, 23, 26, 32, 35, 37, 40\})$ can be interpreted arbitrarily. ♯

3.5 Basic properties of the method

c-resolution together with the disinference rules we have presented in Subsection 3.2 form the core of the inference principle RAMC. Some rules (c-resolution, c-factoring, GPL) are *deductive* rules, i.e. they add new c-clauses without changing the set derived so far. The other rules (the disinference rules) are also deductive, but at the same time *deletion rules*, e.g. the distautology rule creates a new clause and deletes an old one. Note that the distautology rule infers a new object *only* on the general clause level; on the set of all ground instances it is a *pure deletion rule.* The strongest and most important disinference rule is dissubsumption (look at the examples of the last subsection). Indeed some of the other disinference rules can be expressed by dissubsumption and a deduction rule:

PROPOSITION 3.1 *c-Disresolution and c-disfactoring can be expressed by dissubsumption, resolution and factoring, or more formally:*

(a) Let $D =$ *c-Dfact*$(C, L(\bar{t}_1), L(\bar{t}_2))$*. Then either* $D \equiv C$ *or there exists a factor* C' *of* C *with* $D =$ *c-Dsub*(C, C')*.*

(b) Let $C_3 =$ *c-Dres*$(C_2, L(\bar{t}), C_1)$*. Then either* $C_3 \equiv C_1$ *or there exists a resolvent* R *of* C_1 *and* C_2 *with* $C_3 =$ *c-Dsub*(R, C_1)*.*

Proof:

(a) : Let us assume $D \not\equiv C$.
By definition of c-disfactoring C is of the form

$$[\![E \vee P(\bar{t}_1) \vee P(\bar{t}_2) : \mathcal{X}]\!]$$

and

$$D = [\![E \vee P(\bar{t}_1) \vee P(\bar{t}_2) : \mathcal{X} \wedge \bar{t}_1 \neq \bar{t}_2]\!].$$

By $D \not\equiv C$ the equational formula $\mathcal{X} \wedge \bar{t}_1 = \bar{t}_2$ must have a solution. But then the clause C' for

$$C' \quad = \quad [\![E \vee P(\bar{t}_1) : \mathcal{X} \wedge \bar{t}_1 = \bar{t}_2]\!]$$

is a c-factor of C. Clearly $D = c\text{-}Dsub(C, C')$.

(b) : Let

$$\begin{array}{lll} C_1 & : & [\![L(\bar{t}_1) \vee E : \mathcal{X}]\!], \text{ and} \\ C_2 & : & [\![L(\bar{t}_2) : \mathcal{Y}]\!]. \end{array}$$

Again we may assume that $C_3 \not\equiv C_1$. The proof proceeds quite similarly: by $C_3 \not\equiv C_1$ the c-clause

$$R \quad : \quad [\![E : \mathcal{X} \wedge \mathcal{Y} \wedge \bar{t}_1 = \bar{t}_2]\!]$$

is indeed a c-resolvent of C_1 and C_2. That $C_3 = c\text{-}Dsub(R, C_1)$ follows directly from the definition of c-dissubsumption. $\diamond$

Remark: Proposition 3.1 tells us that, in presence of disresolution and the deduction rules we can sacrifice c-disfactoring and c-disresolution without reducing the power of the whole system. Thus we keep only c-distautology and c-dissubsumption as deletion rules. $\sharp$

DEFINITION 4.73 (RAMC-REDUCTION) Let S be a set of c-clauses. Then $S \rhd_1 S'$ if $S' = S \cup \{C\}$ where C is obtained from S in one of the following ways:

a. C is a clause in S or

b. C is a c-factor of a clause C' in S or

c. C is a c-resolvent of variants of clauses $C_1, C_2 \in S$ or

d. C is C' where C' is a result of the GPL-rule on S or

e. C is obtained by simplification of a clause in S.

$S \rhd_2 S'$ if either $S' = simp(S)$ where *simp* is an operator simplifying all clauses in a set, or

$$S' = (S \setminus \{C\}) \cup \{D\},$$

and D is obtained in one of the following ways:

(1) $D = c\text{-}Dsub(C, C')$ for a clause C' which is a variant of a clause in S, or

(2) $D = c\text{-}Dtaut(C)$.

Finally we define $S \rhd_r S'$ if either $S \rhd_1 S'$ or $S \rhd_2 S'$. If $S \rhd_r S'$ then we say that S' is obtained from S via RAMC. ♮

Remark: $\rhd_1$ describes the RAMC-deduction principle without deletion and thus describes the purely deductive part. $\rhd_2$ describes the deletion and simplification part. Note that $\rhd_1$ and $\rhd_2$ differ in the use of simplification. In $\rhd_1$ simplified clauses are merely added, while in $\rhd_2$ clauses in the original set may be deleted. The application of *simp* is required to obtain real deletion, i.e. the removal of clauses with empty constraints. ♮

While $\rhd_1$ is monotone (i.e. $S \subseteq S'$ for $S \rhd_1 S'$), $\rhd_2$ and $\rhd_r$ are not. Thus it does not make sense to define the deductive closure as the union of derivable clauses. In fact, it is more natural to describe the inference via replacement sequences like in [Leitsch, 1997].

DEFINITION 4.74 (RAMC-SEQUENCE) A sequence of sets of c-clauses $\gamma : (S_i)_{i \in \mathbb{N}}$ is called a *RAMC-sequence* on S if $S_0 = S$ and for all i $S_i \rhd_r S_{i+1}$.

γ is called a *refutation sequence* on S if $S_0 = S$ and there exists an S_i with $S_i = \{\Box\}$. γ is called *convergent* on S if $S_0 = S$ and there exists an i with $S_i = S_{i+1}$. We write $S \rhd_r^* S'$ if there exists a RAMC sequence $(S_i)_{i \in \mathbb{N}}$ on S and a $j \in \mathbb{N}$ s.t. $S' = S_j$. ♮

Remark: If a RAMC-sequence γ is a refutation sequence then there exists a converging sequence too: once we have $\Box \in S_i$ then a consecutive application of $\rhd_2$ deletes all other clauses in S_i, i.e. $S_{i+2k} = \{\Box\}$, where k is the number of clauses in S_i. Thus, in some sense, $\Box$ is the *normal form* of unsatisfiable clause sets under $\rhd_r$. The most interesting case is convergence on satisfiable clause sets. Indeed if RAMC converges to a set S' where S' consists of unit clauses only, we have found a representation of a Herbrand model; we may simply interpret all positive instances as true and the other ones as false. If RAMC converges to S' for a set S' containing nonunit clauses we only know that the original set is satisfiable.

Due to the undecidability of c-clause logic there are sets of clauses S s.t. no RAMC-sequence terminates on S; thus nontermination is unavoidable in principle. But RAMC *improves* the termination behavior by the disinference rules and thus gives more *logical information* than the purely deductive part $\rhd_1$. ♯

Due to the presence of the GPL rule in $\rhd_r$, RAMC is *not* sound in the standard sense. Indeed $S \rhd_r^* S'$ does not imply that $S \models S'$. But still soundness holds in weaker sense which is sufficient for model building. This weaker concept also played an important role in model building by resolution in chapter 3. Indeed, for the T-operator in MB, we had $\mathcal{C} \sim_{sat} T(\mathcal{C})$ and $T(\mathcal{C}) \models \mathcal{C}$, but not $\mathcal{C} \models T(\mathcal{C})$; thus T is weakly sound in the sense defined below:

DEFINITION 4.75 (WEAK SOUNDNESS) A binary relation $\rhd$ on sets of c-clauses is called *weakly sound* if, for all sets of c-clauses S and S', $S \rhd S'$ implies

- S and S' are sat-equivalent and
- $S' \models S$. ♯

Remark: Note that weak soundness is more than just sat-equivalence. Indeed a relation $\rhd$ with

$$\{[\![P(a) : \top]\!]\} \rhd \{[\![P(a) \vee \neg P(a) : \top]\!]\}$$

is not weakly sound, although sat-equivalence is preserved. ♯

THEOREM 3.3 (WEAK SOUNDNESS OF RAMC) *RAMC is weakly sound, i.e. the relation $\rhd_r$ is weakly sound.*

Proof: All rules in RAMC, with the exception of GPL, are (strongly) sound. If $S \cup \{C\}$ is derived from S via GPL then, by Lemma 4.70, S is sat-equivalent to $S \cup \{C\}$. $S \cup \{C\} \models S$ is trivial. ◇

THEOREM 3.4 (COMPLETENESS) *RAMC is refutationally complete, i.e. for all unsatisfiable set of c-clauses S, $S \rhd_r^* \{\Box\}$*

Proof: The result follows from the completeness of cR-deduction, i.e. from Theorem 3.2: Indeed if γ is a cR-derivation of a clause C from S then there exists a set S' with $C \in S'$ and $S \rhd_1^* S'$.

So for any unsatisfiable set of c-clauses S we obtain a set S' with $S \rhd_1^* S'$ and $\Box \in S'$.

But, as already mentioned in a remark above, a finite sequence of $\rhd_2$-reductions δ on S' then leads to $\{\Box\}$. Clearly γ, δ is a RAMC-refutation sequence on S and $S \rhd_r^* \{\Box\}$. $\diamond$

The definition of $\rhd_r$ and of RAMC-sequences is quite liberal in the sense that it contains a lot of nondeterminism. Thus it is not made clear when and how often inference- and deletion rules have to be applied. Looking at the proof of Theorem 3.4 we see that deletion rules are only necessary at the end of derivations, but on the other hand they should be applied as soon as possible in order to reduce the search space. We will adress this problem in developing a strong version of RAMC-reduction where the deletion rules are exhaustively applied after any deductive step; this is possible as the deletion rules themselves are terminating.

DEFINITION 4.76 We define the strict version $>_2$ of $\rhd_2$ by

$$S >_2 S' \quad \text{iff} \quad S \rhd_2 S' \text{ and } S' \neq S.$$

$\sharp$

LEMMA 4.77 *$>_2$ is strongly terminating, i.e. there exists no infinite sequence of finite sets of c-clauses $(S_i)_{i \in \mathbb{N}}$ s.t. $S_i >_2 S_{i+1}$ for all $i \in \mathbb{N}$.*

Proof: Let S be an arbitrary finite set of c-clauses. We define define three measures ρ_1, ρ_2, ρ by

$$\begin{aligned}
\rho_1(S) &= |\{(C,D) | C, D \in S, \textit{c-Dsub}(C,D) \not\simeq C\}|, \\
\rho_2(S) &= |\{C | C \in S, \textit{c-Dtaut}(C) \not\simeq C\}|, \\
\rho(S) &= \rho_1(S) + \rho_2(S).
\end{aligned}$$

We show first that there exists no infinite reduction sequence w.r.t. $>_2$ with only dissubsumption and distautology applied (we omit the *simp* reduction for now). Indeed we prove that for $S >_2 S'$ via distautology or dissubsumption we have $\rho(S') < \rho(S)$.

- $S >_2 S'$ via distautology:
  Then, by definition, $S' = (S \setminus \{C\}) \cup \{D\}$ for $D = \textit{c-Dtaut}(C)$ and $C \not\simeq D$.

  Thus clearly $\rho_2(S') < \rho_2(S)$.
  On the other hand $\rho_1(S') \leq \rho_1(S)$: Indeed it is impossible that $\textit{c-Dsub}(C,E) \simeq C$ and $\textit{c-Dsub}(D,E) \not\simeq D$, as $\mathcal{S}(D) \subseteq \mathcal{S}(C)$. Similarly it is impossible that $\textit{c-Dsub}(E,C) \simeq E$ and $\textit{c-Dsub}(E,D) \not\simeq E$: clearly if a ground instance D' of D subsumes a ground instance E' of E then D' is also a ground instance of C.

Thus we obtain $\rho(S') < \rho(S)$.

- $S >_2 S'$ via dissubsumption:
  By definition $S' = (S \setminus \{C\}) \cup \{D\}$ for $D = \textit{c-Dsub}(C, E)$ for some $E \in S$ and $C \not\simeq D$.

  But then $\rho_2(S') < \rho_2(S)$. This is clear by the fact that $D \simeq \textit{c-Dsub}(D, E)$ and because $\mathcal{S}(D) \subset \mathcal{S}(C)$. Clearly also $\rho_1(S') \leq \rho_1(S)$: it is impossible that D contains a tautological instance, but C does not.

  Again we obtain $\rho(S') < \rho(S)$.

Thus all reduction sequences w.r.t. $>_2$ using only distautology and dissubsumption terminate. Let us consider simplification:
Clearly there exists no reduction $S >_2 S' >_2 S''$ via simplification as $\textit{simp}(\textit{simp}(S)) = \textit{simp}(S)$. Thus any infinite $>_2$ sequence must contain infinitely many applications of distautology and dissubsumption. But by definition of *simp*, which does not change the set of represented ground instances (but may delete clauses), $\rho(S') \leq \rho(S)$ for $S >_2 S'$ via *simp*. Therefore at most $2 \cdot \rho(S)$ reduction steps are possible on S. ◇

Below we define δ-RAMC, a version of RAMC with a strong use of the deletion relation $>_2$. Some other operators are used in the analysis and in the completeness proof for δ-RAMC.

DEFINITION 4.78 Let S be a finite set of c-clauses. We define $\rho(S)$: $\textit{simp}(S) \cup \textit{cfact}(S) \cup \textit{cresol}(S)$ where $\textit{cfact}(S)$ denotes the set of all c-factors definable in S and $\textit{cresol}(S)$ denotes the set of all c-resolvents from clauses in S.

- R_{cr} : $R_{cr}(S) = S \cup \rho(S)$.
  R_{cr} describes deductive inference without GPL.

- δ is a normalization operator w.r.t. $>_2$, i.e. $S \rhd_2^* \delta(S)$ and $\delta(S)$ is irreducible under $>_2$. Such operators exist by Lemma 4.77.

- $R_{\rhd_1}$: $R_{\rhd_1}(S) = \bigcup\{S' | S \rhd_1 S'\}$.

- R_δ : $R_\delta(S) = \delta(R_{\rhd_1}(S))$.
  R_δ formalizes δ-RAMC, a version of RAMC applying maximal disinference after any level of inference. ♯

The analysis of the disinference rules requires a subsumption relation $\leq_{sHt}$ which is stronger than our subsumption concepts used so far; $\leq_{sHt}$ is quite close to the relation $\leq_{sH}$ defined in Chapter 5. But $\leq_{sHt}$ is not used algorithmically as a deletion method, it only serves for analyzing the operator R_δ.

DEFINITION 4.79 Let S be a set of c-clauses, C be a c-clause and H be a Herbrand universe. Then $S \leq_{sHt} C$ if for all nontautological clauses C' with $C' \in \mathcal{S}_H(C)$ there exists a D in S s.t. D subsumes C', i.e. $c\text{-}Dsub(C', D) = \emptyset$.

$\leq_{sHt}$ is extended to sets of clauses by

$$S' \leq_{sHt} S \quad \text{iff} \quad \text{for all } C \in S : S' \leq_{sHt} C. \ \sharp$$

EXAMPLE 4.80 Let

$$\begin{aligned} S &: \{D_1 : [\![P(f(f(x))) : \top]\!],\ D_2 : [\![P(f(a)) : \top]\!]\}, \\ C &: [\![\neg P(a) \vee P(x) : \top]\!] \text{ and} \\ H &: \{f^n(a) | n \in \mathbb{N}\}. \end{aligned}$$

Then C is neither subsumed by D_1 nor by D_2. Not even the ground instance $\neg P(a) \vee P(a)$ of C is subsumed by by S, but it is a tautology. The ground instance $\neg P(a) \vee P(f(a))$ is subsumed by D_2 and the ground instances $\neg P(a)) \vee P(f^{n+2}(a))$ by D_1. Thus all nontautological ground instances of C are subsumed by one of the clauses D_1, D_2; but this means $S \leq_{sHt} C$. $\sharp$

LEMMA 4.81 *$\leq_{sHt}$ is reflexive and transitive on sets of c-clauses.*

Proof: Reflexivity is trivial.
Transitivity: Let $S'' \leq_{sHt} S'$ and $S' \leq_{sHt} S$, let C be a clause in S and $C' \in \mathcal{S}(C)$ s.t. C' is not a tautology. By $S' \leq_{sHt} S$ there exists a clause $D \in S'$ s.t. D subsumes C'. But C' is ground and thus there exists a ground instance $D' \in \mathcal{S}(D)$ s.t. D' subsumes C'. Then D' cannot be a tautology (otherwise C' would be a tautology too). By $S'' \leq_{sHt} S'$ there exists a clause $E \in S''$ s.t. E subsumes D'. By transitivity of subsumption, E subsumes C'. Therefore $S'' \leq_{sHt} C$. ◇

LEMMA 4.82 *Let S be a finite set of c-clauses, then $\delta(S) \leq_{sHt} S$.*

Proof: It is enough to show that, for $S >_2 S'$, we have $S' \leq_{sHt} S$ (we have just shown that $\leq_{sHt}$ is transitive).

If $S' = simp(S)$ then $S' \leq_{sHt} S$ is trivial by $\mathcal{S}(S) = \mathcal{S}(S')$.

Let us assume that $S >_2 S'$ via distautology. Then the set $\mathcal{S}(S) \setminus \mathcal{S}(S')$ consists of tautologies only, i.e. all nontautological ground instances of S occur in S' too. Again $S' \leq_{sHt} S$ follows directly from the definition of $\leq_{sHt}$.

Thus it remains to investigate the case $S >_2 S'$ via dissubsumption. Then

$$\begin{aligned} S' &= (S \setminus \{C\}) \cup \{D\} \text{ and} \\ D &= c\text{-}Dsub(C, E) \text{ for some } E \in S \setminus \{C\}. \end{aligned}$$

It is enough to show $S' \leq_{sHt} C$:
Let $C' \in \mathcal{S}(C)$. If $C' \in \mathcal{S}(D)$ then, clearly, D subsumes C'. If $C' \notin \mathcal{S}(D)$ then, by definition of D, E subsumes C'. Therefore $\{D, E\} \leq_{sHt} \{C\}$ and thus $S' \leq_{sHt} C$. $\diamond$

LEMMA 4.83 *Let S, S' be two finite sets of c-clauses and $S' \leq_{sHt} S$. Then*

$$R_\delta(S') \quad \leq_{sHt} \quad R_{cr}(S).$$

Proof: We have to prove that

$$R_\delta(S') \leq_{sHt} S \cup simp(S) \cup cfact(S) \cup cresol(S).$$

$R_\delta(S') \leq_{sHt} S \cup simp(S)$ is easy to show:
First of all $\mathcal{S}(S) = \mathcal{S}(simp(S))$ and the problem can be reduced to $R_\delta(S') \leq_{sHt} S$: But $R_\delta(S') = \delta(R_{\rhd_1}(S'))$ and $S' \subseteq R_{\rhd_1}(S')$. Thus, in particular,

$$R_{\rhd_1}(S') \leq_{sHt} S' \leq_{sHt} S.$$

By transitivity of $\leq_{sHt}$ we conclude $R_{\rhd_1}(S') \leq_{sHt} S$. By Lemma 4.82 we get $\delta(R_{\rhd_1}(S')) \leq_{sHt} S$, i.e. $R_\delta(S') \leq_{sHt} S$.

It remains to show that

(a) $R_\delta(S') \leq_{sHt} cfact(S)$ and

(b) $R_\delta(S') \leq_{sHt} cresol(S)$.

case (a):
Let $C : [\![P(\bar{t}_1) \vee P(\bar{t}_2) \vee E : \mathcal{X}]\!]$ be a clause in S and $C_1 : [\![P(\bar{t}_1) \vee E : \mathcal{X} \wedge \bar{t}_1 = \bar{t}_2]\!]$ be a factor of C.

Now let $C_1' : P(\bar{s}) \vee E'$ be a ground instance of C_1. If C_1' is a tautology there is nothing to show. Thus we may assume that C_1' is nontautological.

By $S' \leq_{sHt} S$ there exists a clause $D \in S'$ s.t. D subsumes $C' : P(\bar{s}) \vee P(\bar{s}) \vee E'$ (which is ground instance of C and also nontautological). By definition of subsumption there exists a ground substitution θ s.t. $D\theta$ (as a multiset) is contained in C'. If $D\theta$ contains $P(\bar{s})$ twice then a factor of $D\theta$ subsumes C_1'; by lifting there exists a factor of D' of D s.t.

D' subsumes C'_1. If $D\theta$ does not contain $P(\bar{s})$ twice then $D\theta$ already subsumes C'_1. In any case

$$S' \cup cfact(S') \quad \leq_{sHt} \quad C'_1.$$

and, as this holds for an arbitrary factor of an arbitrary clause in S we get

$$S' \cup cfact(S') \quad \leq_{sHt} \quad cfact(S).$$

But $S' \cup cfact(S') \subseteq R_{\rhd_1}(S')$ and thus also

$$R_{\rhd_1}(S') \quad \leq_{sHt} \quad cfact(S).$$

By Lemma 4.82 $\delta(R_{\rhd_1}(S')) \leq_{sHt} cfact(S)$, i.e.

$$R_\delta(S') \leq_{sHt} cfact(S).$$

case (b):
Let C be a resolvent of C_1, C_2 for two clauses $C_1, C_2 \in S$. By assumption we have $S' \leq_{sHt} \{C_1, C_2\}$.

Now let C' be a ground instance of C which is not a tautology. Then there are ground instances C'_1, C'_2 of C_1, C_2 s.t. C' is resolvent of C'_1 and C'_2 (this follows directly from the definition of c-resolution).

By $S' \leq_{sHt} \{C_1, C_2\}$ either C'_i is a tautology or there exists a clause $D_i \in S'$ s.t. D_i subsumes C'_i (for $i = 1, 2$). If both C'_1, C'_2 are tautologies then C' is a tautology too, contradicting our assumption on C'. If C'_1 is a tautology then C' is subsumed by C'_2 and so D_2 subsumes C'; the case where C'_2 is a tautology is symmetric.

So we may assume that both, C'_1 and C'_2 are not tautologies and there are clauses D_1, D_2 in S' which subsume C'_1 and C'_2, respectively. In particular there are ground instances D'_1, D'_2 of D_1, D_2 which subsume C'_1, C'_2. By a well-known lemma in resolution theorem proving (see e.g. [Leitsch, 1997]) either one of the clauses D'_1, D'_2 subsumes C' or there exists a resolvent D' of D'_1, D'_2 which subsumes C'; in the last case there exists a resolvent D of D_1, D_2 s.t. D' is an instance of D and so D subsumes C'. Therefore

$$S' \cup cresol(S') \quad \leq_{sHt} \quad C$$

and, as this holds for any resolvent C, we have

$$\begin{aligned} S' \cup cresol(S') &\quad \leq_{sHt} \quad cresol(S), \\ R_{\rhd_1}(S') &\quad \leq_{sHt} \quad cresol(S) \text{ and finally} \\ R_\delta(S') &\quad \leq_{sHt} \quad cresol(S) \end{aligned}$$

by Lemma 4.82. ◇

LEMMA 4.84 *For all* $i \in \mathbb{N}$ *:* $R^i_\delta(S) \leq_{sHt} R^i_{cr}(S)$.

Proof: By induction on i.
For $i = 0$ the proof is trivial. So let us assume that $R^i_\delta(S) \leq_{sHt} R^i_{cr}(S)$. Then, by Lemma 4.82

$$R_\delta R^i_\delta(S) \quad \leq_{sHt} \quad R_{cr} R^i_{cr}(S),$$

i.e. $R^{i+1}_\delta(S) \leq_{sHt} R^{i+1}_{cr}(S)$. ◇

We are now in the position to prove our strong completeness result, namely that δ-RAMC is complete:

THEOREM 3.5 *δ-RAMC is complete, i.e. for all finite unsatisfiable sets of c-clauses S there exists an i s.t.* $R^i_\delta(S) = \{\Box\}$.

Proof: Let S be a finite unsatisfiable set of c-clauses. R_{cr} is complete by the completeness of cR-deduction (see Theorem 3.2) and thus there exists an i s.t. $\Box \in R^i_{cr}(S)$. By Lemma 4.84 we have $R^i_\delta(S) \leq_{sHt} R^i_{cr}(S)$. As $\Box$ is not a tautology (indeed!) there exists a clause $C \in R^i_\delta(S)$ s.t. C subsumes $\Box$; this is only possible if $C \simeq \Box$. But then $\delta(R^i_\delta(S)) = \{\Box\}$ and $R^{i+1}_\delta(S) = \{\Box\}$. ◇

3.6 Comparing MB and RAMC

So far we have presented two methods of model building on clause logic, MB and RAMC. MB is based on hyperresolution and unit reduction, while the characteristic features of RAMC are the disinference rules and GPL. We have already shown that both methods are weakly sound, but not sound in the strong sense. They share the property to iteratively reduce the set of models till a model can be extracted from the derived set (both GPL and the selection rule α in Chapter 3 serve this purpose).

But there also obvious differences and the reader may ask which method is in fact preferable. The analysis below will reveal that RAMC and MB are in fact incomparable (even on sets of standard clauses), but combinations of both methods can lead to considerable improvements. The following table illustrates common and different features:

MB:	RAMC:
weakly sound	weakly sound
hyperr. + subsumption	c-resolution + disrules
selection rule	GPL
model constr. as postprocessing	simultaneous construction
based on inference closure	simple closure not sufficient
success = positive clauses unit	success = all c-clauses unit.

Indeed in the form presented in the chapters of the book, MB and RAMC are not comparable in their potential to express models in clause logic. For instance

- RAMC terminates and constructs models on the Bernays-Schönfinkel class BS, but MB does not terminate and, even if it does on a satisfiable set S then possibly without model representation.
- MB terminates on PVD and constructs models on this class. RAMC fails on PVD, it neither terminates nor can get rid of nonunit clauses.

We illustrate the difference mentioned above in two examples:

EXAMPLE 4.85 In Example 3.17 we have presented an example of a set of clauses in the Bernays-Schönfinkel class where hyperresolution does not terminate (even if we rename the signs of literals). The corresponding set of clauses was $\mathcal{C} : \{C_1, \ldots, C_4\}$ for

$$
\begin{aligned}
C_1 &= P(x,x,a), \\
C_2 &= P(x,z,u) \vee \neg P(x,y,u) \vee \neg P(y,z,u), \\
C_3 &= P(x,y,u) \vee P(y,z,u) \vee \neg P(x,z,u), \\
C_4 &= \neg P(x,x,b).
\end{aligned}
$$

RAMC terminates on $\mathcal{C}$ and constructs a model because in the solution of some constraints the variables are substituted by ground terms. E.g. in applying dissubsumption on C_2 via C_1 we obtain

$$C_5 : \ c\text{-}Dsub(C_2, C_1) = [\![C_2 : x \neq z \vee u \neq a]\!].$$

and C_2 may be deleted. But $u \neq a \equiv u = b$ over $H = \{a, b\}$ and so

$$C_5 \quad = \quad [\![C_2 : x \neq z \vee u = b]\!].$$

As the constraint is an $\vee$-formula we can split and obtain

$$C_6 \quad = \quad [\![C_2 : u = b]\!].$$

But C_6 is equivalent to

$$C_6' : P(x,z,b) \vee \neg P(x,y,b) \vee \neg P(y,z,b).$$

By resolving C_6' and C_4 we get

$$C_7 : \ \neg P(x,y,b) \vee \neg P(y,x,b).$$

Dissubsubsumption with C_4 gives

$$C_8 : \ [\![\neg P(x,y,b) \vee \neg P(y,x,b) : x \neq y]\!].$$

Using dissubsumption with C_4 (twice) and C_8 we can transform C_6' into

$$C_9 : \ [\![P(x,z,b) \vee \neg P(x,y,b) \vee \neg P(y,z,b) : x \neq y \wedge y \neq z \wedge z \neq x]\!].$$

and delete C_6'. But the constraint of C_9 is unsolvable over the domain $\{a,b\}$ and so we delete C_9 altogether. This way we can get rid of all the nonunit clauses, though it requires a lot of steps. Similarly to C_8 we obtain (resolution of C_1 and C_3 and then dissubsumption by C_1):

$$C_{10} : \ [\![P(x,y,a) \vee P(y,x,a) : x \neq y]\!]$$

After some further steps all mixed clauses are deleted (in the same way as C_9) and we are left with C_1, C_4, C_8 and C_{10}; finally two applications of the pure literal rule give the unit clauses:

$$\{P(x,x,a), [\![P(x,y,a) : x \neq y]\!], \neg P(x,x,b), [\![\neg P(x,y,b) : x \neq y]\!]\}.$$

which describe a unique Herbrand model over $\{a,b\}$. $\sharp$

The following example shows the other side of the problem:

EXAMPLE 4.86 Let

$$\mathcal{C} \ = \ \{P(f(a)), \ \neg P(f(x)) \vee P(x), \neg P(b)\}.$$

Then $\mathcal{C} \in \mathcal{PVD}$ and

$$R^*_{H,s}(\mathcal{C}) \ = \ \{P(f(a)), \ P(a), \ \neg P(f(x)) \vee P(x), \neg P(b)\}.$$

Clearly $\mathcal{A} : \{P(f(a)), P(a)\}$ is an ARM of $\mathcal{C}$.

But RAMC does not terminate on $\mathcal{C}$, in fact all clauses $[\![\neg P(f^n(b)) : \top]\!]$ are derivable. In particular the clause $\neg P(f(x)) \vee P(x)$ is never removed, i.e. after n dissubsumption steps we get

$$[\![\neg P(f(x)) \vee P(x) : x \neq t_1 \wedge \ldots x \neq t_n]\!]$$

for ground terms $t_1, \ldots, t_n$. The GPL-rule is ineffective as the second clause is self-resolving and stays so after arbitrary many steps of dissubsumption. $\sharp$

Thus after the two examples above we might ask the following questions:

- which method should be used under which circumstances and
- can RAMC be combined with MB?

The first question is not easy to answer, the exceptions are the syntax classes where the methods are known to be successful.

Concerning the second question we observe that RAMC works on sets of c-clauses, while hyperresolution works on ordinary clauses only. Even if RAMC starts with a set of ordinary clauses, c-clauses are generated by the disinference rules. But this is not a problem at all: hyperresolution can be defined on c-clause logic exactly as on clause logic. In the definition of a clash we only have to replace the term "resolvent" by "c-resolvent" and factor by c-factor. All the logical properties for hyperresolution hold for c-clause logic too, essentially because they depend only on the *set of all ground instances*, and not on the specific form of clauses. The operators R_H and $R_{H,s}$ can be defined in the same way. We have defined ARMs as a set of positive unit clauses, describing the set of all their ground instances over a Herbrand universe. The extension to c-clauses is obvious: A set of c-unit clauses $\mathcal{A}: \{C_1, \ldots, C_n\}$ over a Herbrand universe H represents just the set of ground atoms:

$$\mathcal{S}_H(C_1) \cup \cdots \cup \mathcal{S}_H(C_n).$$

Instead of ARM we call $\mathcal{A}$ a CARM. We will give a detailed analysis of those representations in Chapter 5. With this extension we get a generalization of our crucial lemma (compare to Lemma 3.24):

LEMMA 4.87 *Let S be a finite set of nonpositive c-clauses and $\mathcal{A}$ be a finite set of positive unit c-clauses s.t. $S \cup \{\mathcal{A}\}$ is stable (under hyperresolution). Then $\mathcal{A}$ is a CARM of $S \cup \mathcal{A}$, i.e. the instances of $\mathcal{A}$ define a Herbrand model of $S \cup \mathcal{A}$ over $H(S \cup \mathcal{A})$.*

Proof: Essentially analogous to the proof of Lemma 3.24. $\diamond$

So there is no problem in doing model building on c-clause logic via hyperresolution. Also combining hyperresolution with RAMC in a fully nondeterministic way does neither affect the (weak) soundness nor completeness. But we can do it in a more subtle way. Consider the success conditions for MB and for RAMC: for RAMC, success is achieved if *all nonunit clauses* are deleted, for MB if *all positive* clauses are unit and the set is a fixed point under hyperresolution + subsumption. Looking at the Example 4.86 we see that RAMC + hyperresolution will never delete all ground instances of the second clause! What we can easily do is to combine RAMC with the more liberal success criterion of MB. I.e. we stop if the *positive* c-clauses are unit *and* the set is a fixed point under hyperresolution; there is no need to get rid of mixed and negative nonunit clauses. This results in the following method, where *POS* denotes the subset of positive clauses and *Unit* is a predicate on c-clause sets giving true iff all clauses are unit:

hyper-RAMC:
begin
while $R_{H,s}(S) \neq S \vee \neg Unit(POS(S))$ **do** $S \leftarrow R_\delta(S)$;
$\mathcal{M} \leftarrow POS(S)$.
end

Clearly hyper-RAMC improves the model building capacity of RAMC. Indeed, if we obtain a set of c-unit clauses (the original success criterion for RAMC) and RAMC stops then the set is also a fixed point under hyperresolution (because there are no resolvents at all). On the other hand the more liberal criterion may lead to success where RAMC fails:

EXAMPLE 4.88 Reconsider Example 4.86. We had

$$S = \{P(f(a)),\ \neg P(f(x)) \vee P(x), \neg P(b)\}.$$

We first compute R_δ:

$$\begin{aligned} R_\delta(S) &= \{[\![P(f(a)) : \top]\!], [\![\neg P(b) : \top]\!], [\![P(a) : \top]\!], [\![\neg P(f(b)) : \top]\!], \\ &\quad [\![\neg P(f(x)) \vee P(x) : x \neq f(a)]\!]. \end{aligned}$$

But $R_{H,s}(R_\delta(S)) = R_\delta(S)$ and all positive clauses are unit. Indeed, the only hyperresolvent which is definable is $[\![P(a) : \top]\!]$ which is already in the set. Therefore hyper-RAMC stops with

$$\mathcal{M} : \{[\![P(f(a)) : \top]\!], [\![P(a) : \top]\!]\},$$

which is a CARM of $R_\delta(S)$ and thus of S itself. In contrast RAMC would continue to derive negative clauses infinitely. ♯

The improvement hyper-RAMC is in some sense straightforward. We might also consider more complex combinations where the unit reduction step α of MB is applied together with RAMC. This would lead to an algorithm hyper-RAMC-II below:

hyper-RAMC-II:
begin
while $R_{H,s}(S) \neq S \vee \neg Unit(POS(\mathcal{C}))$
do begin
if $R_{H,s}(S) = S$ **then** $S \leftarrow \alpha(S)$;
$S \leftarrow R_\delta(S)$
end
$\mathcal{M} \leftarrow POS(\mathcal{C})$.
end

Combinations of RAMC with hyperresolution are largely uninvestigated. In particular it is not clear how hyper-RAMC-II actually performs. Additional theoretical and experimental work will be necessary. However there are some results about using semantic clash resolution with constraints; the interested reader should have a look at [Peltier, 1997b].

4. The RAMCET method

In this section, we use the same techniques as in Section 3, namely equational constraints to discard irrelevant instances of formulae, in order to extend semantic tableaux.

It is well known that the tableau method enumerates partial potential models of the considered formula. This enumeration will possibly lead to closed branches when the potential model does *not* exist. Finite failures in the attempt to close branches allow to specify models as a side effect (when these models are finite Herbrand models). The so-called model generation tableaux provers are model generators in this sense. Several works have already incorporated *unification* to semantic tableaux (see for example [Reeves, 1987, Fitting, 1990]). Here, we propose to extend tableaux with *equational constraints* (also taking into account *dis-unification* conditions).

Instead of extracting a model from a *finite* open branch (as a *by-product* of the method), the method tries to build a model *during the search.* The main advantage of this approach is that it can prune the search space of the refutation procedure and may build models for formulae for which the classical approach does not terminate. The properties of the original method (i.e. soundness and refutational completeness) are preserved.

This extension formalizes on an object level reasoning which usually takes place in an ad-hoc manner in the meta-level.

For the sake of simplicity and for consistency with the previous sections, we assume that all the considered formulae are in clausal form (though the method may be extended in order to deal with arbitrary formulae, see for example [Caferra and Zabel, 1993, Peltier, 1999]).

4.1 Definition of tableaux

The tableau method uses two different kinds of variables: universally quantified variables, which correspond to the variables occurring in the original clause sets and "rigid" variables, which correspond to terms introduced by the δ-rule. Since c-clauses will contain both kinds of variables, we need to distinguish them by making the universal quantifier explicit in the c-clauses. Thus, we need to introduce the following definition:

DEFINITION 4.89 A *generalized c-clause* (*gc-clause* for short) is of the form

$$(\forall \overline{x})\{C : \mathcal{X}, \}$$

where C is a clause and $\mathcal{X}$ an equational formula.

A variable is said to be *free* in $(\forall \overline{x})\{C : \mathcal{X}\}$ iff it occurs in C or free in $\mathcal{X}$, but does not occur in $\overline{x}$. A gc-clause without free variables is said to be *closed.* ♯

As usual, a gc-clause of the form $(\forall \overline{x})\{C : \top\}$ will be simply denoted as $(\forall \overline{x})C$. Moreover, a gc-clause of the form $(\forall \overline{x})\{C : \mathcal{X}\}$, where $\overline{x}$ is empty will be more simply denoted by $\{C : \mathcal{X}\}$.

A closed gc-clause $(\forall \overline{x})\{C : \mathcal{X}\}$ is equivalent to the c-clause $[\![C : \mathcal{X}]\!]$, as stated by the following definition:

DEFINITION 4.90 Let $(\forall \overline{x})\{C : \mathcal{X}\}$ be a closed gc-clause. Let $\mathcal{I}$ be an interpretation. We say that $\mathcal{I} \models (\forall \overline{x})\{C : \mathcal{X}\}$ iff $\mathcal{I} \models [\![C : \mathcal{X}]\!]$. ♯

Thus, c-clauses are particular cases of gc-clauses. As we shall see, non closed gc-clauses will be interpreted by instantiating the free variables occurring in it.

For any ground substitution σ, and for any gc-clause $(\forall \overline{x})\{C : \mathcal{X}\}$ we denote by $(\forall \overline{x})\{C : \mathcal{X}\}\sigma$ the gc-clause $(\forall \overline{x})\{C\sigma' : \mathcal{X}\sigma'\}$ where σ' is defined as follows:

- $\sigma'(x) = x$ if x occurs in $\overline{x}$.
- $\sigma'(x) = \sigma(x)$ otherwise.

Note that if σ is ground and if $dom(\sigma)$ contains all the free variables of $(\forall \overline{x})\{C : \mathcal{X}\}$ then $(\forall \overline{x})\{C : \mathcal{X}\}\sigma$ contains no free variable, hence it is corresponds exactly to the c-clause $[\![C : \mathcal{X}]\!]\sigma$ and is interpreted in the same way. However, a c-clause $[\![C : \mathcal{X}]\!]$ should *not* be confused with the *gc-clause* $\{C : \mathcal{X}\}$. Indeed, in the first case, variables occurring in $[\![C : \mathcal{X}]\!]$ are implicitly universally quantified, whereas in the second case, all variables are free. But, since only gc-clauses are considered in the rest of the section, there will be no confusion).

DEFINITION 4.91 (Tableaux) A *tableau* $\mathcal{T}$ is a pair $\{T : \mathcal{P}\}$ where T is a finite set of sets of gc-clauses and $\mathcal{P}$ is an equational formula (the constraint part of $\mathcal{T}$). $\sharp$

Intuitively speaking, $\mathcal{P}$ expresses conditions on the free variables occurring in the tableau. More formally:

DEFINITION 4.92 Let $\mathcal{I}$ be an interpretation. $\mathcal{I}$ is said to be a *model* of a tableau $\mathcal{T} = \{T : \mathcal{P}\}$ iff for all substitutions $\sigma \in \mathcal{S}(\mathcal{P})$ there exists a set $S \in T$ such that $\mathcal{I} \models S\sigma$.

A tableau $\mathcal{T}$ is said to be *satisfiable* iff it has a model, *unsatisfiable* otherwise.

Two tableaux $\mathcal{T}$ and $\mathcal{T}'$ are said to be *equivalent* iff they have exactly the same set of models. $\sharp$

A tableau $\mathcal{T} = (T, \mathcal{X})$ is said to be *closed* if $T = \emptyset$ **and** $\mathcal{X}$ has at least one solution. By definition, any closed tableau is unsatisfiable.

4.2 Tableau construction rules

We now present the rules allowing to construct tableaux.

Their typical form is the following:

$$\frac{E}{E_1 \quad \dots \quad E_n}$$

$$\mathcal{X} \to \mathcal{Y}$$

This notation has the following meaning: Given a tableau $\{T : \mathcal{X}\}$ any set $B \in T$ with $E \subseteq B$ may be *deleted* and *replaced* by the n sets $(B \setminus E) \cup E_1, \dots, (B \setminus E) \cup E_n$, *provided* that the constraint part $\mathcal{X}$ of the tableau is replaced by $\mathcal{Y}$ **and** that $\mathcal{Y}$ is satisfiable (applications of rules that leads to unsatisfiable constraints are forbidden). Thus the new tableau becomes:

$$\{T \setminus B \cup \{(B \setminus E) \cup E_i \mid i \in [1..n]\} : \mathcal{Y}\}$$

The set B is called the *ancestor* of the sets $(B \setminus E) \cup E_i$.

If the condition $\mathcal{X} \to \mathcal{Y}$ is missing, then $\mathcal{X}$ is kept unchanged. n can be equal to 0, in this case E is simply deleted from the tableau (in this case the empty sequence $E_1, \dots, E_n$ is denoted by $\perp$).

Note that the handling of constraints in the extension rules (both into the gc-clauses and at the global level) forms the main new part of our method. This approach allows us to perform meta-level reasoning and to express properties of tableaux that could not be expressed at the object level in classical tableaux.

Initially, the tableau contains only one set, the given set of c-clauses, and the constraint part of the tableau is $\top$.

The Splitting rule The splitting rule allows to generate unit clauses, by "splitting" a set containing a disjunction $C \vee D$ into two sets containing the formulae C and D respectively. Of course, additional conditions are needed on C and D to guarantee the soundness of this rule: essentially C and D must not share variables. However, we also have to take into account the constraint part of the gc-clauses. Indeed, consider for example the gc-clause $(\forall x)\{P(x) \vee Q(x) : x = a\}$. Here $P(x)$ and $Q(x)$ share the variable x. However this variable is instantiated by the ground term a, thus splitting can be performed anyway. On the other hand, consider the gc-clause $(\forall x, y)\{P(x) \vee Q(y) : x = y\}$. $P(x)$ and $Q(y)$ do not share variables, however, due to the constraint x must be equal to y, hence splitting cannot be performed.

The Splitting rule is defined as

$$\frac{\{(\forall \overline{x})\{C \vee D : \mathcal{X}\}\}}{\{(\forall \overline{x})\{C : \mathcal{X}\} \qquad (\forall \overline{x})\{D : \mathcal{X}\}\}}$$

under the conditions (where σ and θ denotes two renamings of $\overline{x}$ such that $dom(\sigma) \cap dom(\theta) = dom(\sigma) \cap rg(\theta) = rg(\sigma) \cap rg(\theta) = \emptyset$):

- the equational formula $\neg(\mathcal{X}\sigma \wedge \mathcal{X}\theta) \vee (\exists \overline{x})(\mathcal{X} \wedge \overline{y}\sigma = \overline{z} \wedge \overline{x}\theta = \overline{x})$ is valid.
- C and D are non-empty disjunctions of literals.
- $\overline{y}$ are the variables occurring in C and $\overline{z}$ are the variables occurring in D.

EXAMPLE 4.93 Let S be the following set of gc-clauses:

$$S = \{(\forall x)P(x), (\forall y_1, y_2)\{\neg P(y_1) \vee Q(y_2) : y_1 \neq a \wedge y_2 \neq a\}\}.$$

We consider the renamings $\sigma = \{y_1 \leftarrow y_1', y_2 \leftarrow y_2'\}$ and $\theta = \{y_1 \leftarrow y_1'', y_2 \leftarrow y_2''\}$. The reader can easily check that the formula

$$\neg(y_1' \neq a \wedge y_2' \neq a \wedge y_1'' \neq a \wedge y_2'' \neq a) \vee$$
$$(\exists y_1, y_2)(y_1 \neq a \wedge y_2 \neq a \wedge y_1' = y_1 \wedge y_1'' = y_1 \wedge y_2'' = y_2)$$

is valid.

Thus S may be replaced by the two following sets:

$$S_1 = \{(\forall x)P(x), (\forall y_1, y_2)\{\neg P(y_1) : y_1 \neq a \wedge y_2 \neq a\}\}$$

and

$$S_2 = \{(\forall x)P(x), (\forall y_1, y_2)\{Q(y_2) : y_1 \neq a \wedge y_2 \neq a\}\}$$

Note that the constraint part of the gc-clauses $\{\neg P(y_1) : y_1 \neq a \wedge y_2 \neq a\}$ and $\{Q(y_2) : y_1 \neq a \wedge y_2 \neq a\}$ may be simplified (since y_2 and y_1 respectively does not occur in the clausal part of the gc-clauses) yielding the gc-clauses $\{\neg P(y_1) : y_1 \neq a\}$ and $\{Q(y_2) : y_2 \neq a\}$. ♯

The Instantiation Rule The instantiation rule corresponds to the γ-rule in the classical presentation of tableaux.

$$\frac{(\forall \overline{x})\{C : \mathcal{X}\}}{(\forall \overline{x})\{C : \mathcal{X} \wedge \neg \mathcal{Y}\},\ (\forall \overline{x})\{C : \mathcal{X} \wedge \mathcal{Y}\}}$$

where $\mathcal{Y}$ is a constraint and $\mathcal{X} \wedge \mathcal{Y}$ has at least one solution.

Remark: No special assumption on $\mathcal{Y}$ is made in order to keep as much flexibility as possible. Additional strategy could be defined for choosing the formula $\mathcal{Y}$ on which the rule is applied (for example it can be controlled by the user). However, for proving refutational completeness, we assume that $\mathcal{Y}$ is of the form $x = y$, where x is a variable occurring in $\overline{x}$ and y a new free variable, not occurring elsewhere in the tableau.

EXAMPLE 4.94 (Continued from Example 4.93) The set

$$S_1 = \{(\forall x)P(x), (\forall y_1)\{\neg P(y_1) : y_1 \neq a\}\}$$

can be replaced by:

$$S_1' = \{(\forall x)P(x), (\forall y_1)\{\neg P(y_1) : y_1 \neq a \wedge y_1 \neq u\}\}$$
$$\cup\{(\forall y_1)\{\neg P(y_1) : y_1 \neq a \wedge y_1 = u\}\}$$

where u is a new free variable.

After simplification (replacement of y_1 by u), the gc-clause

$$(\forall y_1)\{\neg P(y_1) : y_1 \neq a \wedge y_1 = u\}$$

may be replaced by

$$\{\neg P(u) : u \neq a\}$$

(note that u is not universally quantified). ♯

The Simplification rule The simplification rule allows to take advantage of the information deduced so far concerning the model corresponding to the current set, in order to simplify the gc-clauses occurring in the set. This is done in two ways:

- By discarding instances of gc-clauses that are subsumed by unit gc-clauses in the set (the unit dissubsumption rule).
- By removing literals that are false in the considered set (as in unit resolution and in disresolution).

Unit dissubsumption.

$$\frac{(\forall\overline{x}_1)\{L(\overline{t}) : \mathcal{X}\}, (\forall\overline{x}_2)\{L(\overline{s}) \vee C : \mathcal{Y}\}}{(\forall\overline{x}_1)\{L(\overline{t}) : \mathcal{X}\}, (\forall\overline{x}_2)\{L(\overline{s}) \vee C : \mathcal{Y} \wedge \forall\overline{x}_1.[\neg\mathcal{X} \vee \overline{s} \neq \overline{t}]\}}$$

If $\overline{t} = \overline{s} \wedge \mathcal{X} \wedge \mathcal{Y}$ has at least one solution.

EXAMPLE 4.95 Let $S = \{(\forall x, y)(\neg P(x, y) \vee Q(y)), (\forall x')\neg P(x', u)\}$ where u denotes a free variable. The dissubsumption rule leads to the set:

$$S' = \{(\forall x, y)\{\neg P(x, y) \vee Q(y) : \forall x'.(x \neq x' \vee y \neq u)\}, (\forall x')P(x', u)\}.$$

Using constraint simplification, S' may be transformed into:

$$S' = \{(\forall x, y)\{\neg P(x, y) \vee Q(y) : y \neq u\}, (\forall x')P(x', u)\}.$$

♯

Disresolution.

$$\frac{(\forall\overline{x}_1)\{L(\overline{t}) : \mathcal{X}\}, (\forall\overline{x}_2)\{\neg L(\overline{s}) \vee C : \mathcal{Y}\}}{(\forall\overline{x}_1)\{L(\overline{t}) : \mathcal{X}\}, C_{res}, C_{disres}}$$

where:

- $C_{disres} = (\forall\overline{x}_2)\{\neg L(\overline{s}) \vee C : \mathcal{Y} \wedge \forall\overline{x}_1.[\neg\mathcal{X} \vee \overline{s} \neq \overline{t}]\}$;
- $C_{res} = (\forall\overline{x}_2, \overline{x}_1)\{C : \mathcal{X} \wedge \mathcal{Y} \wedge \overline{t} = \overline{s}\}$;
- $\overline{t} = \overline{s} \wedge \mathcal{X} \wedge \mathcal{Y}$ has at least one solution.

EXAMPLE 4.96 (Continued from Example 4.94) The set

$$S_1' = \{(\forall x)P(x), (\forall y_1)\{\neg P(y_1) : y_1 \neq a \wedge y_1 \neq u\}, \{\neg P(u) : u \neq a\}\}.$$

may be replaced by:

$$S_1'' = \{(\forall x)P(x), (\forall y_1)\{\neg P(y_1) : y_1 \neq a \wedge y_1 \neq u\}\}$$

$$\cup\{\{\neg P(u) : u \neq a \wedge \forall x. x \neq u\}, \{\Box : u \neq a \wedge x = u\}\}.$$

Since $(\exists x)x = u$ is valid, S_1'' may be simplified, yielding:

$$S_1'' = \{(\forall x)P(x), (\forall y_1)\{\neg P(y_1) : y_1 \neq a \wedge y_1 \neq u\}, \{\Box : u \neq a\}\}.$$

♯

The Clash Rule The clash rule allows to delete sets that contain a pair of complementary literals. As usual, unification is not performed explicitly: instead new equality constraints are added in the constraint part of the tableau in order to insure that the two literals are complementary.

$$\frac{(\forall \overline{x})\{\Box : \mathcal{X}\}}{\perp}$$

$$\mathcal{P} \to \mathcal{P} \wedge (\exists \overline{x}.\mathcal{X}).$$

Where $\mathcal{P}$ denotes the constraints of the tableau.

EXAMPLE 4.97 (Continued from Example 4.96) The set S_1'' may deleted, by adding the constraint $u \neq a$ into the global constraints of the tableau. ♯

NOTATION 4.1 *We denote by RAMCET the system defined by the rules: splitting, instantiation, clash, simplification. A RAMCET-derivation is a (possibly infinite) sequence of tableaux $(\mathcal{T}_i)_{i \in I}$ (where I is either $\mathbb{N}$ or of the form $\{0..n\}$ for some $n \in \mathbb{N}$) such that for any $i \in I \setminus \{0\}$, $\mathcal{T}_i$ is obtained from $\mathcal{T}_{i-1}$ by a rule of RAMCET.*

Remark: The reader should note that *all* RAMC rules could be extended to gc-clauses and incorporated into RAMCET, simply by considering free variables as constant symbols. Thus RAMCET can be seen as an extension of RAMC, obtained by allowing splitting and introduction of rigid variables. However, we limit ourselves to the rules that are necessary either for refutational completeness, or for building models.

4.3 Basic properties of the method

In this section, we show that RAMCET is sound and refutationally complete.

We need the following lemma:

LEMMA 4.98 *Let $\mathcal{T} = (T, \mathcal{P})$ be a tableau and let $\mathcal{T}' = (T', \mathcal{P}')$ be a tableau deduced from $\mathcal{T}$ by applying Splitting, Instantiation, Clash or Simplification on $\mathcal{T}$. Then*

$$\mathcal{S}(\mathcal{P}') \subseteq \mathcal{S}(\mathcal{P}).$$

Proof: This follows immediately from the definition of RAMCET's rules (see Section 4.2. Indeed, the rules always add new conditions in the constraint part of the tableau and never remove them.

LEMMA 4.99 *(Soundness) Let $\mathcal{I}$ be an interpretation. Let $\mathcal{T}$ be a tableau and $\mathcal{T}'$ be a tableau deduced from $\mathcal{T}$ by applying the Splitting, Instantiation, Clash or Simplification rule on $\mathcal{T}$. If $\mathcal{I} \models \mathcal{T}$ then $\mathcal{I} \models \mathcal{T}'$.*

Proof: Let $\mathcal{T} = (T, \mathcal{P})$ and $\mathcal{T}' = (T', \mathcal{P}')$. We consider each rule separately. Let S denote the set on which the rule is applied. Let $\mathcal{I}$ be a model of $\mathcal{T}$ and let $\sigma \in \mathcal{S}(\mathcal{P}')$. By Lemma 4.98, we have $\mathcal{S}(\mathcal{P}') \subseteq \mathcal{S}(\mathcal{P})$ hence $\sigma \in \mathcal{S}(\mathcal{P})$. This implies that there exists $B \in T$ with $\mathcal{I} \models B\sigma$.

- **Splitting rule.** The set S is replaced by two sets $(S \setminus \{C_1\}) \cup \{C_2\}$ and $(S \setminus \{C_1\}) \cup \{C_3\}$ where:

$$C_1 = (\forall \overline{x})\{C \vee D : \mathcal{X}\}$$

$$C_2 = (\forall \overline{x})\{C : \mathcal{X}\}$$
$$C_3 = (\forall \overline{x})\{D : \mathcal{X}\}$$

  where the following conditions hold (σ and θ denotes two renamings of $\overline{x}$ such that $dom(\sigma) \cap dom(\theta) = dom(\sigma) \cap rg(\theta) = rg(\sigma) \cap rg(\theta) = \emptyset$).

  - the equational formula $\neg(\mathcal{X}\sigma \wedge \mathcal{X}\theta) \vee (\exists \overline{x})(\mathcal{X} \wedge \overline{y}\sigma = \overline{z} \wedge \overline{x}\theta = \overline{x})$ is valid.
  - C and D are non-empty disjunctions of literals.
  - $\overline{y}$ are the variables occurring in C and $\overline{z}$ are the variables occurring in D.

  We show that there exists a $B' \in T'$ such that $\mathcal{I} \models D\sigma$. If $B \neq S$, the proof is immediate, since $B \in T'$. Now, assume that $B = S$. We have $\mathcal{I} \models S\sigma$, thus $\mathcal{I} \models C_1\sigma$. If $\mathcal{I} \not\models (S \setminus \{C_1\} \cup C_2)\sigma$ and $\mathcal{I} \not\models (S \setminus \{C_1\} \cup C_3)\sigma$, we must have $\mathcal{I} \not\models C_2\sigma$ and $\mathcal{I} \not\models C_3\sigma$. Thus, there must exists $\sigma_1 \in \mathcal{S}(\mathcal{X}\sigma)$ and $\sigma_2 \in \mathcal{S}(\mathcal{X}\sigma)$ such that $\mathcal{I} \not\models C\sigma\sigma_1$ and $\mathcal{I} \not\models D\sigma\sigma_2$.

  Let $(\forall \overline{x}')\{C' \vee D' : \mathcal{X}'\}$ be a renaming of C_1. There exists $\sigma_1 \in \mathcal{S}(\mathcal{X}\sigma)$ and $\sigma_2 \in \mathcal{S}(\mathcal{X}'\sigma)$ such that $\mathcal{I} \not\models C\sigma\sigma_1$ and $\mathcal{I} \not\models D'\sigma\sigma_2$. By the application condition of the splitting rule, this implies that there exists a solution σ_3 of $\mathcal{X}\sigma$ such that: $C\sigma\sigma_3 = C\sigma\sigma_1$ and $D\sigma\sigma_3 = D'\sigma\sigma_2$. But then we have $\mathcal{I} \not\models (C \vee D)\sigma\sigma_3$ and $\sigma_3 \in \mathcal{S}(\mathcal{X}\sigma)$, which is impossible, since $\mathcal{I} \models S\sigma$.

- **Instantiation rule.**

Again, we show that there exists a set B' in $\mathcal{T}'$ such that $\mathcal{I} \models B'\sigma$. If $B \neq B'$ then the proof is trivial. Otherwise, the set S is replaced by a set $(S \setminus \{D\}) \cup \{D_1, D_2\}$ where $D_1 = (\forall \overline{x})\{C : \mathcal{X} \wedge \neg \mathcal{Y}\}$ $D_2 = (\forall \overline{x})\{C : \mathcal{X} \wedge \mathcal{Y}\}$ and $D = (\forall \overline{x})\{C : \mathcal{X}\}$. Obviously, we have $\mathcal{S}(D_i\sigma) \subseteq \mathcal{S}(D\sigma)$ $(\forall i \in \{1,2\})$. Therefore $D_i\sigma$ is a logical consequence of $D\sigma$. Thus, we have $\mathcal{I} \models S\sigma \setminus \{D\sigma\} \cup \{D_1\sigma, D_2\sigma\}$.

- **Simplification rules.**
  - **Dissubsumption.** S is replaced by $S' = S \setminus \{C\} \cup \{D\}$, where $C = (\forall \overline{x}_2)\{L(\overline{s}) \vee C : \mathcal{Y}\}$ and
  $$D = (\forall \overline{x}_2)\{L(\overline{s}) \vee C : \mathcal{Y} \wedge (\forall \overline{x}_1[\neg \mathcal{X} \vee \overline{s} \neq \overline{t}])\}$$
  Clearly, $S'\sigma$ may be deduced from $S\sigma$ by applying the (unit) dissubsumption rule. By Lemma 4.64, $S'\sigma \equiv S\sigma$. Thus $\mathcal{T}$ and $\mathcal{T}'$ are equivalent.
  - **Disresolution.** The proof is similar.

- **Clash rule.** A set S containing a gc-clause $(\forall \overline{x})\{\Box : \mathcal{X}\}$ is deleted, and the constraint $(\exists \overline{x})\mathcal{X}$ is added to the constraint part of the tableau. We show that there exists a set $B' \in T'$ such that $\mathcal{I} \models B'\sigma$. If $B = S$ then the proof is immediate, since B occurs in T'. Otherwise, we must have $\mathcal{I} \models (\forall \overline{x})\{\Box : \mathcal{X}\}\sigma$. But $\sigma \in \mathcal{S}((\exists \overline{x})\mathcal{X})$ thus $\Box \in S\sigma$, which is impossible.

◇

COROLLARY 4.1 *Let S be a set of c-clauses. If there exists a RAMCET-derivation $(\mathcal{T}_1, \ldots, \mathcal{T}_n)$ such that $\mathcal{T}_1 = (\{S\}, \top)$ and $\mathcal{T}_n$ is closed, then S is unsatisfiable.*

Proof: By Lemma 4.99, we know that $\mathcal{T}_n$ is satisfiable if $\mathcal{T}_1$ is satisfiable. Since $\mathcal{T}_n$ is unsatisfiable, this implies that $\mathcal{T}_1$ is unsatisfiable. By definition, this implies that S is unsatisfiable (indeed, any model of S would be a model of S). ◇

Now, we prove that RAMCET is refutationally complete. Note that two different kinds of completeness result may be given for tableau calculi:

- A tableau calculus is said to be **weakly complete** iff for any unsatisfiable tableau $\mathcal{T}$, there exists a derivation from $\mathcal{T}$ leading to a closed tableau.

- A tableau calculus is said to be **strongly complete** iff for any unsatisfiable tableau $\mathcal{T}$, *any* derivation from $\mathcal{T}$ must lead to a closed tableau.

Obviously, strong completeness implies weak completeness. Strong completeness ensures that no backtracking is needed during the construction of the tableau. Now, we prove that RAMCET is strongly refutationally complete, provided that the rules are applied in a fair way, i.e. that no rule can be undefinitely delayed.

The proof of the completeness of semantic tableaux uses *Hintikka sets* [Fitting, 1990, Smullyan, 1968]. There are several significant differences between our method and the classical tableau method, especially due to the use of constraints and simplification rules. Hence, we have to adapt the definition of Hintikka sets to the particular features of our (tableau-based) method. We first need a few preliminar definitions.

DEFINITION 4.100 Let C be a gc-clause. We denote by $deg(C)$ the pair (l, ξ) where:

- l is the number of literals in the clausal part of C.
- $\xi = \min\{n \mid \exists$ a gc-clause $D : (\forall x_1, \ldots, x_n)\{C' : \mathcal{X}\}$ s.t. $D \cong C\}$ (i.e. ξ denotes the minimal number of bound variables in C).

$deg(C)$ is called the *degree* of C. ♯

EXAMPLE 4.101 The gc-clause $(\forall x, y)\{P(x) \vee Q(x, y) : x = y\}$ is of degree $(2, 1)$. Indeed, its clausal part contains 2 literals and it is equivalent to the gc-clause $(\forall x)\{P(x) \vee Q(x, x) : \top\}$ which contain only one bound variable.

Similarly, $(\forall x)\{P(x, u) : u = x\}$ is of degree $(1, 0)$ (u is a free variable). ♯

The above definition induces a partial ordering on gc-clauses, by comparing the degrees of the gc-clauses using the lexicographic extension of the usual ordering on natural number. More formally:

DEFINITION 4.102 Let C_1 and C_2 be two gc-clauses. We write $C_1 <_d C_2$ iff $deg(C_1) = (l_1, x_1)$, $deg(C_2) = (l_2, x_2)$ and either $l_1 < l_2$ or: $l_1 = l_2$ and $x_1 < x_2$. ♯

DEFINITION 4.103 A (possibly infinite) set of closed gc-clauses S is said to be a *c-Hintikka set* iff the following conditions holds:

1 S contains no pair of gc-clauses $(\forall \overline{x})\{P(\overline{s}) : \mathcal{X}\}$ and $(\forall \overline{x}')\{\neg P(\overline{t}) : \mathcal{Y}\}$ such that $(\exists \overline{x}, \overline{x}')\mathcal{X} \wedge \mathcal{Y} \wedge \overline{t} = \overline{s}$ is satisfiable.

2 For any gc-clause $C \in S$, either C is a unit gc-clause, or there exists a set of gc-clauses S' such that $S' \models \mathcal{S}(C)$ and for all $C' \in S'$, $C' <_d C$.

♯

LEMMA 4.104 *C-Hintikka sets are satisfiable.*

Proof: Let $\mathcal{I}$ be the Herbrand interpretation defined as follows: for any ground atom $P(\bar{t})$, $\mathcal{I} \models P(\bar{t})$ iff there exists a gc-clause $(\forall \bar{x})\{P(\bar{s}) : \mathcal{X}\} \in S$ and a solution $\sigma \in \mathcal{S}(\mathcal{X})$ such that $\bar{s}\sigma = \bar{t}$. We prove that for all $C \in S$, $\mathcal{I} \models C$. The proof is by induction on $<_d$.

- **Base case.** C is a unit gc-clause. If C is positive, then the proof is immediate, by definition of $\mathcal{I}$. If C is of the form $(\forall \bar{y})\{\neg P(\bar{t}) : \mathcal{Y}\}$, and if $\mathcal{I} \not\models C$ then this means that there exists $\theta \in \mathcal{S}(\mathcal{Y})$ such that $\mathcal{I} \models P(\bar{t})\theta$. By definition of $\mathcal{I}$, this means that there exists a gc-clause $(\forall \bar{x})\{P(\bar{s}) : \mathcal{X}\} \in S$ and a solution $\sigma \in \mathcal{S}(\mathcal{X})$ such that $\bar{s}\sigma = \bar{t}\theta$. Thus $\mathcal{X} \wedge \mathcal{Y} \wedge \bar{t} = \bar{s}$ is satisfiable. By Condition 1 in the definition of c-Hintikka set, this is impossible.

- **Induction case.** C is non unit. By Condition 2 in the definition of c-Hintikka set, there must exists a set $S' \subset S$ such that: $S' \models \mathcal{S}(C)$ and for any $C' \in S'$, $C' <_d C$. By the induction hypothesis, we have $\mathcal{I} \models C'$. Thus $\mathcal{I} \models \mathcal{S}(C)$. Hence $\mathcal{I} \models C$.

◇

NOTATION 4.2 *Let S be a set of gc-clauses and let C be a gc-clause. We write $S \prec_s^d C$ if there exists $S' \subseteq S$ such that:*

- *$DSub(S', C) = \emptyset$ (i.e. any clause in $\mathcal{S}(C)$ is subsumed by a clause in $\mathcal{S}(S')$).*

- *For any $C' \in S'$, $C' <_d C$.*

  We write $S \prec_s^d S'$ iff for any $C' \in S'$, $S \prec_s^d C'$.

DEFINITION 4.105 Let $\rho = (\mathcal{T}_i)_{i \in I}$ be a RAMCET-derivation. A *branch* through ρ is a set of gc-clauses of the form $S = \bigcup_{i \in I} S_i$, where:

- For all $i \in I$, S_i occurs in $\mathcal{T}_i$.

- For all $i \in I \setminus \{0\}$, S_{i-1} is an ancestor of S_i.

♯

We now give a formal definition of the notion of fairness:

DEFINITION 4.106 Let $\mathcal{T}$ be a tableau. A RAMCET-derivation $\rho = (\mathcal{T}_i)_{i \in \mathbb{N}}$ is said to be *fair* iff there exists a substitution σ such that:

- for all $i \in \mathbb{N}$, σ is a solution of the constraints of $\mathcal{T}_i$.
- for any branch S through ρ:
  - The clash rule is not applicable on $S\sigma$.
  - If a RAMCET rule is applicable to $S\sigma$ yielding n branches $S_1\sigma, \ldots, S_n\sigma$, then there exists $i \in \{1..n\}$ such that $S\sigma \prec_s^d S_i\sigma$.
- For any ground term t, there exists a variable x such that $x\sigma = t$.

♯

The second condition ensures that no rule can be infinitely delayed and the third one expresses that the instantiation of variables must be done in a fair way (i.e. we cannot instantiate repeatedly all rigid variables with the same ground term, because this would lead to incompleteness).

LEMMA 4.107 *Let S be a set of closed gc-clauses. Let $\rho = (\mathcal{T}_i)_{i \in I}$ be a fair derivation. If S is unsatisfiable then there exists an $i \in I$ such that $\mathcal{T}_i$ is closed.*

Proof: Assume that there is no $i \in I$ such that $\mathcal{T}_i$ is closed. We show that there exists a branch S^* through ρ. Since ρ is fair, there exists a (possibly infinite) substitution σ such that:

- σ is a solution of $\mathcal{T}_i$, $\forall i \in I$.
- The clash rule is not applicable on $S^*\sigma$.
- If a RAMCET rule is applicable on $S^*\sigma$ yielding n branches $S_1\sigma, \ldots, S_n\sigma$, then there exists $i \in \{1 \ldots n\}$ such that $S^* \prec_s^d S_i$.
- For any ground term t, there exists a variable x such that $x\sigma = t$.

We are going to prove that $S^*\sigma$ is a c-Hintikka set.

- Assume that there exist two unit gc-clauses $(\forall \overline{x})\{P(\overline{t}) : \mathcal{X}\}$ and $(\forall \overline{y})\{\neg P(\overline{s}) : \mathcal{Y}\}$ in S^* such that $\mathcal{X}\sigma \wedge \mathcal{Y}\sigma \wedge \overline{t}\sigma = \overline{s}\sigma$ is satisfiable. Then the unit disresolution rule would be applicable to $S^*\sigma$. Thus, $S^*\sigma \prec_s^d \{P(\overline{t}) : \mathcal{X} \wedge \mathcal{Y} \wedge \overline{t} = \overline{s}\}\sigma$. Therefore $S^*\sigma$ contains $\sqcup$, which is impossible since the Clash rule would be applicable.
- Let C be a non unit gc-clause in S^*. We have to show that there exists a set of gc-clauses $S' \subseteq S^*\sigma$ such that $S' \models \mathcal{S}(C\sigma)$ and any gc-clause in S' is smaller than $C\sigma$ according to $<_d$.

If the splitting rule is applicable on $C\sigma$, then $C\sigma$ is of the form $(\forall\overline{x})\{A \vee B : \mathcal{X}\}\sigma$. Since ρ is fair, we must have either $S^*\sigma \prec_s^d (\forall\overline{x})\{A : \mathcal{X}\}\sigma$ or $S^*\sigma \prec_s^d (\forall\overline{x})\{B : \mathcal{X}\}\sigma$. W.l.o.g. we assume that $S^* \prec_s^d (\forall\overline{x})\{A : \mathcal{X}\}\sigma$. By definition, this means that for all $\theta \in \mathcal{S}(\mathcal{X})$, there exists $C_\theta \in S^*$ such that $C_\theta\sigma \prec_s^d A\theta\sigma$. Let $S' = \{C_\theta\sigma \mid \theta \in \mathcal{S}(\mathcal{X})\}$. We have $S' \models A\theta\sigma$ for all $\theta \in \mathcal{S}(\mathcal{X})$, hence $S' \models \mathcal{S}(C)$. Moreover, for all $C' \in S'$, $C' <_d C$ (since it contains strictly smaller literals).

Thus, the splitting rule cannot be applied on $C\sigma$. Let $C = \{R : \mathcal{X}\}$. This means that the minimal number of bound variables in $C\sigma$ (hence in C) is greater than 0 (else $\mathcal{X}\sigma$ would have only one solution and the splitting rule would be applicable). Therefore, there exists a bound variable x in C such that the instantiation rule is applicable on C with a constraint of the form $\mathcal{Y} : (x = y)$. Let θ be a solution of $\mathcal{X}\sigma$. Since ρ is fair, there exists a variable y such that $y\sigma = x\theta$. Moreover, $S^*\sigma \prec_s^d \{R : \mathcal{X} \wedge x = y\}$, i.e. there exists $S_\theta \subseteq S^*$ such that: $S_\theta\sigma \models \mathcal{S}(\{R : \mathcal{X} \wedge x = y\}\sigma)$ and for all $C' \in S_\theta$, $C' <_d C$. Let $S' = \bigcup_{\theta \in \mathcal{S}(\mathcal{X}_\sigma)} S_\theta$ We have $S' \models \{\{R : \mathcal{X} \wedge x = x\theta\} \mid \theta \in \mathcal{S}(\mathcal{X}\sigma)\}$ hence $S' \models C$. Moreover for all $C' \in S'$, $C' <_d C$.

Therefore $S^*\sigma$ is a c-Hintikka set. By Lemma 4.104 we deduce that $S^*\sigma$ is satisfiable. But by definition of ρ, $S^*\sigma$ must contain $S\sigma$. But S is closed and so $S\sigma = S$. Thus S is satisfiable. ◇

4.4 The Model Building algorithm

Below, we present the algorithm extracting models from open branches.

The method is based on generalization techniques. Given an open branch S, the set of unit gc-clauses in S may be seen as a representation of the (partial) model corresponding to the branch. Thus, we may construct a model of the whole branch by generalizing this set of unit gc-clauses. To this purpose, we eliminate free variables by considering the universal closure of the gc-clauses (constrained by the equational formula corresponding to the global tableau). This set can still contain inconsistencies, hence, we have to consider the maximal subset E' of E that is a Hintikka set. This set is obtained by removing from E all these (ground) literals of E whose complement is also in E. This set can be expressed by a *finite set* of constrained literals. Now we can check whether all input formulae are true in E'.

For example, given the branch:

$$\{(\forall x)P(f(x)), \neg P(u)\}$$

where u is a free variable, we may compute the eq-interpretation:

$$\{(\forall x)P(f(x)), (\forall u)\{\neg P(u) : (\forall y)u \neq f(y)\}\}.$$

It should be stressed that during our whole task, we try to build a model *not* by *using induction*, but rather by computing *symbolically* all the ground substitutions of the (constrained) free object variables of the formulae on the branch at hand, which make this formulae true under the partial interpretation considered.

procedure *Model*
input: A set of gc-clauses S and a constraint $\mathcal{P}$
output: An eq-interpretation (represented by a set of unit c-clauses)
begin
  Let $E = \mathbf{Unit}(S)$
  $E' = \emptyset$
  for each $(\forall \overline{x})\{P(\overline{t}) : \mathcal{X}\} \in E$
    Let $X_{clash} = \bigvee_{[\![\neg P(\overline{s}):\mathcal{Y}]\!]\in E'} (\exists \overline{u})(\overline{s} = \overline{t} \wedge \mathcal{X} \wedge \mathcal{Y})$
    (where $\overline{u} = \mathcal{V}ar([\![\neg P(\overline{s}) : \mathcal{Y}]\!])$)
    $C_{pure} = [\![P(\overline{t}) : \mathcal{P} \wedge \mathcal{X} \wedge \wedge\neg X_{clash}]\!]$
    $E' := E' \cup \{C_{pure}\}$
  end for each
  return(E')
end

By construction, *Model*(S) is free from contradiction, hence represents a eq-interpretation. Then, we only have to check whether this interpretation validates the initial formula (which can be done using the techniques presented in Chapter 5).

4.5 Further improvements of the method

Several improvement have been considered in order to enrich RAMCET.

- First, the method may be extended (as the standard tableaux) in order to deal with any first-order formula instead of clausal forms. This may be done by encoding equational constraint into the formulae. To this purpose, new equational atoms are considered, interpreted as syntactic equality on the Herbrand universe. This general calculus allows to avoid transformation into clausal form and offers more flexibility.
- **Semantic strategies** can be added in order to prune the search space and improve the efficiency of the method. The idea is to re-

strict the application of the instantiation and splitting rules to gc-clauses that are false in the considered interpretation. In order to preserve refutational completeness, the unit dis-resolution rule (see Section 3) must be used in order to remove literals that are false in the considered branch from gc-clauses that are satisfied by the underlying interpretation. It may be proven that this strategy preserves refutational completeness [Peltier, 1999].

It is worth mentioning that model generation theorem provers, such as positive unit hyperresolution tableaux (PUHR-tableaux) [Bry and Yahya, 1996], SATCHMO [Manthey and Bry, 1988], hypertableaux [Baumgartner et al., 1996] etc. can be considered as particular cases of this approach, using an interpretation maping each positive atoms to **false**.

- Other model extraction techniques may be defined in order to increase the capabilities of the method. In [Peltier, 1999] a more powerful (but more costly) method is proposed. Its basic principle is to use, as a starting point for generalizing partial interpretations, an (arbitrarily chosen) *partition* of the Herbrand base via constraints, and to *refine* the partition during the construction of the tableau, until a eq-model can be extracted.

  Using this method, it may be proven that RAMCET is *complete* w.r.t. the class of formulae having an eq-model. More formally, we have the following theorem:

THEOREM 4.1 *Let S be a set of clauses. If S has an eq-model, then for any fair derivation $(\mathcal{T}_i)_{i \in \mathbb{N}}$ starting from $(\{S\}, \top)$, there exists $i \in \mathbb{N}$ and a branch B in $\mathcal{T}_i$ such that $B\sigma$ is the representation of a eq-model of S.*

Thus the obtained method is a decision procedure for any class of formula $\mathcal{C}$ having the following property: any satisfiable formula in $\mathcal{C}$ has a eq-model. Note that many interesting classes of formulae satisfy this property: Bernays–Schönfinkel class [Dreben and Goldfarb, 1979, Börger et al., 1997], PVD and OCCIN [Fermüller et al., 1993].

Chapter 5

MODEL REPRESENTATION AND EVALUATION

1. The Problem of Model Representation

The model building algorithms presented so far either produce finite models or finite representations of infinite models (in particular Herbrand models). In case of finite models the distinction between the model (as an abstract mathematical object) and its syntactic representation, although necessary from a logical point of view, is rather pointless and sophistic; clearly every finite model is uniquely determined by the enumeration of its domain elements and by the "multiplication"-tables for functions and predicates. The situation becomes less trivial for infinite structures. As every satisfiable formula F of first-order logic has a Herbrand model (defined via the skolemized form), Herbrand interpretations are the most natural candidates. In order to specify a specific Herbrand model of a formula we may always order all models of a formula and take the smallest one, let us call it $\mathcal{M}$; this is possible by the well-ordering theorem (which is equivalent to the axiom of choice). Such a specification is clearly *noncomputational* in a strong sense, e.g. we have no means to evaluate arbitrary ground atoms over $\mathcal{M}$; note that $\mathcal{M}$ is model of F and also a Herbrand interpretation of all skolemized formulae over the signature of F. Leaving aside such pathological representations, we still have the choice between two forms of "computational" specifications of Herbrand models: 1: we define a procedure which enumerates all true ground atoms over the Herbrand universe. 2: we define a formula which (in a sense to be defined) has only one Herbrand model. Both 1 and 2 are syntactic (in 1 the procedure is the syntactic object) but 1 is operational, while 2 is declarative. An example of 1 is the specification of the leftmost infinite branch of a semantic tree for a clause set which is a fixed point under a resolution operator. Examples of 2 are atomic- and constrained atomic representations defined in the chapters 3 and 4. An obvious advantage of declarative interpretations is that they are more informative semantically, but we will also show that they may be superior from a purely computational point of view.

In order to give a more precise meaning to the term "model building" we have to define some general conditions on the output of model building procedures. In particular we have to exclude that a mere proof of satisfiability (e.g. by resolution decision procedures) is also considered as the construction of a model. Let $X(F)$ be a class of syntactical objects representing interpretations of a formula F (yet the term "represents" is informal). Below we list the *postulates* defined in [Fermüller and Leitsch, 1996] with some minor modifications:

- **uniqueness:** each element of $X(F)$ specifies a single interpretation of F.

- **atom test:** there exists a fast (e.g. polynomial) procedure to evaluate arbitrary ground atoms (over the signature of F) in the interpretation defined by a ξ in $X(F)$.

- **formula evaluation:** there exists an algorithm deciding the truth values of arbitrary formulae (having the same syntax type as F) in interpretations defined by $\xi \in X(F)$.

- **equivalence test:** there exist an algorithm which decides whether two representations ξ_1 and ξ_2 in $X(F)$ describe the same interpretation.

Considering the four postulates, automated model building on a set of clauses F simply means the algorithmic construction of an object $\xi \in X(F)$ together with a proof that F is true in the interpretation described by ξ. Clearly all the postulates are fulfilled by interpretations over finite domains. In procedural specifications, like algorithms producing the set of true atoms in a Herbrand interpretation, things become more problematic: still uniqueness and (at least an algorithmic) truth evaluation of ground atoms can be achieved, but there is no direct way to handle clause evaluation and equivalence. In particular it is a nontrivial question whether the four postulates can be fulfilled in a formalism which describes Herbrand models over infinite universes.

In this chapter we present several representation mechanisms fulfilling the four postulates and also locate the limits. Clearly, if we choose stronger representation mechanisms, the algorithmic problems (evaluation and equivalence) become harder; eventually they may become undecidable and we cannot speak any longer of *computational* model representations.

Let us try to test all clausal representations investigated so far according to the postulates :

1 *finite interpretations*: postulates trivially fulfilled.

2 *linear atomic representations*:
Uniqueness is trivial by definition. By Theorem 1.8 every Herbrand model $\mathcal{M}$, described by a linear atomic representation $\mathcal{A}$, has a finite realization $\mathcal{F}(\mathcal{M})$. Thus the conditions on atom test, clause evaluation and equivalence all hold by reduction to finite models.

3 *atomic representations*:
Uniqueness is trivial and the atom test can be performed via matching in linear time. But there are atomic representations of models without finite realizations (see propositions 1.1 and 1.4). Thus clause test and equivalence cannot be reduced to finite interpretations.

4 *constrained atomic representations*:
Again, uniqueness is trivial by definition. An atom test consists of matching and subsequent constraint-solving; the atom test, though computable, may be much more expensive than for atomic representations (by the complexity of constraint-solving itself). As atomic representations are a subset, the problems of clause evaluation and equivalence are at least as hard.

5 *ground equational atomic representations*:
Uniqueness is trivial. Like for linear atomic representations there exist finite realizations (see Theorem 3.1). Thus evaluation and equivalence problem both become decidable. Note that the atom test is more expensive than for linear atomic representations: either we apply Knuth-Bendix completion and normalization or we transform into the finite model. But both transformations can be exponential.

6 *equational atomic representations*:
Uniqueness is trivial, but is the only one which is fulfilled. As the general word problem is undecidable, there exists no computable atom test at all. As a trivial consequence there exist no algorithm to evaluate the truth values of clauses. The equivalence test is undecidable as well, due to the fact that it is undecidable whether the set of ground equations derivable from two different sets of equations are equal.

In the Ph.D. thesis of Robert Matzinger several other representation mechanisms are investigated (see [Matzinger, 2000] and [Matzinger,

1997]); the most natural and prominent one is given by regular term grammars. The Herbrand models described by regular term grammars are the same as those representable by ground equational representations; in both cases the Herbrand models have finite realizations and every finite interpretation is covered by these formalisms. The strongest formalism known so far fulfilling all four postulates is given by atoms with regular and equational constraints [Peltier, 1997d]. A thorough comparison and analysis of algebraic closure properties for all these formalisms is given in [Matzinger, 2000]. Generally it turns out that clause evaluation is harder than equivalence. There are formalisms where the former is undecidable but the latter is decidable [Matzinger, 2000]; if both are decidable then clause evaluation is generally harder than equivalence ([Pichler, 2000]).

All the formalisms mentioned above define recursive Herbrand models. Clearly such models can be obtained whenever a computational proof procedure terminates and yields satisfiability of the set of input clauses. Sometimes it might be enough to have a procedural description and computational atom test only. If, on the other hand, the models themselves (i.e. their representations) are used for semantic inference then clause evaluation is of central importance.

In the classification of the representation mechanisms above, linear atomic and ground equational atomic representations are the most unproblematic ones (due to the realizability by finite models). On the other hand, general equational atomic representations (failing on three postulates) are too strong. The case of atomic and constrained atomic representations remains particularly interesting and will be discussed in subsequent sections.

2. Clause Evaluation over CARMs

In chapter 4 we have demonstrated how models are constructed via c-clause calculi; the representations obtained this way are sets of constrained unit clauses which typically describe a set of models. Indeed, if RAMC constructs such a set of unit c-clauses $\mathcal{D}$ every interpretation extending the partial interpretation given by $\mathcal{D}$ is a model of the set of input clauses. As we are dealing with evaluation and equivalence we have to work with a *single* uniquely specified model. To this aim we extend the semantics of atomic representations to this of constrained atomic representations.

DEFINITION 5.1 (CARM) Let $\mathcal{A}$ be a finite set of c-atoms and H be a Herbrand universe with $H(\mathcal{A}) \subseteq H$. Then $\mathcal{A}$ is called a *con-*

strained atomic representation (*CARM*) of the Herbrand interpretation $\mathcal{M} : (H, \Phi)$ with

$$\Phi(P)(h_1, \ldots, h_k) = \textbf{true} \text{ iff } P(h_1, \ldots, h_n) \in \mathcal{S}_H(\mathcal{A}).$$

$\mathcal{M}$ is denoted by $\mathcal{M}(\mathcal{A}, H)$. ♯

Remark:
Restricting representations by unit c-clauses to CARMs does not result in a loss of expressive power: let $\mathcal{B}$:

$$\{[\![P_1(\bar{s}_1) : \mathcal{X}_1]\!], \ldots, [\![P_n(\bar{s}_n) : \mathcal{X}_n]\!], [\![\neg Q_1(\bar{t}_1) : \mathcal{Y}_1]\!], \ldots, [\![\neg Q_m(\bar{t}_m) : \mathcal{Y}_m]\!]\}$$

be a set of unit c-clauses representing a nonempty set of H-intepretations $\mathcal{I}$ over a Herbrand universe H. Then the interpretation $\mathcal{M}(\mathcal{A}, H)$ for

$$\mathcal{A} : \{[\![P_1(\bar{s}_1) : \mathcal{X}_1]\!], \ldots, [\![P_n(\bar{s}_n) : \mathcal{X}_n]\!]\}$$

is in $\mathcal{I}$. Note that all atoms which are not instances of the $[\![P_i(\bar{s}_i) : \mathcal{X}_i]\!]$ are defined as false; in particular all the literals $[\![\neg Q_i(\bar{t}_i) : \mathcal{Y}_i]\!]$ are verified in $\mathcal{M}(\mathcal{A}, H)$. ♯

DEFINITION 5.2 (CEV-CARM) Let C be a c-clause and $\mathcal{A}$ be a CARM of $\mathcal{M}(\mathcal{A}, H)$ for a Herbrand universe H with $H(\{C\} \cup \mathcal{A}) \subseteq H$. The problem of deciding whether $\mathcal{M}(\mathcal{A}, H) \models C$ is called a *clause evaluation problem* for CARMs. More formally the corresponding decision problem is defined by the set CEV-CARM:

$$\{(\mathcal{A}, C, H) \mid \mathcal{A} \in CARM, C \text{ a c-clause }, H(\mathcal{A} \cup \{C\}) \subseteq H, \mathcal{M}(\mathcal{A}, H) \models C\}.$$

♯

Our main aim is to show that CEV-CARM is decidable and to define a decision algorithm. The decidability of clause evaluation for atomic representations then is an immediate consequence. The key concept in clause evaluation is a kind of subsumption which is defined relative to a fixed Herbrand universe. Subsumption concepts of this type have been studied in [Caferra and Zabel, 1992] (dissubsumption) or in [Fermüller and Leitsch, 1996] (H-subsumption).

DEFINITION 5.3 Let $[\![C : \mathcal{X}]\!]$ be a c-clause and D be a ground clause.Then $[\![C : X]\!]$ *subsumes* D (notation $[\![C : \mathcal{X}]\!] \leq_{sub} D$) if there exists a $\sigma \in \mathcal{S}(C : \mathcal{X})$ over $H(\{[\![C : \mathcal{X}]\!], D\})$ s.t. $\mathrm{LIT}(C\sigma) \subseteq \mathrm{LIT}(D)$.

Let $\mathcal{C}$ be a set of c-clauses, $[\![D : \mathcal{Y}]\!]$ be a c-clause and H be a Herbrand universe with $H(\mathcal{C} \cup \{D\}) \subseteq H$. Then $\mathcal{C}$ *H-subsumes* D (notation $\mathcal{C} \leq_{sH} D$) if for every $\sigma \in \mathcal{S}([\![D : \mathcal{Y}]\!])$ over H there exists a $C \in \mathcal{C}$ with $C \leq_{sub} D\sigma$.

We extend $\leq_{sH}$ to sets of c-clauses $\mathcal{D}$ via

$\mathcal{C} \leq_{sH} \mathcal{D}$ iff for all $D \in \mathcal{D}$: $\mathcal{C} \leq_{sH} D$.

♯

EXAMPLE 5.4 Let

$$\begin{aligned} \mathcal{C} &= \{[\![P(x,y) : x \neq y]\!],\ [\![P(f(x),f(x)) : \top]\!]\}, \\ D &= [\![P(z,y) : z \neq a \wedge y \neq a]\!], \\ H &= \{f^n(a) \mid n \geq 0\}. \end{aligned}$$

We show $\mathcal{C} \leq_{sH} D$:
Let σ be a solution of $z \neq a \wedge y \neq a$ over H.

(a) $z\sigma = y\sigma$:
Then there exists a term $t \in H$ with $z\sigma = y\sigma = f(t)$.
In this case $[\![P(f(x),f(x)) : \top]\!] \leq_{sub} P(f(t),f(t))$.

(b) $z\sigma \neq y\sigma$:
Then $P(z,y)\sigma = P(s,t)$ for $s \neq t$. Here we have $[\![P(x,y) : x \neq y]\!] \leq_{sub} P(s,t)$.

Thus for all solutions σ of $z \neq a \wedge y \neq a$ over H with $dom(\sigma) = \{z,y\}$ there is a clause $C \in \mathcal{C}$ s.t. $C \leq_{sub} D\sigma$.

Note that no single clause in $\mathcal{C}$ H-subsumes D:

$$\begin{aligned} [\![P(x,y) : x \neq y]\!] &\not\leq_{sub} P(f(a),f(a)) \text{ and} \\ [\![P(f(x),f(x)) : \top]\!] &\not\leq_{sub} P(f(f(a)),f(a)). \end{aligned}$$

Moreover $\mathcal{C}$ does not H-subsume D for all Herbrand universes H. Just take

$$H' = \{f^n(a) \mid n \geq 0\} \cup \{f^n(b) \mid n \geq 0\}.$$

Then $\sigma = \{z \leftarrow b, y \leftarrow b\}$ is a solution of the constraint of D over H and

$$\begin{aligned} [\![P(x,y) : x \neq y]\!] &\not\leq_{sub} P(b,b), \\ [\![P(f(x),f(x)) : \top]\!] &\not\leq_{sub} P(b,b). \end{aligned}$$

♯

Remark:
If $\mathcal{C}$ and $\mathcal{D}$ are sets of clauses without constraints then $\mathcal{C} \leq_{sub} \mathcal{D}$ iff $\mathcal{C} \leq_{sH} \mathcal{D}$ for all Herbrand universes H with $H(\mathcal{C} \cup \mathcal{D}) \subseteq H$. ♯

The $\leq_{sH}$-relation directly determines the truth values of positive clauses over CARMs.

LEMMA 5.5 *Let $\mathcal{A}$ be a CARM, $[\![C : \mathcal{Y}]\!]$ be a positive c-clause and H be a universe with $H(\{[\![C : \mathcal{Y}]\!]\} \cup \mathcal{A}) \subseteq H$. Then $\mathcal{C}$ is true in $\mathcal{M}(\mathcal{A}, H)$ iff $\mathcal{A} \leq_{sH} C$.*

Proof: $[\![C : \mathcal{Y}]\!]$ is true in $\mathcal{M}(\mathcal{A}, H) \Rightarrow$:
By Herbrand semantics $C\sigma$ is true in $\mathcal{M}(\mathcal{A}, H)$ for all $\sigma \in \mathcal{S}([\![C : \mathcal{Y}]\!])$. Thus for every such σ there exists an $A \in C\sigma$ s.t. A is true in $\mathcal{M}(\mathcal{A}, H)$. By definition of $\mathcal{M}(\mathcal{A}, H)$ we get $A \in \mathcal{S}_H(\mathcal{A})$; so $\mathcal{A} \leq_{sub} A$ and $\mathcal{A} \leq_{sub} C\sigma$. As this last relation holds for arbitrary $\sigma \in \mathcal{S}([\![C : \mathcal{Y}]\!])$ we have $\mathcal{A} \leq_{sH} C$.

$\mathcal{A} \leq_{sH} [\![C : \mathcal{Y}]\!] \Rightarrow$:
By definition of $\leq_{sH}$ $\mathcal{A} \leq_{sub} C\sigma$ for all $\sigma \in \mathcal{S}([\![C : \mathcal{Y}]\!])$. Thus, for every such $C\sigma$, there exists an $A \in \mathcal{S}_H(\mathcal{A})$ with $A \in \mathrm{LIT}(C\sigma)$. By definition of $\mathcal{M}(\mathcal{A}, H)$ this A is true in $\mathcal{M}(\mathcal{A}, H)$. Thus $C\sigma$ is true for all $\sigma \in \mathcal{S}([\![C : \mathcal{Y}]\!])$ and so $C : \mathcal{Y}$ itself is true in $\mathcal{M}(\mathcal{A}, H)$. $\diamond$

Lemma 5.5 yields a truth criterion for positive clauses only; for negative ones the situation is slightly more difficult.

EXAMPLE 5.6 Let

$$\begin{aligned}
\mathcal{A} &= \{[\![P(a) : \top]\!],\ [\![P(f(f(x))) : \top]\!]\},\\
C &= [\![\neg P(x) \vee P(f(x)) : \top]\!],\\
D &= [\![P(x) \vee P(f(x)) : \top]\!] \text{ and}\\
H &= \{f^n(a) \mid n \geq 0\}.
\end{aligned}$$

Then D is true in $\mathcal{A}$, but C is false in $\mathcal{A}$; note that the instance $\neg P(a) \vee P(f(a))$ is false in $\mathcal{M}(\mathcal{A}, H)$. Indeed we can resolve $[\![P(a) : \top]\!]$ and C to obtain $P(f(a))$, which is false in $\mathcal{M}(\mathcal{A}, H)$. $\sharp$

Example 5.6 suggests that a truth criterion for mixed clauses can be reduced to one for positive clauses via resolution. We show below that such a reduction exists in general.

DEFINITION 5.7 Let C be a c-clause and $\mathcal{A}$ be a CARM. We define

$$\begin{aligned}
HYP(\mathcal{A}, C) &= \{C\} \text{ if } C \text{ is positive},\\
&= \varrho_H(\mathcal{A} \cup \{C\}) \text{ otherwise}.
\end{aligned}$$

$\sharp$

For mixed clauses C, $HYP(\mathcal{A}, C)$ defines the set of all hyperresolvents definable in the set of clauses $\mathcal{A} \cup \{C\}$. Thus $HYP(\mathcal{A}, C)$ is always a finite set of positive clauses. For Example 5.6 we obtain

$$HYP(\mathcal{A}, C) = \{[\![P(f(a)) : \top]\!], [\![P(f^3(x)) : \top]\!]\}.$$

Here the first clause in $HYP(\mathcal{A}, C)$ is false in $\mathcal{M}(\mathcal{A}, H)$, but the second one is true. The following theorem tells us that a mixed clause is true iff all hyperresolvents in $HYP(\mathcal{A}, C)$ are true.

THEOREM 2.1 *Let $\mathcal{A}$ be a CARM, C be a c-clause and H be a Herbrand universe with $H(\mathcal{A} \cup \{C\}) \subseteq H$. Then C is true in $\mathcal{M}(\mathcal{A}, H)$ iff $\mathcal{A} \leq_{sH} HYP(\mathcal{A}, C)$.*

Proof: If C is positive then, by definition, $HYP(\mathcal{A}, C) = \{C\}$ and the result follows from Lemma 5.5. Thus we may assume that C contains negative literals. So let

$$\begin{aligned} C &= [\![C' : \mathcal{Y}]\!] \text{ for} \\ C' &= \neg P_1(\bar{t}_1) \vee \cdots \vee \neg P_n(\bar{t}_n) \vee Q_1(\bar{s}_1) \vee \cdots \vee Q_m(\bar{s}_m) \end{aligned}$$

where $n > 0$ and $m \geq 0$.

C is true in $\mathcal{M}(\mathcal{A}, H)$:
By definition of Herbrand interpretations $C'\sigma$ is true in $\mathcal{M}(\mathcal{A}, H)$ for all $\sigma \in \mathcal{S}(C)$. We distinguish two cases:

(a) $P_i(\bar{t}_i)\sigma$ is false in $\mathcal{M}(\mathcal{A}, H)$ for some $i \in \{1, \ldots, n\}$:
Then, by definition of $\mathcal{M}(\mathcal{A}, H)$, $\mathcal{A} \not\leq_{sub} P_i(\bar{t}_i)\sigma$ and, in particular, there is no $A \in \mathcal{A}$ s.t. $\{A, P_i(\bar{t}_i)\sigma\}$ is a unifiable set. But this implies that there are no hyperresolvents definable in the set $\mathcal{S}_H(\mathcal{A}) \cup \{C'\sigma\}$. Clearly also

$$\varrho_H(\mathcal{A} \cup \{C'\sigma\}) = \emptyset.$$

As $C'\sigma$ is nonpositive we have $HYP(\mathcal{A}, C'\sigma) = \emptyset$ and, trivially, $\mathcal{A} \leq_{sub} HYP(\mathcal{A}, C'\sigma)$.

(b) All $P_i(\bar{t}_i)\sigma$ are true in $\mathcal{M}(\mathcal{A}, H)$:
Then, by definition of $\mathcal{M}(\mathcal{A}, H)$, $\mathcal{A} \leq_{sub} P_i(\bar{t}_i)\sigma$ for $i = 1, \ldots, n$. In particular there exist $[\![A_1 : \mathcal{X}_1]\!], \ldots, [\![A_n : \mathcal{X}_n]\!]$ in $\mathcal{A}$ s.t. the tuple-set

$$\{([\![A_1 : \mathcal{X}_1]\!], \ldots, [\![A_n : \mathcal{X}_n]\!]), (P_1(\bar{t}_1)\sigma, \ldots, P_n(\bar{t}_n)\sigma)\}$$

is unifiable (we may assume that the $[\![A_i : \mathcal{X}_i]\!]$ are variable-disjoint). But then

$$E : Q_1(\bar{s}_1)\sigma \vee \cdots \vee Q_m(\bar{s}_m)\sigma$$

is a hyperresolvent from the set $\mathcal{A} \cup \{C'\sigma\}$ and, by definition of $HYP(,)$, $E \in HYP(\mathcal{A}, C'\sigma)$. Because $C'\sigma$ is true in $\mathcal{M}(\mathcal{A}, H)$, but $\neg P_1(\bar{t}_1)\sigma \vee \cdots \vee \neg P_n(\bar{t}_n)\sigma$ is false, the clause E is true in $\mathcal{M}(\mathcal{A}, H)$. Therefore, by definition of $\mathcal{M}(\mathcal{A}, H)$, $\mathcal{A} \leq_{sub} E$. As this holds for any adequate selection of atoms $[\![A_i : \mathcal{X}_i]\!]$ leading to a unification we obtain

$$\mathcal{A} \leq_{sub} HYP(\mathcal{A}, C'\sigma).$$

Putting (a) and (b) together we find that for all ground instances $C'\sigma$ of $[\![C : \mathcal{Y}]\!]$ we have $\mathcal{A} \leq_{sub} HYP(\mathcal{A}, C'\sigma)$.

Now let $E \in HYP(\mathcal{A}, C)$ and let $E\tau$ be a ground instance of E. Then there exists a ground instance $C'\sigma$ of C s.t. $E\tau$ is a hyperresolvent of $\mathcal{A} \cup \{C'\sigma\}$, i.e. there exists a substitution σ s.t. $E\tau \in HYP(\mathcal{A}, C'\sigma)$.

Therefore $\mathcal{A} \leq_{sub} E\tau$ for all $E \in HYP(\mathcal{A}, C)$ and all ground instances $E\tau$ of E. But this means just

$$\mathcal{A} \leq_{sH} HYP(\mathcal{A}, C).$$

C is false in $\mathcal{M}(\mathcal{A}, H) \Rightarrow$:
Then there exists a ground instance $C'\sigma$ of C s.t. $C'\sigma$ is false in $\mathcal{M}(\mathcal{A}, H)$. But then

$P_1(\bar{t}_1)\sigma, \ldots, P_n(\bar{t}_n)\sigma$ are true in $\mathcal{M}(\mathcal{A}, H)$ and

$Q_1(\bar{s}_1)\sigma, \ldots, Q_m(\bar{s}_m)\sigma$ are false in $\mathcal{M}(\mathcal{A}, H)$.

By definition of $\mathcal{M}(\mathcal{A}, H)$ we thus obtain

(I) $\mathcal{A} \leq_{sub} P_i(\bar{t}_i)\sigma$ for $i = 1, \ldots, n$,

(II) $\mathcal{A} \not\leq_{sub} Q_j(\bar{s}_j)\sigma$ for $j = 1, \ldots, m$.

By (I) there exists a unifier for the set

$$\{([\![A_1 : \mathcal{X}_1]\!], \ldots, [\![A_n : \mathcal{X}_n]\!]), (P_1(\bar{t}_1)\sigma, \ldots, P_n(\bar{t}_n)\sigma)\}$$

for some $[\![A_i : \mathcal{X}_i]\!] \in \mathcal{A}$. This, in turn, yields the hyperresolvent

$E : Q_1(\bar{s}_1)\sigma \vee \cdots \vee Q_m(\bar{s}_m)\sigma$

from $\mathcal{A} \cup \{C'\sigma\}$ and so $E \in HYP(\mathcal{A}, C'\sigma)$.

By the lifting property there exists an $E_0 \in HYP(\mathcal{A}, C)$ and a ground instance $E_0\tau$ with $E_0\tau = E$.

By (II) $\mathcal{A} \not\leq_{sub} E$, i.e. there exists an $E_0 \in HYP(\mathcal{A}, C)$ and a ground instance $E_0\tau$ with $\mathcal{A} \not\leq_{sub} E_0\tau$. By definition of $\leq_{sH}$ this gives $\mathcal{A} \not\leq_{sH} E_0$ and, eventually,

$$\mathcal{A} \not\leq_{sH} HYP(\mathcal{A}, C).$$

◇

COROLLARY 2.1 *Let $\mathcal{A}$ be a CARM, C be a c-clause and H be a Herbrand universe with $H(\mathcal{A} \cup \{C\}) \subseteq H$. Then C is true in $\mathcal{M}(\mathcal{A}, H)$ iff all clauses in $HYP(\mathcal{A}, C)$ are true in $\mathcal{M}(\mathcal{A}, H)$.*

Proof: By Theorem 2.1 C is true in $\mathcal{M}(\mathcal{A}, H)$ iff for all $D \in HYP(\mathcal{A}, C)$ $\mathcal{A} \leq_{sH} D$. But by Lemma 5.5 $\mathcal{A} \leq_{sH} D$ iff D is true in $\mathcal{M}(\mathcal{A}, H)$ for positive clauses D. $\diamond$

Theorem 2.1 gives us a truth criterion for CE-CARM. As $HYP(\mathcal{A}, C)$ is a finite computable set, the decidability of CE-CARM can be reduced to this of $\leq_{sH}$. In order to show decidability of $\leq_{sH}$ we construct equational formulae representing the relation $\mathcal{A} \leq_{sH} C$ for CARMs $\mathcal{A}$ and positive c-clauses C. As equational logic is decidable such a transformation yields the decidability of $\leq_{sH}$ and eventually of CE-CARM.

DEFINITION 5.8 Let $\mathcal{A}$ be a CARM and $C : [\![C' : \mathcal{Y}]\!]$ be a positive c-clause. Let us assume that $V(\mathcal{A}) \cap V(C) = \emptyset$. Then for every formula $B : [\![P(\bar{s}) : \mathcal{X}]\!]$ in $\mathcal{A}$ we define a formula $F(B, C)$ in the following way:

Let $C_P : \{P(\bar{t}_1), \ldots, P(\bar{t}_n)\}$ be the set of all P-atoms in C. If $C_P = \emptyset$ then we define

- $F(B, C) \equiv \bot$;

otherwise we define

$$F(B, C) \equiv (\exists \bar{x})(\mathcal{X} \wedge (\bar{s} = \bar{t}_1 \vee \cdots \vee \bar{s} = \bar{t}_n)),$$

where $\bar{x}$ is a variable vector representing $V(B)$. Finally we define

$$eval(\mathcal{A}, C) \equiv (\forall \bar{y})(\mathcal{Y} \rightarrow \bigvee_{B \in \mathcal{A}} F(B, C)),$$

where $\bar{y}$ is a variable vector representing $V(C)$. $\sharp$

EXAMPLE 5.9 Let

$$\begin{aligned} \mathcal{A} &= \{B_1 : [\![P(x, y) : x \neq y]\!],\ B_2 : [\![P(f(x), f(x)) : \top]\!]\} \text{ and} \\ C &= [\![P(u, a) \vee P(u, f(a)) : u \neq a]\!]. \end{aligned}$$

We define $F(B_1, C)$ and $F(B_2, C)$:

$$\begin{aligned} F(B_1, C) &\equiv (\exists x, y)[x \neq y \wedge ((x = u \wedge y = a) \vee (x = u \wedge y = f(a)))], \\ F(B_2, C) &\equiv (\exists x)[(f(x) = u \wedge f(x) = a) \vee (f(x) = u \wedge f(x) = f(a))]. \end{aligned}$$

Finally we have

$$eval(\mathcal{A}, C) = (\forall u)[u \neq a \rightarrow (F(B_1, C) \vee F(B_2, C))].$$

It is easy to show that $eval(\mathcal{A}, C)$ is true in equality logic over the Herbrand universe $H(\mathcal{A} \cup \{C\})$. Consider a ground term different from a; this term must be of the form $f(t)$; afterwards define the substitution $\sigma = \{y \leftarrow a, x \leftarrow f(t)\}$. Clearly σ verifies the matrix of $F(B_1, C)$. ♯

PROPOSITION 2.1 *Let $\mathcal{A}$ be a CARM, C be a positive c-clause and H be a Herbrand universe with $H(\mathcal{A} \cup \{C\}) \subseteq H$. Then $\mathcal{A} \leq_{sH} C$ iff $eval(\mathcal{A}, C)$ is true (over H).*

Proof: Let $C \equiv [\![C' : \mathcal{Y}]\!]$ for $C' \equiv P_1(\bar{s}_1) \vee \cdots \vee P_n(\bar{s}_n)$.

$\mathcal{A} \leq_{sH} C \Rightarrow$:
Let $C'\sigma$ be a ground instance of C with $\sigma \in \mathcal{S}(C)$. Then, by definition of $\leq_{sH}$, we have $\mathcal{A} \leq_{sub} C'\sigma$. That means there exists an $[\![A : \mathcal{X}]\!] \in \mathcal{A}$ s.t.

$$[\![A : \mathcal{X}]\!] \leq_{sub} P_1(\bar{s}_1)\sigma \vee \cdots \vee P_n(\bar{s}_n)\sigma.$$

But then there exist a $B \in \mathcal{A}$ and $i, \bar{t}, \mathcal{X}$ with $B = [\![P_i(\bar{t}) : \mathcal{X}]\!]$ and

$$B \quad \leq_{sub} \quad P_i(\bar{s}_i)\sigma.$$

This implies that $(\exists \bar{x})(\mathcal{X} \wedge \bar{t} = \bar{s}_i\sigma)$ is true for a variable vector $\bar{x}$ for $V(B)$. By Definition 5.8 this implies the truth of $F(B, C'\sigma)$.

But for all ground instances $C'\sigma$ of C with $\sigma \in \mathcal{S}(C)$ such a $B \in \mathcal{A}$ can be found, i.e. for all such σ
the equational formula

$$\bigvee_{B \in \mathcal{A}} F(B, C'\sigma)$$

is true. Therefore, by the semantics of equational logic, the formula

$$eval(\mathcal{A}, C) \quad : \quad (\forall \bar{y})(\mathcal{Y} \rightarrow \bigvee_{B \in \mathcal{A}} F(B, C))$$

is true (for a variable vector $\bar{y}$ for $V(C)$).

$\mathcal{A} \not\leq_{sH} C \Rightarrow$:
Then, by definition of $\leq_{sH}$, there exists a ground instance $C'\sigma$ of C with $\sigma \in \mathcal{S}(C)$ s.t.

$$\mathcal{A} \not\leq_{sub} P_1(\bar{s}_1)\sigma \vee \cdots \vee P_n(\bar{s}_n)\sigma.$$

Now let $B : [\![Q(\bar{t}) : \mathcal{X}]\!]$ be an arbitrary element in $\mathcal{A}$. Then either Q does not appear in $C'\sigma$ (i.e. $C'\sigma_Q = \emptyset$) or there are indices $i_1, \ldots, i_k$

s.t. $C'\sigma_Q = \{Q(\bar{s}_{i_1})\sigma, \ldots, Q(\bar{s}_{i_k})\sigma\}$ and, as $\mathcal{A} \not\leq_{sub} C'\sigma$, the equational formula

$$(\exists \bar{x})(\mathcal{X} \wedge (\bar{t} = \bar{s}_{i_1}\sigma \vee \cdots \vee \bar{t} = \bar{s}_{i_k}\sigma))$$

is false. In both cases the corresponding formula $F(B, C'\sigma)$ is false (in the first case $F(B, C'\sigma) = \bot$). As $\mathcal{A} \not\leq_{sub} C'\sigma$ this holds for all $B \in \mathcal{A}$ and so

$$\bigvee_{B \in \mathcal{A}} F(B, C'\sigma)$$

is false. As σ is a solution of $\mathcal{Y}$ the equational formula

$$(\exists \bar{y})(\mathcal{Y} \wedge \neg \bigvee_{B \in \mathcal{A}} F(B, C))$$

is true. But the last formula is equivalent to $\neg\, eval(\mathcal{A}, C)$; therefore $eval(\mathcal{A}, C)$ is false. $\diamond$

COROLLARY 2.2 *Let $\mathcal{A}$ be a CARM, C be a positive c-clause and H be a Herbrand universe with $H(\mathcal{A} \cup \{C\}) \subseteq H$. Then C is true in $\mathcal{M}(\mathcal{A}, H)$ iff $eval(\mathcal{A}, C)$ is true (over H).*

Proof: By Proposition 2.1 and Lemma 5.5. $\diamond$

THEOREM 2.2 *Let $\mathcal{A}$ be a CARM, C be a c-clause and H be a Herbrand universe with $H(\mathcal{A} \cup \{C\}) \subseteq H$. Then C is true in $\mathcal{M}(\mathcal{A}, H)$ iff*

$$eval(\mathcal{A}, C) \quad : \quad \bigwedge_{D \in HYP(\mathcal{A}, D)} eval(\mathcal{A}, D)$$

is true over H.

Proof: By Theorem 2.1 C is true in $\mathcal{M}(\mathcal{A}, H)$ iff $\mathcal{A} \leq_{sH} D$ for all $D \in HYP(\mathcal{A}, C)$. By Proposition 2.1 $\mathcal{A} \leq_{sH} D$ iff $eval(\mathcal{A}, D)$ is true (for positive clauses D). Therefore all formulae $eval(\mathcal{A}, D)$ for $D \in HYP(\mathcal{A}, C)$ are true over H. But then also the conjunction $eval(\mathcal{A}, C)$ is true over H. $\diamond$

COROLLARY 2.3 *CE-CARM, the clause evaluation problem over constrained atomic representations, is decidable.*

Proof: Let $\mathcal{A}$ be a CARM, C be a c-clause and H be a Herbrand universe with $H(\mathcal{A} \cup \{C\}) \subseteq H$. By Theorem 2.2 there exists an equational formula $\mathcal{X}$, which can be algorithmically constructed from $(\mathcal{A}, C, H)$, s.t. $\mathcal{X}$ is true over H iff $\mathcal{M}(\mathcal{A}, H) \models C$. Thus CE-CARM can be reduced to the decision problem for equational logic. But equational logic is decidable by the algorithm defined in Subsection 1.2 of Chapter 4. $\diamond$

EXAMPLE 5.10 Let

$$\mathcal{A} = \{[\![P(x,y) : x \neq y]\!],\ [\![P(f(x), f(x)) : \top]\!]\}$$

be a CARM, $H = H(\Sigma)$ for $\Sigma = \{P, a, f\}$ and

$$D = [\![\neg P(u,v) \lor P(u, f(v)) : u \neq a]\!].$$

Let $\mathcal{M}(\mathcal{A}, H)$ be the Herbrand model represented by $\mathcal{A}$ over H. Our aim is to determine whether D is true in $\mathcal{M}(\mathcal{A}, H)$.

By Corollary 2.1 we know that D is true in $\mathcal{M}(\mathcal{A}, H)$ iff all clauses in $HYP(\mathcal{A}, D)$ are true in $\mathcal{M}(\mathcal{A}, H)$. By definition $HYP(\mathcal{A}, D) = \varrho_H(\mathcal{A} \cup \{D\})$ and thus

$$HYP(\mathcal{A}, D) = \{[\![P(u, f(u)) : u \neq a \land u \neq v]\!], [\![P(f(z), f(f(z))) : f(z) \neq a]\!]\}$$

which simplifies to $HYP(\mathcal{A}, D) = \{D_1, D_2\}$ for

$$\begin{aligned} D_1 &= [\![P(u, f(v)) : u \neq a \land u \neq v]\!], \\ D_2 &= [\![P(f(z), f(f(z))) : \top]\!]. \end{aligned}$$

By Lemma 5.5 $\mathcal{A} \leq_{sH} \{D_1, D_2\}$ iff both D_1 and D_2 are true in $\mathcal{M}(\mathcal{A}, H)$. So it remains to test $\mathcal{A} \leq_{sH} D_i$ for $i = 1, 2$.

By Proposition 2.1 $\mathcal{A} \leq_{sH} D_i$ iff $eval(\mathcal{A}, D_i)$ is true as equational formula over A. By Definition 5.8 we get

$$eval(\mathcal{A}, D) = (\forall u, v)[(u \neq a \land u \neq v) \rightarrow \bigvee_{B \in \mathcal{A}} F(B, D_1).$$

Where $\bigvee_{B \in \mathcal{A}} F(B, D_1)$ is

$$(\exists x, y)(x \neq y \land x = u \land y = f(v)) \lor (\exists x)(f(x) = u \land f(x) = f(v))$$

which can be simplified to $u \neq f(v) \lor u = f(v)$ which, in turn, is equivalent to $\top$. So $\bigvee_{B \in \mathcal{A}} F(B, D_1)$ is equivalent to true (over H) and thus $eval(\mathcal{A}, D_1)$ is true.

It remains to evaluate $eval(\mathcal{A}, D_2)$. Now

$$eval(\mathcal{A}, D_2) = (\forall z) \bigvee_{B \in \mathcal{A}} F(B, D_2)$$

where $\bigvee_{B \in \mathcal{A}} F(B, D_2)$ is

$$(\exists x, y)(x \neq y \land x = f(z) \land y = f(f(z))) \lor (\exists x)(f(x) = f(z) \land f(x) = f(f(z)))$$

which simplifies to $f(z) \neq f(f(z)) \lor f(z) = f(f(z))$. This proves that also $eval(\mathcal{A}, D_2)$ is true over H. To sum up we have shown that $\mathcal{A} \leq_{sH} \{D_1, D_2\}$ and thus the clause D is true in $\mathcal{M}(\mathcal{A}, H)$. ♯

Remark: In order to compute the truth values of c-clauses over CARMS efficiently, good decision algorithms for $\leq_{sH}$ play the central role. Clearly it is not wise to use constraint solving under any circumstances. So for ordinary clauses and ARMs there are more efficient algorithms based on the term tupel cover problem developed by R. Pichler [Pichler, 1998]. But Pichler also shows that $\mathcal{A} \leq_{sH} D$, for an ARM $\mathcal{A}$ and a clause D, is co-NP complete; thus, not even in this more restricted case, we cannot hope for very fast algorithms.

Note that the worst-case of solving unrestricted constraints in c-clauses is nonelementary. Thus full CARMs violate the second postulate defined in Section 1. Indeed model representations requiring a nonelementary expense for ground atom evaluation make hardly sense in practice. A viable solution is to onsider only CARMS with constraints in the form of equational problems or even in solved forms (see Chapter 4). This does not decrease the expressivity of the formalism and admits a feasible truth test for ground atoms. ♯

It remains to show that also the equivalence. problem of CARMs is decidable.

DEFINITION 5.11 (EQ-CARM) Let $\mathcal{A}$ and $\mathcal{B}$ two CARMs and H be a Herbrand universe with $H(\mathcal{A} \cup \mathcal{B}) \subseteq H$. Then $\mathcal{A}$ is *equivalent* to $\mathcal{B}$ (notation $\mathcal{A} \sim_H \mathcal{B}$) if $\mathcal{M}(\mathcal{A}, H) = \mathcal{M}(\mathcal{B}, H)$. The *equivalence problem* for constrained atomic representations is given by the set

$$\text{EQ-CARM} = \{(\mathcal{A}, \mathcal{B}, H) \mid \mathcal{A}, \mathcal{B} \in \text{CARM}, H(\mathcal{A} \cup \mathcal{B}) \subseteq H, \mathcal{A} \sim_H \mathcal{B}\}.$$

EXAMPLE 5.12 Let

$$\begin{aligned}
\mathcal{A} &= \{[\![P(a) : \top]\!],\ [\![P(f(a)) : \top]\!],\ [\![P(f(x)) : x \neq a]\!]\},\\
\mathcal{B} &= \{[\![P(x) : \top]\!]\},\\
H &= \{f^n(a) \mid n \geq 0\}.
\end{aligned}$$

Then $\mathcal{A} \sim_H \mathcal{B}$ as $\mathcal{S}_H(\mathcal{A}) = \mathcal{S}_H(\mathcal{B})$. Indeed it is easily verified that

$$\mathcal{S}_H(\mathcal{A}) = \mathcal{S}_H(\mathcal{B}) = \{P(t) \mid t \in H\}.$$

Therefore $\mathcal{M}(\mathcal{A}, H)$ and $\mathcal{M}(\mathcal{B}, H)$ both describe the model in which the predicate corresponding to P is **true** everywhere.

Note that $\mathcal{A}$ and $\mathcal{B}$, as sets of unit c-clauses, are *not logically equivalent.* In particular

$$\mathcal{A} \not\sim_{H'} \mathcal{B} \quad \text{for} \quad H' = \{f^n(a) \mid n \geq 0\} \cup \{f^n(b) \mid n \geq 0\}.$$

This is easy to prove: $P(b) \in I_{H'}(\mathcal{B}) - I_{H'}(\mathcal{A})$; thus $P(b)$ is true in $\mathcal{M}(\mathcal{B}, H')$, but false in $\mathcal{M}(\mathcal{A}, H')$. In particular $(\mathcal{A}, \mathcal{B}, H) \in$ EQ-CARM, but $(\mathcal{A}, \mathcal{B}, H') \notin$ EQ-CARM. ♯

THEOREM 2.3 *EQ-CARM, the equivalence problem for constrained atomic representations, is decidable.*

Proof: Let $\mathcal{A}$ and $\mathcal{B}$ two CARMs and H be a Herbrand universe with $H(\mathcal{A} \cup \mathcal{B}) \subseteq H$. Then $(\mathcal{A}, \mathcal{B}, H) \in$ EQ-CARM iff $\mathcal{A} \sim_H \mathcal{B}$. But $\mathcal{A} \sim_H \mathcal{B}$ iff $\mathcal{M}(\mathcal{A}, H) = \mathcal{M}(\mathcal{B}, H)$ iff $\mathcal{S}_H(\mathcal{A}) = \mathcal{S}_H(\mathcal{B})$. But the last equation is equivalent to

$$\mathcal{A} \leq_{sH} \mathcal{B} \quad \text{and} \quad \mathcal{B} \leq_{sH} \mathcal{A}.$$

Thus it is enough to show how to decide $\mathcal{A} \leq_{sH} \mathcal{B}$. But

$$\mathcal{A} \leq_{sH} \mathcal{B} \text{ iff } \mathcal{A} \leq_{sH} B \ \text{ for all } B \in \mathcal{B}.$$

So the problem is further reduced to decide $\mathcal{A} \leq_{sH} B$ for a CARM $\mathcal{A}$ and some c-atom B. By Proposition 2.1 $\mathcal{A} \leq_{sH} B$ iff the equational formula $eval(\mathcal{A}, B)$ is true over H. By the decidability of equational logic the result follows. $\diamond$

3. Representing Models by Equational Constraints

In the previous section, we investigated the use of sets of unit c-clauses for representing Herbrand models (CARM). We provided algorithms for solving clausal evaluation problems (CE-CARM) and equivalence problems (EQ-CARM).

In this section, we extend these results into two directions.

- First, we show how to solve evaluation problems for first-order formulae. This problem is more general – and harder – than clausal evaluation, due to the presence of existential quantifiers. Note that this problem cannot be handled by using existing algorithm for clausal transformation, since skolemization does not preserve logical equivalence.

- Second, we show how to extend the expressive power of equational constraints by using other, more complex, constraint languages.

Any decidable first-order theory can be used for representing Herbrand models of first-order formulae. In particular, we show how to solve the formula evaluation problem in interpretations represented in this way. We give examples of decidable extensions of equational formulae that can be used for representing models and we compare their expressive powers.

3.1 $\mathcal{T}$-interpretations

In the previous section, we represented Herbrand models by sets of constrained atoms. A ground atom is true in the model represented by a set $\mathcal{A}$ iff it belongs to the set of ground atoms denoted by $\mathcal{A}$. Here, we will use a different point of view: we represent the set of atoms $P(\overline{s})$ that are true in the model as the *solutions* of a formula $\mathcal{F}$, interpreted in a particular constraint language $\mathcal{T}$. More precisely, the interpretation of any predicate P, i.e. the set of tuples $(t_1, \ldots, t_n)$ such that $P(t_1, \ldots, t_n)$ is true in the model, will be denoted by a first-order formula $\mathcal{F}_P$ containing n free variables $x_1, \ldots, x_n$, such that $\mathcal{F}_P\{x_i \leftarrow t_i \mid 1 \leq i \leq n\}$ is true in $\mathcal{T}$ iff $P(t_1, \ldots, t_n)$ is true in the model.

If we want this model representation mechanism to be convenient and usable for applications, the constraint language $\mathcal{T}$ must of course be decidable, i.e. there must exist an algorithm deciding whether a given first-order formula $\mathcal{F}$ is true or false in $\mathcal{T}$; this is necessary for finding the truth values of atoms in interpretations. As we shall see, one of the properties of this kind of representation is that the set of solutions of a first-order formula in an interpretation $\mathcal{I}$, represented by a constraint on $\mathcal{T}$, can itself be represented by a constraint on $\mathcal{T}$. Therefore, numerous problems concerning the interpretations represented in this way can be reduced to the problem of deciding whether a given constraint has a solution or not. Consequently, it suffices to have a (hopefully efficient) algorithm for finding the truth value of a formula in $\mathcal{T}$ for solving these problems in a very natural and *uniform* way.

Indeed, constraints provide a very natural and convenient framework for model representation. Comparisons with the other model representation formalisms introduced in Chapter 3 will be provided later.

DEFINITION 5.13 A *constraint language* $\mathcal{T}$ is a pair $(\mathcal{C}, \mathcal{I})$ such that:

- $\mathcal{C}$ is a set of formulae, closed under negation, universal and existential quantification, conjunction and disjunction, and strictly containing all purely equational formulae.

- $\mathcal{I}$ is an *equational interpretation* of formulae in $\mathcal{C}$ (i.e. an interpretation which interprets $=$ as equality).

A constraint language $\mathcal{T} = (\mathcal{C}, \mathcal{I})$ is said to be *decidable* iff there exists an algorithm deciding whether a closed first-order formula $\mathcal{F} \in \mathcal{C}$ is true or false in $\mathcal{I}$. ♯

For the remaining part of this section, we assume that $\mathcal{T} = (\mathcal{C}, \mathcal{I})$ denotes a given decidable constraint language.

DEFINITION 5.14 Let $\mathcal{T} = (\mathcal{C}, \mathcal{I})$ be a constraint language. Let $\mathcal{F}$ be a first-order formula in $\mathcal{C}$. The *set of solutions* of $\mathcal{F}$ in $\mathcal{T}$ (denoted by $\mathcal{S}_{\mathcal{T}}(\mathcal{F})$) is the set of ground substitutions σ such that $\mathcal{I} \models \mathcal{F}\sigma$.

If $\mathcal{F}_1$ and $\mathcal{F}_2$ have the same set of solutions in $\mathcal{T}$, then $\mathcal{F}_1$ and $\mathcal{F}_2$ are said to be $\mathcal{T}$*-equivalent*, and we write $\mathcal{F}_1 \equiv_{\mathcal{T}} \mathcal{F}_2$. ♯

Constrained clauses on $\mathcal{T}$. The definition of c-clauses (syntax and semantics) can be extended in a straightforward way to clauses with constraints in $\mathcal{C}$ - instead of purely equational constraints on the finite tree algebra (as in Chapter 4). The interpretation of a c-clause $[\![C : \mathcal{X}]\!]$ will be defined as usual, i.e. $[\![C : \mathcal{X}]\!]$ denotes the set of ground c-clauses $C\sigma$ such that $\sigma \in \mathcal{S}_{\mathcal{T}}(\mathcal{X})$. For any c-clause $[\![C : \mathcal{X}]\!]$ we denote by $\mathcal{S}_{\mathcal{T}}([\![C : \mathcal{X}]\!])$ the set:

$$\{C\sigma \mid \sigma \in \mathcal{S}_{\mathcal{T}}(\mathcal{X})\}.$$

Similarly, for any set of c-clauses S, we write $\mathcal{S}_{\mathcal{T}}(S)$ for the set

$$\{\mathcal{S}_{\mathcal{T}}(C) \mid C \in S\}.$$

The notion of satisfiability, unsatisfiability, etc. as well as the inference and disinference rules described in the Section 3 of Chapter 4 can be extended to c-clauses on $\mathcal{T}$. The soundness and completeness results of the method are preserved (provided that $\mathcal{T}$ is decidable).

$\mathcal{T}$-representable interpretations. The notion of $\mathcal{T}$*-representable sets* is defined as follows:

DEFINITION 5.15 Let $n \in \mathbb{N}$ and $\mathcal{H} \subseteq H(\Sigma)^n$. $\mathcal{H}$ is said to be $\mathcal{T}$-*representable* iff there exists a first-order formula $\mathit{Form}(\mathcal{H})(x_1, \ldots, x_n)$ containing at most n free variables $x_1, \ldots, x_n$ such that for all tuples $(t_1, \ldots, t_n) \in H(\Sigma)^n$ we have

$$(t_1, \ldots, t_n) \in \mathcal{H} \Leftrightarrow \mathit{Form}(\mathcal{H})(x_1, \ldots, x_n)\{x_1 \leftarrow t_1, \ldots, x_n \leftarrow t_n\} \equiv_{\mathcal{T}} \top.$$

♯

THEOREM 3.1 *The set of $\mathcal{T}$-representable sets is closed under complement, union, intersection and projection, i.e, for any $n \in \mathbb{N}$ and $\mathcal{H} \subseteq H(\Sigma)^n$:*

- *if $\mathcal{H}$ is $\mathcal{T}$-representable, then $H(\Sigma) \setminus \mathcal{H}$ is $\mathcal{T}$-representable;*
- *if $\mathcal{H}$ and $\mathcal{H}'$ are $\mathcal{T}$-representable, then $\mathcal{H} \cup \mathcal{H}'$ is $\mathcal{T}$-representable;*
- *if $\mathcal{H}$ and $\mathcal{H}'$ are $\mathcal{T}$-representable, then $\mathcal{H} \cap \mathcal{H}'$ is $\mathcal{T}$-representable.*

- *If $\mathcal{H}$ is $\mathcal{T}$-representable, then the set*

$$\mathcal{H}_{|i} = \{(t_1, \ldots, t_{i-1}, t_{i+1}, \ldots, t_n) \mid \exists t_i.(t_1, \ldots, t_n) \in \mathcal{H}\}$$

is $\mathcal{T}$-representable.

Proof: The proof is immediate, since we have:

- $Form(H(\Sigma) \setminus \mathcal{H})(x_1, \ldots, x_n) = \neg Form(\mathcal{H})(x_1, \ldots, x_n)$,
- $Form(\mathcal{H} \cup \mathcal{H}')(x_1, \ldots, x_n) =$
  $Form(\mathcal{H})(x_1, \ldots, x_n) \vee Form(\mathcal{H}')(x_1, \ldots, x_n)$,
- $Form(\mathcal{H} \cap \mathcal{H}')(x_1, \ldots, x_n) =$
  $Form(\mathcal{H})(x_1, \ldots, x_n) \wedge Form(\mathcal{H}')(x_1, \ldots, x_n)$,
- $Form(\mathcal{H}_{|i})(x_1, \ldots, x_{i-1}, x_{i+1}, \ldots, x_n) = \exists x_i.Form(\mathcal{H})(x_1, \ldots, x_n)$.

◇

NOTATION 3.1 *Let $\mathcal{I}$ be an equational Herbrand interpretation, and P be a n-ary predicate symbol. We denote:*

$$\mathcal{I}(P) = \{t \in H(\Sigma) \mid \mathcal{I} \models P(t)\}.$$

DEFINITION 5.16 An equational Herbrand interpretation $\mathcal{I}$ is said to be $\mathcal{T}$-representable iff for all predicate symbols P, the set $\mathcal{I}(P)$ is $\mathcal{T}$-representable. If $\mathcal{I}$ is $\mathcal{T}$-representable it is called a $\mathcal{T}$-interpretation. ♯

$\mathcal{T}$-interpretations correspond exactly to those interpretations that can be described by a finite set of constrained atoms, as evidenced by the following theorem, which proves the equivalence between constraint representation and CARMs.

DEFINITION 5.17 Let S be a set of c-atoms. For all n-place predicate symbols $P \in \mathrm{PS}(S)$ we define

$$S(P) \quad = \quad \{(t_1, \ldots, t_n) \mid (t_1, \ldots, t_n) \in H(\Sigma)^n, P(t_1, \ldots, t_n) \in \mathcal{S}_{\mathcal{T}}(S)\}. \;\sharp$$

The next theorem shows that $\mathcal{T}$-representable Herbrand interpretations are exactly those that can be represented by CARMs.

THEOREM 3.2 *Let S be a CARM. Then for all predicate symbols $P \in \mathrm{PS}(S)$, the set $S(P)$ is $\mathcal{T}$-representable. Conversely, for all $\mathcal{T}$-representable set of ground atoms $\mathcal{I}$ there exists a CARM S such that $\mathcal{S}_{\mathcal{T}}(S) = \mathcal{I}$.*

Proof:

- We have to show that $S(P)$ is $\mathcal{T}$-representable. We consider the following equational formula on $\mathcal{T}$.

$$\mathcal{F}^+ = \bigvee_{[\![P(\bar{t}):\mathcal{X}]\!]\in S} \exists\overline{y}.\overline{x} = \bar{t} \wedge \mathcal{X}$$

  where $\overline{y} = \mathcal{V}ar([\![P(\bar{t}) : \mathcal{X}]\!])$.

  We have by definition : $\sigma \in \mathcal{S}_{\mathcal{T}}(\mathcal{F}^+)$ iff $P(\overline{x})\sigma \in \mathcal{S}_{\mathcal{T}}(S)$, i.e. $\overline{x}\sigma \in S(P)$. Hence $S(P)$ is $\mathcal{T}$-representable.

- Conversely, let $\mathcal{I}$ be a $\mathcal{T}$-representable set of ground atoms. Then we define a CARM S by:

$$S = \{[\![P(\overline{x}) : \mathit{Form}(\mathcal{I}(P))(\overline{x})]\!] \mid P \in \mathrm{PS}(S)\}.$$

  By definition of S we obtain for all ground atoms A:

$$A \in \mathcal{S}_{\mathcal{T}}(S) \text{ iff } \mathcal{I} \models A.$$

  Hence $\mathcal{S}_{\mathcal{T}}(S) = \mathcal{I}$. ◇

EXAMPLE 5.18 We denote by $\mathcal{A}$ the following CARM:

$$\mathcal{A} = \{[\![P(x,y) : x \neq y]\!], [\![Q(a) : \top]\!], [\![Q(b) : \top]\!], [\![R(f(x)) : \top]\!]\}.$$

The model represented by $\mathcal{A}$ can be described by the following equational formulae:

$$\mathit{Form}(S(P))(x,y) = x \neq y$$
$$\mathit{Form}(S(Q))(x) = (x = a) \vee (x = b)$$
$$\mathit{Form}(S(R))(x) = \exists u.x = f(u).$$

♯

3.2 The Formula Evaluation Problem

Many problems of $\mathcal{T}$-interpretations can be reduced to the solution of formulae in $\mathcal{T}$ in a simple way. One of them is the equivalence problem, i.e. the problem to decide whether two given representations $\mathcal{A}$ and $\mathcal{B}$ are equivalent. Indeed, the equivalence problem can be reduced to the problem of deciding whether for all $P \in \mathrm{PS}(\mathcal{A} \cup \mathcal{B})$, the sets $\mathcal{A}(P)$ and $\mathcal{B}(P)$ are equal i.e. whether the corresponding formulae $\mathit{Form}(\mathcal{A}(P))(\overline{x})$ and $\mathit{Form}(\mathcal{B}(P))(\overline{x})$ are $\mathcal{T}$-equivalent. The equivalence problem for $\mathcal{T}$ is therefore reduced to equivalence problem for first-order formulae in $\mathcal{T}$, which is decidable since $\mathcal{T}$ is decidable.

In this section, we show how to solve the formula evaluation problem, i.e. the problem of finding the truth value of a formula in a $\mathcal{T}$-interpretation $\mathcal{I}$. To this aim we define an algorithm computing the solutions of a formula in a $\mathcal{T}$-interpretation. Its principle is to translate the formula $\mathcal{F}$ into a first-order formula $\mathcal{F}'$ such that the set of solutions of $\mathcal{F}$ in $\mathcal{I}$ is the same as the set of solutions of $\mathcal{F}'$ in $\mathcal{T}$. More precisely, the definition 5.19 below introduces two formulae $\phi^+_{\mathcal{M}}(\mathcal{F})$ and $\phi^-_{\mathcal{M}}(\mathcal{F})$ stating conditions for a formula $\mathcal{F}$ to be true (resp. false) in $\mathcal{M}$.

DEFINITION 5.19 Let $\mathcal{M}$ be a $\mathcal{T}$-interpretation and $\mathcal{F}$ be a formula. Let $\phi^+_{\mathcal{M}}(\mathcal{F})$ and $\phi^-_{\mathcal{M}}(\mathcal{F})$ be equational formulae defined as follows:

- If $\mathcal{F}$ is of the form $P(\bar{t})$, then

$$\phi^+_{\mathcal{M}}(\mathcal{F}) = \exists \bar{x}.\bar{t} = \bar{x} \wedge \mathit{Form}(\mathcal{I}(P))(\bar{x})$$

$$\phi^-_{\mathcal{M}}(\mathcal{F}) = \exists \bar{x}.\bar{t} = \bar{x} \wedge \neg \mathit{Form}(\mathcal{I}(P))(\bar{x})$$

- If $\mathcal{F} = \mathcal{F}_1 \vee \mathcal{F}_2$:

$$\phi^+_{\mathcal{M}}(\mathcal{F}) = \phi^+_{\mathcal{M}}(\mathcal{F}_1) \vee \phi^+_{\mathcal{M}}(\mathcal{F}_2)$$

$$\phi^-_{\mathcal{M}}(\mathcal{F}) = \phi^-_{\mathcal{M}}(\mathcal{F}_1) \wedge \phi^-_{\mathcal{M}}(\mathcal{F}_2)$$

- If $\mathcal{F} = \mathcal{F}_1 \wedge \mathcal{F}_2$:

$$\phi^+_{\mathcal{M}}(\mathcal{F}) = \phi^+_{\mathcal{M}}(\mathcal{F}_1) \wedge \phi^+_{\mathcal{M}}(\mathcal{F}_2)$$

$$\phi^-_{\mathcal{M}}(\mathcal{F}) = \phi^-_{\mathcal{M}}(\mathcal{F}_1) \vee \phi^-_{\mathcal{M}}(\mathcal{F}_2)$$

- If $\mathcal{F} = \exists x.\mathcal{F}_1$:

$$\phi^+_{\mathcal{M}}(\mathcal{F}) = \exists x.\phi^+_{\mathcal{M}}(\mathcal{F}_1)$$

$$\phi^-_{\mathcal{M}}(\mathcal{F}) = \forall x.\phi^-_{\mathcal{M}}(\mathcal{F}_1)$$

- if $\mathcal{F} = \forall x.\mathcal{F}_1$:

$$\phi^+_{\mathcal{M}}(\mathcal{F}) = \forall x.\phi^+_{\mathcal{M}}(\mathcal{F}_1)$$

$$\phi^-_{\mathcal{M}}(\mathcal{F}) = \exists x.\phi^-_{\mathcal{M}}(\mathcal{F}_1)$$

- If $\mathcal{F} = \neg \mathcal{F}_1$:

$$\phi^+_{\mathcal{M}}(\mathcal{F}) = \phi^-_{\mathcal{M}}(\mathcal{F}_1)$$

$$\phi^-_{\mathcal{M}}(\mathcal{F}) = \phi^+_{\mathcal{M}}(\mathcal{F}_1)$$

♯

THEOREM 3.3 *Let $\mathcal{M}$ be a $\mathcal{T}$-interpretation and $\mathcal{F}$ be a first-order formula. Let σ be a substitution. If $\sigma \in \mathcal{S}_{\mathcal{T}}(\phi^+_{\mathcal{M}}(\mathcal{F}))$, then $\mathcal{M} \models \mathcal{F}\sigma$. If $\sigma \in \mathcal{S}_{\mathcal{T}}(\phi^-_{\mathcal{M}}(\mathcal{F}))$, then $\mathcal{M} \models \neg\mathcal{F}\sigma$.*

Proof: By structural induction on the set of formulae.

- Let $\mathcal{F} = P(\bar{t})$. Assume that $\sigma \in \mathcal{S}_{\mathcal{T}}(\phi^+_{\mathcal{M}}(\mathcal{F}))$. Then, by definition, $\sigma \in \mathcal{S}_{\mathcal{T}}(\exists\bar{x}.\bar{t} = \bar{s} \wedge Form(\mathcal{I}(P))(\bar{x})$; hence $\{\bar{x} \to \bar{t}\sigma\}$ is a solution of $Form(\mathcal{I}(P))(\bar{x})$, therefore $P(\bar{t})\sigma \in \mathcal{I}(P)$. Hence $\mathcal{M} \models \mathcal{F}\sigma$. The proof is similar for $\mathcal{S}_{\mathcal{T}}(\phi^-_{\mathcal{M}}(\mathcal{F}))$.
- $\mathcal{F} = \mathcal{F}_1 \vee \mathcal{F}_2$: if $\sigma \in \mathcal{S}_{\mathcal{T}}(\phi^+_{\mathcal{M}}(\mathcal{F}))$, then either $\sigma \in \mathcal{S}_{\mathcal{T}}(\phi^+_{\mathcal{M}}(\mathcal{F}_1))$ or $\sigma \in \mathcal{S}_{\mathcal{T}}(\phi^+_{\mathcal{M}}(\mathcal{F}_2))$. By induction hypothesis, either $\mathcal{M} \models \mathcal{F}_1\sigma$ or $\mathcal{M} \models \mathcal{F}_2\sigma$. Hence, $\mathcal{M} \models \mathcal{F}_1\sigma \vee \mathcal{F}_2\sigma$, i.e. $\mathcal{M} \models \mathcal{F}\sigma$. If $\sigma \in \mathcal{S}_{\mathcal{T}}(\phi^-_{\mathcal{M}}(\mathcal{F}))$, then $\sigma \in \mathcal{S}_{\mathcal{T}}(\phi^-_{\mathcal{M}}(\mathcal{F}_1))$ and $\sigma \in \mathcal{S}_{\mathcal{T}}(\phi^-_{\mathcal{M}}(\mathcal{F}_2))$. By induction hypothesis, $\mathcal{M} \models \neg\mathcal{F}_1\sigma$ and $\mathcal{M} \models \neg\mathcal{F}_2\sigma$. Hence $\mathcal{M} \models \neg\mathcal{F}_1\sigma \wedge \neg\mathcal{F}_2\sigma$, i.e. $\mathcal{M} \models \neg(\mathcal{F}_1 \vee \mathcal{F}_2)\sigma$, and $\mathcal{M} \models \neg\mathcal{F}\sigma$.
- The proof is similar for $\mathcal{F} = \mathcal{F}_1 \wedge \mathcal{F}_2$.
- $\mathcal{F} = \neg\mathcal{F}_1$: if $\sigma \in \mathcal{S}_{\mathcal{T}}(\phi^+_{\mathcal{M}}(\mathcal{F}))$, then $\sigma \in \mathcal{S}_{\mathcal{T}}(\phi^-_{\mathcal{M}}(\mathcal{F}_1))$, hence $\mathcal{M} \models \neg\mathcal{F}_1\sigma$, i.e. $\mathcal{M} \models \mathcal{F}\sigma$. If $\sigma \in \mathcal{S}_{\mathcal{T}}(\phi^-_{\mathcal{M}}(\mathcal{F}))$, then $\sigma \in \mathcal{S}_{\mathcal{T}}(\phi^+_{\mathcal{M}}(\mathcal{F}_1))$, hence $\mathcal{M} \models \mathcal{F}_1\sigma$, i.e. $\mathcal{M} \models \neg\neg\mathcal{F}_1\sigma$.
- $\mathcal{F} = \exists x.\mathcal{F}_1$: if $\sigma \in \mathcal{S}_{\mathcal{T}}(\phi^+_{\mathcal{M}}(\mathcal{F}))$, then $\sigma \in \mathcal{S}_{\mathcal{T}}(\exists x.\phi^+(\mathcal{F}_1))$, i.e. there exists a closed term t such that $\sigma \cup \{x \leftarrow t\} \in \mathcal{S}_{\mathcal{T}}(\phi^+_{\mathcal{M}}(\mathcal{F}_1))$. Therefore $\mathcal{M} \models \mathcal{F}_1\{x \leftarrow t\}\sigma$. Hence $\mathcal{M} \models (\exists x.\mathcal{F}_1)\sigma$. If $\sigma \in \mathcal{S}_{\mathcal{T}}(\phi^-_{\mathcal{M}}(\mathcal{F}))$, then $\sigma \in \mathcal{S}_{\mathcal{T}}(\forall x.\phi^-(\mathcal{F}_1))$, i.e. for all ground terms t, $\sigma\{x \leftarrow t\} \in \mathcal{S}_{\mathcal{T}}(\phi^-(\mathcal{F}_1))$. Hence $\forall t.\mathcal{M} \models \neg\mathcal{F}_1\{x \leftarrow t\}\sigma$ and $\mathcal{M} \models (\forall x.\neg\mathcal{F}_1)\sigma$. i.e. $\mathcal{M} \models \neg\mathcal{F}\sigma$.
- The proof is similar for $\mathcal{F} = \forall x.\mathcal{F}_1$. ◇

THEOREM 3.4 *Let $\mathcal{F}$ be a first-order formula and let $\mathcal{I}$ be a $\mathcal{T}$-interpretation. Then $\phi^+_{\mathcal{I}}(\mathcal{F}) \vee \phi^-_{\mathcal{I}}(\mathcal{F}) \equiv \top$.*

Proof: By structural induction on $\mathcal{F}$.

- $\mathcal{F}$ is an atomic formula $P(\bar{t})$ for $P \in \mathrm{PS}_n$: By definition $\sigma \in \mathcal{S}_{\mathcal{T}}(\phi^+_{\mathcal{I}}(\mathcal{F}))$ iff $\bar{t}\sigma \in \mathcal{I}(P)$. But this is equivalent to $\bar{t}\sigma \notin H^n \setminus \mathcal{I}(P)$, hence to $\sigma \notin \mathcal{S}_{\mathcal{T}}(\phi^-_{\mathcal{I}}(\mathcal{F}))$.
- $\mathcal{F} = \mathcal{F}_1 \vee \mathcal{F}_2$: if $\sigma \notin \mathcal{S}_{\mathcal{T}}(\phi^+_{\mathcal{I}}(\mathcal{F}))$, then $\sigma \notin \mathcal{S}_{\mathcal{T}}(\phi^+_{\mathcal{I}}(\mathcal{F}_1))$ and $\sigma \notin \mathcal{S}_{\mathcal{T}}(\phi^+_{\mathcal{I}}(\mathcal{F}_2))$. By induction hypothesis, it implies that $\sigma \in \mathcal{S}_{\mathcal{T}}(\phi^-_{\mathcal{I}}(\mathcal{F}_1))$ and $\sigma \in \mathcal{S}_{\mathcal{T}}(\phi^-_{\mathcal{I}}(\mathcal{F}_2))$, i.e. $\sigma \in \mathcal{S}_{\mathcal{T}}(\phi^-_{\mathcal{I}}(\mathcal{F}))$.

- The proof is similar for $\mathcal{F} = \mathcal{F}_1 \wedge \mathcal{F}_2$.
- $\mathcal{F} = \neg\mathcal{F}'$: then $\sigma \notin \mathcal{S}_{\mathcal{T}}(\phi^+_{\mathcal{I}}(\mathcal{F}))$ implies $\sigma \notin \mathcal{S}_{\mathcal{T}}(\phi^-_{\mathcal{I}}(\mathcal{F}'))$ hence (by induction hypothesis) $\sigma \in \mathcal{S}_{\mathcal{T}}(\phi^+_{\mathcal{I}}(\mathcal{F}'))$ and $\sigma \in \mathcal{S}_{\mathcal{T}}(\phi^-_{\mathcal{I}}(\mathcal{F}))$.
- $\mathcal{F} = \exists x.\mathcal{F}'$: if $\sigma \notin \mathcal{S}_{\mathcal{T}}(\phi^+_{\mathcal{I}}(\mathcal{F}))$ then, by definition of terms accepted in $H(\Sigma)$, we have $\sigma \notin \mathcal{S}_{\mathcal{T}}(\phi^+_{\mathcal{I}}(\mathcal{F}'\{x \leftarrow t\}))$ for all $t \in H(\Sigma)$. Hence by induction hypothesis for all $t \in H(\Sigma)$: $\sigma \in \mathcal{S}_{\mathcal{T}}(\phi^-_{\mathcal{I}}(\mathcal{F}\{x \leftarrow t\}))$, i.e. $\sigma \in \mathcal{S}_{\mathcal{T}}(\phi^-_{\mathcal{I}}(\mathcal{F}))$.
- The proof is similar for $\mathcal{F} = \forall x.\mathcal{F}'$. ◇

EXAMPLE 5.20 Let us consider the formula

$$\mathcal{F} : P(x) \vee \neg Q(x, y)$$

and the equational Herbrand interpretation $\mathcal{I}$ defined by:

- $P(x)$ is true in $\mathcal{I}$ if $x = a$ (and false otherwise).
- $Q(x, y)$ is false in $\mathcal{I}$ if $x = y$ (and true otherwise).

Then

$$\phi^+_{\mathcal{I}}(\mathcal{F}) \equiv (x = a \vee x = y)$$

and

$$\phi^-_{\mathcal{I}}(\mathcal{F}) \equiv (x \neq a \wedge x \neq y).$$

But this implies

$$\phi^+_{\mathcal{I}}(\mathcal{F}) \vee \phi^-_{\mathcal{I}}(\mathcal{F}) \equiv \top.$$

♯

Finally we obtain the following result, that allows us to reduce the formula evaluation problem to the solution of first-order constraints in $\mathcal{T}$.

COROLLARY 3.1 *Let $\mathcal{I}$ be an equational Herbrand interpretation and $\mathcal{F}$ a formula. Moreover let σ be a substitution on the domain $V(\mathcal{F})$. Then*

$$\mathcal{I} \models \mathcal{F}\sigma \textit{ iff } \sigma \in \mathcal{S}_{\mathcal{T}}(\phi^+_{\mathcal{I}}(\mathcal{F})).$$

Proof: It suffices to show that $\mathcal{I} \models \mathcal{F}\sigma$ implies $\sigma \in \mathcal{S}_{\mathcal{T}}(\phi^+_{\mathcal{I}}(\mathcal{F}))$ (the converse is a consequence of Theorem 3.3). Assume that $\mathcal{I} \models \mathcal{F}\sigma$ and that $\sigma \notin \mathcal{S}_{\mathcal{T}}(\phi^+_{\mathcal{I}}(\mathcal{F}))$. Then, by Theorem 3.4, we have $\sigma \in \mathcal{S}_{\mathcal{T}}(\phi^-_{\mathcal{I}}(\mathcal{F}))$. Therefore $\mathcal{I} \models \neg\mathcal{F}\sigma$ by Theorem 3.3. Hence $\mathcal{I}$ validates simultaneously $\mathcal{F}\sigma$ and $\neg\mathcal{F}\sigma$, which is impossible. ◇

3.3 Examples of Decidable Constraint Languages

In this section, we present some examples of constraint languages that may be used for model representation and model building.

3.3.1 Finite tree algebra

In chapter 4, we used equational formulae, interpreted on the algebra of finite trees to represent Herbrand models. This corresponds to the constraint language

$$\mathcal{T}_{\emptyset} = (\mathcal{C}, \mathcal{I}),$$

where $\mathcal{C}$ denotes the set of equational formulae and $\mathcal{I}$ is the finite tree algebra (i.e. $\mathcal{I}$ is the Herbrand interpretation). In Section 1.2 we have shown that $\mathcal{T}_{\emptyset}$ is decidable.

The constraint languages that we consider in the rest of this section are non-trivial extensions of the language of equational formulae. They have been designed to express infinite sequences of terms, with a "regular" structure. We do not give any proof of the decidability of the theories nor any description of the corresponding solution algorithms, because this is beyond the scope of this book, but we provide definitions of the syntax and semantics of the considered languages, examples of applications and comparisons of their expressive powers w.r.t. model building. References to papers containing detailed proofs are given.

3.3.2 Equational constraints with equations: $\mathcal{T}_E$

A first, quite natural, extension of $\mathcal{T}_{\emptyset}$ is to interpret formulae on more complex structures than Herbrand universes. An interesting idea is to consider a set of equations E and to interpret equalities modulo this set of equations (thus, if E contains the equations $f(a) = g(b)$ and $b = c$, then $f(a)$ and $g(c)$ must be equal). This corresponds to interpretations of equational formulae over a quotient of the Herbrand universe defined by some congruence $=_E$.

More formally, let E be a set of equations. Given two terms t, s, we write $t =_E s$ iff $E \models (t = s)$. Obviously, $=_E$ is a congruence on the considered signature Σ. For any term t, we denote by $[t]_E$ the equivalence class of t modulo $=_E$.

DEFINITION 5.21 We denote by $\mathcal{T}_E$ the constraint language $(\mathcal{C}, \mathcal{I}_E)$, where $\mathcal{C}$ denotes the set of equational formulae and $\mathcal{I}_E$ is the interpretation defined as follows:

- The domain of $\mathcal{I}_E$ is the set $\{[t]_E \mid t \in H\}$.
- For all $f \in \Sigma$, $t_{\mathcal{I}_E}(f(t_1, \dots, t_n)) = [f(t_1, \dots, t_n)]_E$. ♯

Note that $\mathcal{I}_E$ is well-defined, since $=_E$ is a congruence.

Of course, $\mathcal{T}_E$ is undecidable in general but there are some particular cases in which decision algorithms do exist. For example, algorithms have been provided for solving equational problems in the case in which E is a *ground* equational theory [Comon, 1993] or a *shallow* theory (a theory is said to be *shallow* iff variables occurs at depth at most one) [Comon et al., 1994]. Thus $\mathcal{T}_E$ is decidable, if E is ground or shallow.

Note that interpretations represented by GEARMs (see Section 2) are particular cases of $\mathcal{T}_E$-interpretation, where E is a set of ground equations and where, for all predicate symbols P, the set $\mathcal{I}(P)^+$ is represented by a formula of the form $\bigwedge_{i=1}^{n} x_i = t_i$, for ground terms t_i.

3.3.3 Equational formulae with membership constraints: $\mathcal{T}_\in$

The idea is to enrich the language of equational first-order formulae by adding *membership constraints*, i.e. atomic formulae of the form

$$t \in S$$

where t is a term (possibly containing variables) and S is a set of terms. We restrict our investigation to first-order constraints, hence no quantification over sets of terms is possible. Therefore, any set of terms S occurring in the formula must be a *constant*. Its value (i.e. the explicit subset of $H(\Sigma)$ denoted by S) is given by a tree automaton $\mathcal{A}_S$ (or equivalently by a regular term grammar), which must be specified together with the formula. Informally speaking, a tree automaton $\mathcal{A}$ is a set of rewriting rules. The set of ground terms that are denoted by a given set symbol S is the set of terms that can be rewritten to S using rules in $\mathcal{A}$.

EXAMPLE 5.22 The following formula is an equational formula with membership constraints.

$$\mathcal{F} : \exists y, z.(x = y + z \wedge y \in S \wedge z \in S)$$

where S is defined by the following automaton.

$$\begin{array}{lcl} succ(S) & \rightarrow & S \\ 0 & \rightarrow & S \end{array}$$

S denotes the set of natural number, i.e. the terms of the form $succ^n(0)$. The solutions of $\mathcal{F}$ is the set of substitution $\sigma : x \rightarrow t + s$, where t, s are natural numbers. ♯

Syntax and semantics. In this section, we give a formal definition of the language of equational formulae with membership constraints. First we introduce a formal definition of the notion of *tree automata*, which can be considered as a generalization of standard finite word automata.

DEFINITION 5.23 Let Σ be a set of function symbols and let $\mathcal{S}$ be a set of symbols of arity 0, called *set symbols*. A *tree automaton* $\mathcal{A}$ is a set of *production rules* that are either of the form

- $f(S_1, \ldots, S_n) \rightarrow S$, where $(S_1, \ldots, S_n, S) \in \mathcal{S}^{n+1}$ and f is a function symbol of arity n,
- or of the form $S' \rightarrow S$ where $(S', S) \in \mathcal{S}^2$.

The terms $f(S_1, \ldots, S_n)$ and S' are called *production terms*. ♯

DEFINITION 5.24 Let $\mathcal{A}$ be a tree automaton. An $\mathcal{A}$-*derivation* is a sequence of terms $(t_1, \ldots, t_n)$ over the signature $\Sigma \cup \mathcal{S}$ such that for all $i \in \{1, \ldots, n-1\}$, t_{i+1} is obtained from t_i by replacing one occurrence of a term t by a set symbol S provided there exist in a rule of the form $t \rightarrow S$ in $\mathcal{A}$.

A ground term t is said to be *accepted* by $\mathcal{A}$ at S, iff there exists an $\mathcal{A}$-derivation of the form $(t, \ldots, S)$. ♯

EXAMPLE 5.25 Let $\mathcal{A}$ be the following tree automaton:

$$\begin{array}{lcl} N + N & \rightarrow & N \\ 0 & \rightarrow & N \\ \mathit{succ}(N) & \rightarrow & N \\ \mathit{or}(B, B) & \rightarrow & B \\ \mathit{and}(B, B) & \rightarrow & B \\ \mathit{not}(B) & \rightarrow & B \\ \mathit{true} & \rightarrow & B \\ \mathit{false} & \rightarrow & B \end{array}$$

The set $\mathit{succ}(0) + 0$ is accepted by $\mathcal{A}$ at N. Indeed, we have the following $\mathcal{A}$-derivation:

$$\begin{array}{lcl} \mathit{succ}(0) + 0 & \rightarrow & \mathit{succ}(N) + 0 \\ & \rightarrow & N + 0 \\ & \rightarrow & N + N \\ & \rightarrow & N. \end{array}$$

The set $\mathit{and}(\mathit{true}, \mathit{or}(\mathit{false}, \mathit{true}))$ is accepted by $\mathcal{A}$ at B.

$$\begin{array}{lcl} \mathit{and}(\mathit{true}, \mathit{or}(\mathit{false}, \mathit{true})) & \rightarrow & \mathit{and}(B, \mathit{or}(\mathit{false}, \mathit{true})) \\ & & \mathit{and}(B, \mathit{or}(B, \mathit{true})) \\ & & \mathit{and}(B, \mathit{or}(B, B)) \\ & & \mathit{and}(B, B) \\ & & B. \end{array}$$

The set of terms accepted by $\mathcal{A}$ at N is the set of arithmetic expressions over $\{0, succ, +\}$, whereas the set of terms accepted by $\mathcal{A}$ at B is the set of Boolean expressions over $\{and, or, neg, true, false\}$. ♯

It is known that tree automata are essentially equivalent to regular term grammars, i.e. they have *exactly* the same expressive power. Moreover, tree automata share all the nice properties of word automata, i.e. they are closed under the usual boolean operations (complement, union, intersection), and the emptiness problem (i.e. the problem of deciding whether the set of terms accepted by a given automata is empty) is decidable.

DEFINITION 5.26 Let Σ be a set of function symbols and let $\mathcal{S}$ be a set of set symbols. The set of *equational formulae with membership constraints* $F_{\in}$ is the least set satisfying the following properties.

- $\top$ (true) and $\bot$ (false) are in $F_{\in}$.
- If s and t are two terms, then $t = s$ is in $F_{\in}$.
- If t is a term and S is a set symbol, then $t \in S$ is in $F_{\in}$.
- If $\mathcal{F}$ is in $F_{\in}$, then $\neg\mathcal{F} \in F_{\in}$.
- If $\mathcal{F}$ is in $F_{\in}$, and x is a variable, then $\exists x.\mathcal{F}$ and $\forall x.\mathcal{F}$ are in $F_{\in}$.
- If $\mathcal{F}$ and $\mathcal{G}$ are in $F_{\in}$, then $\mathcal{F} \vee \mathcal{G}$ and $\mathcal{F} \wedge \mathcal{G}$ are in $F_{\in}$. ♯

DEFINITION 5.27 Let Σ be a set of function symbols and let $\mathcal{S}$ be a set of symbols, called *states*. Let $\mathcal{F} \in F_{\in}$ and let $\mathcal{A}$ be a tree automaton.

We denote by $\mathcal{T}_{\in}$ the constraint language $(F_{\in}, \mathcal{I}_{\in})$ where $\mathcal{I}_{\in}$ is the interpretation defined as follows:

- The domain of $\mathcal{I}_{\in}$ is the set of ground terms (i.e. the Herbrand universe).
- For any function symbol f,

$$t_{\mathcal{I}_{\in}}(f(t_1, \ldots, t_n)) = f(t_{\mathcal{I}_{in}}(t_1), \ldots, t_{\mathcal{I}_{\in}}(t_n)).$$

- For any term t and for any state symbol S, $t \in S$ is true in $\mathcal{I}_{\in}$ iff t belongs to the set of terms accepted by the tree automaton $\mathcal{A}$ at S.

♯

EXAMPLE 5.28 Let

$$\mathcal{A} = \{a \rightarrow S,\ f(f(S)) \rightarrow S)\}$$

be a tree automaton. The set of terms accepted by $\mathcal{A}$ at S is

$$\{f^{2n}(a) \mid n \in \mathbb{N}\}$$

The formula

$$\mathcal{F} : x = y \vee (x = f(y) \wedge y \in S)$$

is in $F_{\in}$ (x and y are free variables).

The set of solutions of $\mathcal{F}$ in $\mathcal{T}_{\in}$ is the set of ground substitutions σ such that

- either $x\sigma = y\sigma$;
- or $x\sigma = f(y\sigma)$ and $y\sigma$ is accepted by $\mathcal{A}$ at S.

For example

$$\{x \to f(f(f(a))), y \to f(f(a))\}$$

is a solution of $\mathcal{F}$ but $\{x \to f(f(a)), y \to f(a)\}$ is not. ♯

Next theorem states that $\mathcal{T}_{\in}$ is decidable.

THEOREM 3.5 *Given a signature Σ and a finite tree automaton $\mathcal{A}$, there exists an algorithm to decide whether a given equational formula with membership constraint $\mathcal{F}$ has a solution or not w.r.t. $\Sigma, \mathcal{A}$.*

Proof: See [Comon, 1990, Comon, 1988] or [Comon and Delor, 1994]. ◇

$\mathcal{T}_{\in}$ is strictly more expressive than $\mathcal{T}_{\emptyset}$, as we have seen before. Moreover, $\mathcal{T}_{\in}$ is also strictly more expressive than tree automata. Take, for example, the interpretation

$$\mathcal{I}: P(x, y) \text{ is true iff } x \neq y.$$

$\mathcal{I}$ can be represented by an equational formula but *not* by a tree automaton. As we have seen before, interpretations representable by term grammars correspond exactly to interpretations that have a finite realization (in the sense that for any $\mathcal{T}_{\in}$-representable interpretation $\mathcal{I}$, there exists a finite interpretation validating exactly the same non-equational atoms as $\mathcal{I}$, see Chapter 3 for more details); however this is not the case for equational formulae. But equational formulae with membership constraints are also more expressive than the *union* of tree automata and equational formulae, i.e. there exists interpretations that can be represented by an equational formulae with membership constraints but can neither be represented by equational formulae nor by tree automata, as evidenced by the following example.

EXAMPLE 5.29 Let us consider the following interpretation $\mathcal{I}$ defined on $H(\Sigma)$, where $\Sigma = \{0, succ\}$. $\mathcal{I} \models P(x, y, z)$ iff:

- x and y are the same ground terms and z is of the form $succ^{2n}(0)$.
- x and y are distinct ground terms and z is of the form $succ^{2n+1}(0)$.

Clearly, $\mathcal{I}$ has no finite model, hence cannot be represented by a term grammar. Moreover, it cannot be represented by an equational formula. The equational formula with membership constraints representing this interpretation is:

$$(x = y \wedge x \in S) \vee (x \neq y \wedge x \notin S)$$

where S is defined as in Example 5.22. ♯

Extensions of tree automata. Several extensions of tree automata have been considered. For example, it is possible to extend this notion by allowing equality or disequality tests between the subterms of a production term. This defines a new class of tree automata, the class of *tree automata with equality tests* [Mongy, 1981]. Unfortunately, the emptiness problem - i.e. the problem of deciding whether the set of ground terms accepted by a given automaton is empty or not - has been shown to be undecidable for this class of automata. As a consequence, the corresponding first-order theory of membership constraints defined on these automata is *undecidable* (indeed the emptiness problem can be expressed by a first-order formula, for example $\neg\exists x.x \in S$). However, it is possible to restrict the class of equality tests in order to preserve the properties of standard tree automata. In particular, if only equality or disequality tests between *brothers* (i.e. between terms occurring at depth at most 1 in production terms) are allowed, then the emptiness problem becomes decidable. Moreover the obtained language is closed under all Boolean transformations (union, intersection, complement) [Bogaert and Tison, 1992]. In [Comon and Delor, 1994], it is proven that $\mathcal{T}_\in$ is decidable.

3.3.4 Equational formulae with terms having integer exponents: $\mathcal{T}_I$

Since the 90's (and independently of model building or model representation), many different formalisms have been proposed in order to give a finite representation of infinite sequences of structurally similar terms: recurrence terms [Chen et al., 1990], ω-terms [Chen and Hsiang, 1991], I-terms [Comon, 1992], R-terms [Salzer, 1992], and primal grammars [Hermann, 1992, Hermann, 1994]. These formalisms have been used in order to study infinite loops and divergence, for example in theorem proving [Salzer, 1994] or in term rewriting. For some of these languages, unification is known to be decidable, and algorithms have been proposed for solving *unification* problems. A very large bibliography can be found in [Hermann, 1994, Salzer, 1994].

Term schematization can be used to enrich the language of equational formulae. As an example, we will choose the language of *terms with integer exponents* (or *I-terms* for short), which has been introduced by H. Comon [Comon, 1995] and then extended in [Peltier, 1997a]. The theory of I-terms is the most powerful term schematization theory that is currently known to be decidable.

Informal presentation and example. The *terms with integer exponents* (or *I-terms*) have been defined in [Comon, 1995]. Roughly speaking, these terms are of the form:

$$t^n.u$$

where n is an integer variable, u is an I-term, and t is a term containing at a position p a symbol $\diamond$, called a "hole". Terms can be iterated n times along the path p, i.e. the interpretation of a term $t^n.u$ is inductively defined as follows: either $t^n.u = u$ if $n = 0$, or $t^n.u = t[t^{n-1}.u]_p$ if $n > 0$.

EXAMPLE 5.30 (I-terms)
The term $t = f(x, g(\diamond))^n.a$ is an I-term. Instances of t are:

$f(x, g(a))$	for $n = 1$
$f(x, g(f(x, g(a))))$	for $n = 2$
$f(x, g(f(x, g(f(x, g(a))))))$	for $n = 3$
etc.	

♯

In [Comon, 1995], only I-terms with one hole are considered. Nevertheless, it is possible to allow several occurrences of $\diamond$ in the context. The expressive power of the extended formalism is strictly greater than the initial one. It allows for example to express sequences of terms such as the set of all complete binary trees:

$$t = f(\diamond, \diamond)^n.a$$

Instances of t are:

$f(a, a)$
$f(f(a, a), f(a, a))$
$f(f(f(a, a), f(a, a)), f(f(a, a), f(a, a)))$
etc.

Remark: The obtained formalism is in fact strictly equivalent to the R-terms of [Salzer, 1994] (with a slightly different syntax); but we keep the notation above that appears to be more readable. Using the syntax of R-terms, an I-term $t^n.s$ is denoted by $\Phi(t, n, s)$, with exactly the same semantics. ♯

Equational Formulae with I-Terms.

DEFINITION 5.31 Let V_N be an infinite set of symbols of arity 0 denoting *integer variables.* The *set of arithmetic terms* T_N is the least set of terms built as usual on the set of variables V_N and on the set of function symbols $\{0, succ, +\}$, of arity $0, 1, 2$ respectively. ♯

Remark: According to the standard usage, we denote by $1, 2, 3, \ldots$ arithmetic terms of the form $succ(0), succ(succ(0)), succ^3(0), \ldots$. ♯

DEFINITION 5.32 Let $\diamond$ be a symbol of arity 0, called the *hole* symbol. The set T_I of *terms with integer exponents (I-terms)* and the set $T_\diamond$ of *terms with multiple holes* are the least sets that satisfy the following properties:

$$\begin{array}{lll} f(\overline{s}) \in T_I & \text{if} & \overline{s} \in T_I^{a(f)}, \\ \mathcal{V} \subset T_I, & & \\ \diamond \in T_\diamond, & & \\ f(s_1, \ldots, s_n) \in T_\diamond & \text{if} & \text{for all } i \in \{1, \ldots, n\},\ s_i \in T_\diamond \wedge n = a(f), \\ s^n.t \in T_I & \text{if} & s \in T_\diamond,\ t \in T_I,\ n \in T_N \text{ and } s \neq \diamond. \end{array}$$

♯

I-terms of the form $s^n.u$ are called *N-terms.*

DEFINITION 5.33 The set of *equational formulae with I-terms* F_I is the subset of first-order formulae inductively defined as follows.

- $\top$ (true) and $\bot$ (false) are in F_I.
- If s and t are in T_I, then $t = s \in F_I$.
- If t and s are two arithmetic terms, then $t < s \in F_I$ and $t = s \in F_I$.
- If $\mathcal{F} \in F_I$ then $\neg\mathcal{F} \in F_I$.
- If $\mathcal{F} \in F_I$ and x is a variable, then $\exists x.\mathcal{F} \in F_I$ and $\forall x.\mathcal{F} \in F_I$.
- If $\mathcal{F}$ and $\mathcal{G}$ are in F_I, then $\mathcal{F} \vee \mathcal{G} \in F_I$ and $\mathcal{F} \wedge \mathcal{G} \in F_I$. ♯

The notion of first-order formula, clause, constraint clauses etc. can be extended to formulae containing I-terms in a straightforward way.

DEFINITION 5.34 We denote by $\mathcal{T}_I$ the constraint language $(F_I, \mathcal{I}_I)$, where $\mathcal{I}_I$ is the interpretation defined below.

- The domain of $\mathcal{I}_I$ is the set of ground terms on the signature Σ and the set of natural numbers. Variables in V_N are interpreted on $\mathbb{N}$ and variables in $\mathcal{V}$ are interpreted as ground terms.

- For any function symbol f distinct from $0, succ, +$, we have

$$t_{\mathcal{I}_I}(f(t_1, \ldots, t_n)) = f(\mathcal{I}_I(t_1), \ldots, \mathcal{I}_I(t_n)).$$

- The semantics of the symbols $\{0, succ, +, <\}$ is defined as usual:

  - $t_{\mathcal{I}_I}(0) = 0.$
  - $t_{\mathcal{I}_I}(succ(t)) = t_{\mathcal{I}_I}(t) + 1.$
  - $t_{\mathcal{I}_I}(t + s) = t_{\mathcal{I}_I}(t) + t_{\mathcal{I}_I}(s).$
  - $\mathcal{I}_I \models t < s$ iff $t_{\mathcal{I}_I}(t) < t_{\mathcal{I}_I}(s).$

- The semantics of I-terms is defined as follows:

  - $t_{\mathcal{I}_I}(t^n.u) = t_{\mathcal{I}_I}(u)$ if $t_{\mathcal{I}_I}(n) = 0.$
  - $t_{\mathcal{I}_I}(t^n.u) = t_{\mathcal{I}_I}(t\{\diamond \rightarrow t^m.u\})$ if $t_{\mathcal{I}_I}(n) > 0$, where $m = t_{\mathcal{I}_I}(n) - 1.$ ♯

The following properties are straightforward consequences of the definition of I-terms.

- If $t \in T$, then the term $t^n.u$ is equivalent to t.

- If $t \equiv \diamond$, then the term $t^n.u$ is equivalent to u.

- If $n = 0$, then the term $t^n.u$ is equivalent to u.

- If $n = 1$, then the term $t^n.u$ is equivalent to $t\{\diamond \leftarrow u\}$.

- Let $t \in T_\diamond$, $s \in t$ and n, m be arithmetic ground terms; then

$$t^n.(t^m.s) \equiv t^{m+n}.s$$

THEOREM 3.6 *$\mathcal{T}_I$ is decidable, i.e. given a signature Σ, there exists an algorithm deciding whether a given formula in F_I has solution in $\mathcal{I}_I$.*

Proof: The proof of this assertion and the corresponding solution algorithm can be found in [Peltier, 1997a]. ◇

$\mathcal{T}_I$ is strictly more expressive than $\mathcal{T}_\emptyset$. For example, the interpretation $\{P(x)/x = succ^{2\times n}(0)\}$, that cannot be represented by equational constraints on finite trees, is simply represented by the formula:

$$P(x)\colon \exists m.n = m + m \wedge x = succ(\diamond)^n.$$

Moreover, the expressive power of $\mathcal{T}_I$ is incomparable to that of $\mathcal{T}_\in$, in the sense that there exist interpretations representable by using I-terms and *not* by using membership constraints, and vice versa.

EXAMPLE 5.35 Let $\Sigma = \{a, f\}$ and $\mathcal{I}$ be the Herbrand interpretation defined as:

$$\mathcal{I} : \{P(f^n(a), f^m(a))/n < m\}.$$

Then $\mathcal{I}$ is representable by the formula $P(x, y)$ in F_I for

$$P(x, y) = \exists n.n > 0 \wedge y = f(\diamond)^n(x).$$

But $\mathcal{I}$ is *not* representable by a formula with membership constraints. ♯

EXAMPLE 5.36 Consider the following interpretation $\mathcal{I}$, defined on the signature $\Sigma = \{a, g, f\}$ and let P be the only predicate symbol of a formula. We define

$P(x)$ is true iff one of the following conditions holds:

- $x = g(y)$.
- $x = f(y, z)$ and $\mathcal{I} \models (P(y) \wedge P(z))$.

$\mathcal{I}$ can be represented by the following formula in $F_\in$:

$$P(x) \text{ is true iff } x \in S$$

where S is defined by the following rules:

$$\begin{array}{lcl} f(S, H) & \to & S \\ g(H) & \to & S \end{array}$$

where H denotes the set of all terms. But $\mathcal{I}$ cannot be represented by using I-terms (see for example [Amaniss et al., 1993] for details). ♯

EXAMPLE 5.37 Let $\Sigma = \{a, b, f, g\}$, where a, b are of arity 0, f is of arity 1 and g is of arity 2. We define

$$\forall x.(P(x) \Leftrightarrow (x = a \vee \exists y.(x = f(y) \wedge P(y)) \vee \exists y.(x = g(y) \wedge P(y)))).$$

The only Herbrand model of this formula is the set $\mathcal{M}$ of atoms defined inductively:

- $P(a) \in \mathcal{M}$;
- if $P(t) \in \mathcal{M}$ then $P(f(t)), P(g(t)) \in \mathcal{M}$.

This set of terms *cannot be expressed by an equational formula on I-terms* since at each inductive step any term t in $\mathcal{M}$ may produce two new distinct term $f(t)$ and $g(t)$.

However, $\mathcal{M}$ can be easily described by the formula

$$x \in S$$

with the following tree automaton:

$$\begin{array}{lcl} f(S) & \rightarrow & S, \\ g(S) & \rightarrow & S, \\ a & \rightarrow & S. \ \sharp \end{array}$$

Remark: An essential difference between I-terms and tree automata lies in the fact that in tree automata the variables occurring at each level in the iterative context ($f(x, y)$ in the previous example), as well as the variables occurring in the base term ($g(y)$) *must* be independent, whereas in I-terms these variables *cannot* be independent (otherwise the theory would become undecidable). Moreover, with I-terms, only *one* possible context is allowed, whereas several distinct ones are possible in the case of tree automata (see Example 5.37). $\sharp$

Computing Model Representations Using I-Terms. I-terms are a powerful tool for representing interpretations, but it remains to show how such representations are generated from first-order formulae, or sets of first-order clauses. Of course, I-terms can be simply added in the language in order to enlarge the expressive power of first-order logic, allowing for example the user to replace sets of clauses of the form

$$\{even(0), \neg even(x) \vee even(succ(succ(x)))\}$$

by

$$\{even(succ^{2 \times n}(0))\}.$$

However, I-terms can also by generated automatically from standard clauses, by adding inference rules simulating in only one step several applications of the resolution principle. The idea is to use I-terms to specify all the resolvents of some kinds of self-resolving clauses. For example, it is possible to deduce from a clause $C : \neg P(x) \vee P(f(x))$ a new clause $C' : P(x) \vee P(f^n(\diamond).x)$, where C' is obtained by n applications of the resolution rule on C. By this rule, it is possible to discard potentially infinite derivations and to improve the termination behavior of resolution.

Such clauses can be computed using the following inductive rule (see also [Peltier, 1997a], a similar - but less general - version of the rule is defined in [Salzer, 1994]):

DEFINITION 5.38 **Inductive Generalization rule** (IG)**:**

$$\frac{\neg P\sigma \vee P\theta \vee R}{\neg P\sigma_n \vee P\theta_n \vee R}$$

if:

- σ is a substitution of the form $\{x_i \leftarrow t_i \mid i = 1, \ldots, k\}$.
- θ is a substitution of the form $\{y_i \leftarrow s_i \mid j = 1, \ldots, m\}$.
- For all $i \in \{1, \ldots, k\}$, $V(t_i) \cap \{x_i, y_j \mid 1 \leq i \leq k, 1 \leq j \leq m\} \subseteq \{x_i\}$.
- For all $j \in \{1, \ldots, m\}$, $V(s_j) \cap \{x_i, y_j \mid 1 \leq i \leq k, 1 \leq j \leq m\} \subseteq \{y_j\}$.
- $V(R) \cap \{x_1, \ldots, x_k, y_1, \ldots, y_m\} = \emptyset$.
- $\{x_1, \ldots, x_k\} \cap \{y_1, \ldots, y_m\} = \emptyset$.

and where :

- n is a new integer variable.
- σ_n is the substitution mapping each variable x_i $(1 \leq i \leq k)$ to $t_i'^n.x_i$, where t_i' is obtained from t_i by replacing each occurrence of x_i by the hole symbol "$\diamond$".
- θ_n is the substitution mapping each variable y_j (for $1 \leq j \leq m$) to $s_j'^n.y_j$, where s_j' is obtained from s_j by replacing each occurrence of y_j by the hole symbol "$\diamond$". ♯

EXAMPLE 5.39 Consider the clause:

$$R(u) \vee \neg P(x, y) \vee P(f(x), z)$$

It is of the form:

$$[\![R(u) \vee \neg P(x, y)\sigma \vee P(x, y)\theta : \top]\!]$$

where σ is the identity and $\theta = \{x \rightarrow f(x), y \rightarrow z\}$.

These clause satisfies the conditions of Definition 5.38, hence we can deduce the following clause:

$$C' : R(u) \vee \neg P(x, y) \vee P(f(\diamond)^n.x, z).$$

It is easy to see that C' is a logical consequence of C. ♯

THEOREM 3.7 *The inductive generalization rule is sound.*

Proof: Let $\mathcal{I}$ be an interpretation satisfying $\neg P\sigma \vee P\theta \vee R$. We first show - by induction on $n \in \mathbb{N}$ - that $\mathcal{I}$ must satisfy the clause $\neg P\sigma^n \vee P\theta^n \vee R$.

- The proof is immediate for $n = 0$.
- Assume that $\mathcal{I} \models \neg P\sigma^n \vee P\theta^n \vee R$. $\mathcal{I}$ also models $\neg P\sigma \vee P\theta \vee R$. Then it must satisfy also $(\neg P\sigma^n \vee P\theta^n \vee R)\sigma$ and $(\neg P\sigma \vee P\theta \vee R)\theta^n$, i.e. $\neg P\sigma^{n+1} \vee P\theta^n\sigma \vee R\sigma$ and $\neg P\sigma\theta^n \vee P\theta^{n+1} \vee R\theta^n$.

  Since the domains of θ and σ are disjoint, $\sigma\theta = \theta\sigma$, hence $\sigma\theta^n = \theta^n\sigma$. Therefore $\mathcal{I}$ validates $\neg P\sigma^{n+1} \vee P\sigma\theta^n \vee R\sigma$ and $\neg P\sigma\theta^n \vee P\theta^{n+1} \vee R\theta^n$. By applying the resolution rule on the literal $P\sigma\theta^n$, we get the clause: $\neg P\sigma^{n+1} \vee P\theta^{n+1} \vee R\sigma \vee R\theta^n$. This clause must be true in $\mathcal{I}$, by soundness of the resolution rule. Moreover, R does not contain any variable occurring in the domains of σ and θ, hence $R\sigma \equiv R$ and $R\theta^n \equiv R$.

  Therefore we have:

  $$\mathcal{I} \models \neg P\sigma^{n+1} \vee P\theta^{n+1} \vee R.$$

  It only remains to show that $P\sigma^{n+1} \equiv P\sigma_{n+1}$ and $P\theta^{n+1} \equiv P\theta_{n+1}$. Without loss of generality, we give only the proof of $P\sigma^{n+1} \equiv P\sigma_n$ (the other part of the proof is analogous).

  Let σ be a substitution fulfilling the conditions of Definition 5.38 and let $i \in \{1, \ldots, k\}$. We show, by induction on n, that $x_i\sigma^n = x_i\sigma_n$.

  - $n = 0$. The proof is immediate, since $x_i\sigma^n = x_i$ and $x_i\sigma_n = x_i$.
  - Assume that $n > 0$ and $x_i\sigma^n = x_i\sigma_n$.
    We have: $x_i\sigma^{n+1} = x_i\sigma\sigma^n$. By definition, we have $x_i\sigma = t_i$, hence $x_i\sigma^{n+1} = t_i\sigma^n$. Since t_i contains no variable in $dom(\sigma)$ except x_i, we have:

    $$x_i\sigma^{n+1} = t_i\{x_i \leftarrow x_i\sigma^n\}.$$

    But by induction hypothesis $x_i\sigma^n = x_i\sigma_n$, hence:

    $$x_i\sigma^{n+1} = t_i\{x_i \leftarrow x_i\sigma_n\}$$

    i.e. (by definition of σ_n):

    $$x_i\sigma^{n+1} = t_i\{x_i \leftarrow t_i'^n.x_i\}.$$

By definition $t_i\{x_i \leftarrow t\} \equiv t_i'^1.t$ for any term t, hence

$$t_i\{x_i \leftarrow t_i'^n.x_i\} \equiv t_i^1.t_i'^n.x_i \equiv t_i^{n+1}.x_i$$

and therefore

$$x_i\sigma^{n+1} \equiv x_i\sigma_{n+1}.$$

◇

This feature considerably improves the termination behavior of the resolution method (since potentially infinite derivations can be replaced by finite ones) and allows to generate saturated (c-)clauses sets in cases where the resolution rule does not terminate. This is especially interesting for model building, but also for theorem proving, since it may prune the search space and greatly reduce the length of the proofs, as evidenced by the following example.

EXAMPLE 5.40 Let $S =$

1: $\neg P(x) \vee P(f(x))$
2: $P(a)$
3: $\neg P(f^k(a))$

It is well known that the shortest resolution proof of S is of length $o(ln(k))$. Using I-terms we can obtain a 3-step proof:

4:	$\neg P(x) \vee P(f^n(x))$	(IG,1)
5:	$P(f^n(a))$	(resolution, 4, 1)
6:	$\square$	(resolution, 5, 3)

An interesting feature is that the length of the proof is *independent from the value of the integer k.* ♯

The Inductive Generalization Rule in Model Building. The following example illustrates the impact of the inductive generalization rule on model building.

EXAMPLE 5.41 Let S be the following set of clauses over $\Sigma = \{a, f\}$:

1: $\neg P(x, y) \vee \neg P(y, z) \vee P(x, z)$
2: $P(x, f(x))$
3: $\neg P(x, x)$

S states that P is an irreflexive and transitive relation satisfying $P(x, f(x))$, i.e. a monotonic ordering. It is clear that the only Herbrand model of S is:

$$P(x, y) \text{ is true iff } x = f^n(0), y = f^m(0) \text{ and } m > n.$$

This model cannot be expressed by a first-order equational formula not even with additional use of membership constraints. Using the resolution rule, we derive:

$$4: \quad \neg P(x, y) \vee P(x, f(y)) \quad \text{(resolution, 1,2)}$$

At this point, we can apply the inductive generalization rule instead of continuing with the resolution rule. Then, we obtain:

$$5: \quad \neg P(x, y) \vee P(x, f(\diamond)^n.y) \quad \text{(IG, 4)}$$

Then, applying the resolution rule on 5 and 2, we obtain:

$$6: \quad P(x, f(\diamond)^n.f(x)) \quad \text{(resolution, 5, 2)}$$

On this point, the set of Horn clauses $1-6$ is saturated under hyperresolution. Indeed, the only way of applying the hyperresolution rule on this set of clauses would be:

- Either to unify $P(x, f(\diamond)^n.f(x))$ and $\neg P(x, x)$ (which would allow to derive a contradiction by applying the resolution principle on $6, 3$). However, it is clear that these two literals are not unifiable.
- Or to apply the hyperresolution rule on 1 and 6 which gives:

  $$P(x', f(\diamond)^{n'}.f(f(\diamond)^{n'}.f(x'))).$$

  Note that 2 is subsumed by 6 and thus can be deleted immediately after production of 6. Thus only 6 is left for hyperresolution with 1.

  However also the hyperresolvent above is *subsumed* by the clause 6. Indeed, the term
  $$f(\diamond)^{n'}.f(f(\diamond)^{n'}.f(x))$$
  is an instance of the term $f(\diamond)^n.f(x)$ (it suffices to take the substitution $n = m + m + 1$). The subsumption test can be done using constraint solving, i.e. it suffices to check whether the formula:

  $$\forall x', n'(\exists x, n.x' = x \wedge f(\diamond)^{n'}.f(f(\diamond)^{n'}.f(x')) = f(\diamond)^n.f(x))$$

  is true or false.

Therefore (using the techniques described in the Section 1 of Chapter 3), we obtain immediately the finite description of a Herbrand model of the clause set:

$$\{P(x, f(\diamond)^n(f(x)))\}.$$

♯

The inductive generalization rule and semantic refinements of the resolution method. An important feature of all semantic refinements of the resolution principle (such as positive resolution, or hyperresolution) is that they prevent application of the resolution rule between some clauses, thus leading in several cases to a significant pruning of the search space, and to an improvement of the termination behavior of the method (i.e. the method terminates on more sets of clauses). The prize to be payed of course is that the proofs may become longer (for the set of clauses in Example 5.40, it is well known that the shortest hyperresolution proof is exponentially longer than the shortest resolution proof). In particular, the application of the resolution rule between self-resolving clauses is always prevented.

If I-terms are added to the method, then the situation concerning termination is very different. The inductive generalization rule allows to detect cycles in resolution derivation and to remove these cycles by providing a finite description of the set of resolvents. But, since the rule relies on the presence of self-resolvent clauses to detect such cycles, this feature is useful *only* if these clauses are explicitly generated by the resolution rule. However, the use of resolution refinement can *prevent* the generation of such self-resolving clauses, thus also preventing the detection and deletion of infinite cycles in the derivation. Therefore, in some cases the use of such refinements can lead to non-termination of the resolution process.

EXAMPLE 5.42 Consider for example the following clause set:

$$\begin{array}{ll} 1: & \neg P(x) \vee Q(f(x)) \\ 2: & \neg Q(x) \vee P(f(x)) \\ 3: & P(a) \end{array}$$

Using the resolution calculus enriched by the inductive generalization rule it is easy to get a saturated clause set S:

$$\begin{array}{lll} 4: & \neg P(x) \vee P(f(f(x))) & (resolution, 1, 2) \\ 5: & \neg P(x) \vee P(f(f(\diamond))^n.x) & (IG, 4) \\ 6: & P(f(f(\diamond))^n.a) & (resolution, 3, 5) \\ 7: & Q(f(f(f(\diamond))^n).a) & (resolution, 1, 6) \\ 8: & \neg Q(x) \vee Q(f(f(x))) & (resolution, 1, 2) \\ 9: & \neg Q(x) \vee Q(f(f(\diamond))^n.x) & (IG, 8) \end{array}$$

S is saturated under the resolution rule, i.e. all derivable clauses are subsumed by existing ones (of course under the subsumption principle extended to I-terms). From S, an atomic representation of a model of S can be extracted. The representation is the following one:

$$\{P(f(f(\diamond))^n.a), Q(f(f(f(\diamond))^n).a)\}.$$

However, if we use hyperresolution instead of resolution, then the clauses

$$\neg P(x) \vee P(f(f(x)))$$

and

$$\neg Q(x) \vee Q(f(f(x)))$$

will *never* be generated. Instead, hyperresolution will generate the following infinite set of resolvents:

4:	$Q(f(a))$	(hyperresolution, 1, 3)
5:	$P(f(f(a)))$	(hyperresolution, 4, 1)
6:	$Q(f(f(f(a))))$	(hyperresolution, 6, 3)
7:	$P(f(f(f(f(a)))))$	(hyperresolution, 7, 1)
...		

All these clauses are ground and the inductive generalization rule can never be applied. Therefore hyperresolution does not terminate[1]. ♯

The previous example shows that the explicit generation of cycles allowed by the resolution rule is necessary for termination in some cases. Using the hyperresolution rule - or similar semantic refinements - the generation of such cycles is hidden, hence the inductive generalization rule cannot be applied. In order to apply the inductive generalization rule with hyperresolution, we would have to add a mechanism for explicitly generating such cycles. Otherwise, the application of the rule will be limited to clauses that are already contained in the initial clauses set. The situation is similar in the tableau method, where cycles are never made explicit, due to the lack of the lemma building property of the resolution method. If I-terms are to be applied into tableaux-based procedures, an additional mechanism must be added to make the construction of self-resolvent clauses explicit. In [Stolzenburg, 1999, Peltier, 2001a], techniques are proposed for detecting cycles occurring in the proof procedure and for generating the corresponding term scheme automatically.

RAMC and I-terms. Since the first-order theory of I-terms is decidable, I-terms can be added to RAMC without losing the properties of the method. Below, we give an example of an application of *RAMC* (see Section 3 of Chapter 4) enriched by the use of I-terms and by the inductive generalization rule.

[1]Of course, in this simple case the method would terminate if the appropriate setting is chosen.

EXAMPLE 5.43 Consider the set of the four following clauses:

1	$[\![\neg Q(x, f(x)) : \top]\!]$
2	$[\![P(x, f(x)) : \top]\!]$
3	$[\![\neg P(x, y) \vee Q(x, f(y)) : \top]\!]$
4	$[\![\neg Q(x, y) \vee Q(x, f(y)) : \top]\!]$

This set of clauses is satisfiable and has no finite models [Dreben and Goldfarb, 1979].

5	$[\![\neg P(x, x) : \top]\!]$	(c-resolution 3,1)
6	$[\![\neg Q(x, x) : \top]\!]$	(c-resolution 4,1)
7	$[\![\neg P(x, y) \vee Q(x, f(y)) : x \neq y]\!]$	(c-dissubsumption 3,5)
8	$[\![\neg P(x, y) : y \neq f(x)]\!]$	(GPL)
9	$[\![\neg Q(x, y) \vee Q(x, f^n(\diamond).y) : \top]\!]$	(I-term introduction 4)
10	$[\![\neg Q(x, y) \vee Q(x, f^n(\diamond).y) : x \neq y]\!]$	(c-dissubsumption, 6, 9)
11	$[\![Q(x, f^n(\diamond).y) : x \neq y]\!]$	(GPL)

This gives us the following CARM on $\mathcal{T}_I$:

$$\{[\![P(x, f(x)) : \top]\!],\\ [\![Q(x, f^n(\diamond).y) : x \neq y]\!]\}.$$

♯

Extension of I-terms. Several languages have been proposed in order to extend the theory of I-terms. For example, Hermann introduced [Hermann, 1994] the notion of a *primal grammar*, which is strictly more expressive. However, the decidability of the first-order theory of primal grammars is not known. Only the decidability of the less general unification problem has been established [Hermann and Galbavý, 1997].

Chapter 6

FINITE MODEL BUILDING

In the first part of this chapter the main conceptual differences between finite and infinite models are pointed out, the corresponding theoretical results are recalled and representative methods and systems for finite model building are presented.

In the second part we illustrate the importance of finite model building to various fields. Some results obtained with running systems give a short overview of the present capabilities of automated model building in practical applications.

The chapter is divided in 5 sections:

1 Models and finite models

2 The fundamental theorems

3 Some representative methods for finite model building

4 Finite models in different fields

5 Practical applications of model building: some results

1. Models and finite models

Though at first glance finite model building may appear as a "simple" case of model building in general, it deserves a treatment on its own. There are theoretical and practical reasons justifying a clear separation in the treatment of such models[1].

[1]Of course, this separation does not invalidate the usefulness of practical systems offering a unified framework able to build finite and infinite models.

The theoretical ones are the deep differences between the general case and the finite one for key properties and fundamental theorems. Most of the classical properties of first-order logic fail on finite structures. In the next section we recall those which are most relevant to automated model building.

Finite models play a key role in the study of one of the capital problems in logic, i.e. in the study of decidable classes in first-order logic: if a class $\mathcal{C}$ of formulae has the finite model property[2] it is decidable (see for ex. [Börger et al., 1997]). It follows from the fact that it is possible to recursively enumerate finite structures over a finite vocabulary and that it is decidable whether a given structure satisfies a given first-order formula. A particular case that deserves to be mentioned here is Ramsey's theorem (see [Boolos and Jeffrey, 1974]), that has been proven in order to show the decidability of a certain fragment of first-order logic (see [Börger et al., 1997]). Ramsey has proven the finite version of his theorem in order to obtain a method to build finite models.

Within Computer Science, the idea of restricting the study of models to finite ones has been expressed by Hájek in 1977 (see [Fagin, 1993]) and the interest in finite structures initiated fundamental developments leading to what is known today as finite model theory (see the seminal works of Gurevich [Gurevich, 1984, Gurevich, 1988] or [Fagin, 1993, Ebbinghaus and Flum, 1999]). The practical reasons are of psychological (it is in general much more comfortable for a human being to deal with finite structures) and of technical nature: finite models can always be *finitely represented* and, as pointed out above, it is decidable whether a first-order formula F is satisfied in a finite interpretation (the intuition is that it is possible to check propositional instances of F in each point of the – finite – domain).

2. The fundamental theorems

For a long time finite models did not have a status similar to that of infinite ones in mathematical logic. In reference books in the domain such as [Chang and Keisler, 1990, Hodges, 1993] there are no special sections devoted to finite models. Things have changed with the interplay of logic with computer science (complexity theory, databases, logic programming,...) and, as just mentioned in the previous section, *finite* model theory is now a well established research domain.

We recall the main theoretical results necessary to a good understanding of the field. We mainly follow the presentation of [Fagin, 1993].

[2] A class $\mathcal{C}$ has the *finite model property* iff every satisfiable formula in $\mathcal{C}$ has a finite model. This property is called *finite controllability* in [Dreben and Goldfarb, 1979].

[Ebbinghaus and Flum, 1999] provides an exhaustive treatment of the subject (see also [Trahtenbrot, 1963, Börger et al., 1997]).

The first result to be reminded is that of the undecidability of the validity problem over finite domains, i.e. given a first-order formula decide whether it is *valid* over (all) finite structures.

THEOREM 2.1 (TRAHTENBROT) *Let $\mathcal{L}$ be a first-order logic language containing non-unary predicate symbols (possibly with equality -see [Trahtenbrot, 1963]). Then the set of sentences of $\mathcal{L}$ which are valid over finite domains is co-recursively enumerable but* **not recursively enumerable**.

In [Smoryński, 1991] it is stated as:

THEOREM 2.2 (TRAHTENBROT) *First-order logic has the finite inseparability property.*

A first-order theory $\mathcal{T}$ has the *finite inseparability* property if the sets of finitely refutable sentences (these are sentences with finite countermodels) over $\mathcal{T}$ and of theorems of $\mathcal{T}$ are *effectively (or recursively) inseparable.*

Two disjoint sets A, B are *recursively separable* if there is a recursive set C such that $A \subseteq C$ and $B \subseteq \overline{C}$. They are *recursively inseparable* if they are not recursively separable.

A set A is recursive if and only if A and $\overline{A}$ are recursively separable.

The proof of Trahtenbrot's theorem can be found in [Trahtenbrot, 1963, Ebbinghaus and Flum, 1999, Smoryński, 1991].

As Fagin remarks the "easy part" of the proof (that also suggests an obvious automated procedure) is to prove that the set of sentences valid over finite structures is co-r.e. This is simply because in order to prove that a sentence F is not valid in a finite domain, it is enough to enumerate the finite structures $\mathcal{M}$ over the language of $\Sigma(F)$ and perform the (decidable) test: $\mathcal{M} \models F$? If there exists an $\mathcal{M}$ with $\mathcal{M} \not\models F$ this $\mathcal{M}$ is eventually found (thus this test corresponds to the search for finite counter-examples).

The same argument (but replacing the search for $\mathcal{M} \not\models F$ by this for $\mathcal{M} \models F$) can be used to prove the following theorem which serves as a theoretical basis to the very successful model building approaches based on enumeration such as FINDER (see 3.6), SEM (see 3.7), MACE [McCune, 1998, Rose and Wilkinson, 2001] or FMC (see section 3).

THEOREM 2.3 *The set of finitely satisfiable sentences of first-order logic is recursively enumerable.*

Then, the crucial point in designing systems of finite model building *in practice* is to invent clever ways of producing and testing candidate structures (for example avoiding testing isomorphic structures, using symmetries,...).

It is worth noticing that logicians have proposed a calculus corresponding to (the co-recursively enumerable part of) Trahtenbrot's theorem [Hailperin, 1961] and to the theorem above [Bullock and Schneider, 1973].

We finish this section by comparing the fundamentals results in ordinary first-order logic with those corresponding to first-order logic over finite structures. They should enlighten the importance of studying finite models.

DEFINITION 6.1 A proof system $\mathcal{P}$ (the derivability by $\mathcal{P}$ is denoted by $\vdash$) is called *sound* if, for all formulae F, $\vdash F$ implies $\models F$; it is called *complete* if, for all formulae F, $\models F$ implies $\vdash F$. ♯

The theorem below, proven by Gödel in his famous Ph.D. thesis, justifies the central role first-order logic is playing within logic and computer science.

THEOREM 2.4 (COMPLETENESS OF FIRST-ORDER LOGIC) *There are proof systems for first-order logic which are sound and complete*[3].

As proofs can be effectively generated (this follows immediately from the definition of a proof system) we obtain as an immediate consequence:

THEOREM 2.5 *The set of valid sentences of first-order logic is recursively enumerable.*

But then, as the existence of a proof system entails that the set of valid sentences is r.e., it follows from Trahtenbrot's theorem that there are no proof systems for first-order logic over finite structures:

THEOREM 2.6 (NON-COMPLETENESS FOR FINITE STRUCTURES) *There is no sound and complete calculus for validity over finite structures.*

Another central result in first-order logic is:

THEOREM 2.7 (COMPACTNESS OF FIRST-ORDER LOGIC) *Let $\mathcal{F}$ be a set of first-order sentences. If every finite subset of $\mathcal{F}$ is satisfiable then $\mathcal{F}$ is satisfiable.*

[3]The term "completeness of first-order logic" is actually an abuse of language. Not the logic but the calculus is sound and complete.

On the other hand we have

THEOREM 2.8 (NON-COMPACTNESS FOR FINITE STRUCTURES) *The compactness theorem does not hold for satisfiability over finite structures.*

Proof: The (counter-) example used to prove the theorem is the same as that for proving the non-compactness of second order logic (see for example [Enderton, 1972]).

The intended meaning of the following sentence is

"there are at least n things."

Let

$$G_n(n \geq 2)\colon\ \exists x_1 \exists x_2, \ldots, \exists x_n (x_1 \neq x_2 \wedge x_1 \neq x_3 \wedge, \ldots, \wedge\, x_{n-1} \neq x_n)$$

where the conjunction contains $n(n-1)/2$ disequations.

It is obvious, from the intended interpretation, that each G_n is satisfiable in models with universes D s.t. $card(D) \geq n$ and unsatisfiable in universes D with $card(D) < n$.

Now, let us consider the infinite set $\mathcal{F}$: $\{G_i \mid i \geq 2\}$ and all the finite intervals, i.e. the sets

$$\mathcal{F}_n\colon\ \{G_i \mid 2 \leq i \leq n\}.$$

Clearly, all sets $\mathcal{F}_n$ are satisfiable over universes D with $card(D) \geq n$, thus they are **finitely satisfiable**

But $\mathcal{F}$ **is not** finitely satisfiable. Otherwise, let D_f a universe s.t. $card(D_f) = n$ in which $\mathcal{F}$ is satisfied, i.e. *all* formulae in $\mathcal{F}$ are satisfied. But there are infinitely many formulae in $\mathcal{F}$ that are falsified in D_f, namely all formulae $\mathcal{F}_{n+j}$ for $(j \geq 1)$. We obtain a contradiction. ◇

3. Some representative methods for finite model building

As mentioned in Chapter 1, since the beginning of logic in occidental civilization the study of *particular* cases of arguments (i.e. models, counter-examples) has been intimately related to the specific relationship holding between premises and conclusion in *all* possible situations (i.e. deductive reasoning, proof), as well as to other forms of reasoning[4]. This close interplay formally appears in the completeness theorem for first-order logic: given a first-order formula F the proof of the theorem

[4] Aristotle said: "The example is similar to induction and induction is a form of reasoning".

either yields a cut-free proof or a counter-model of F (see for example [Hodges, 1983]).

Concerning methods in automated deduction, simultaneous search for a proof (contradiction) and counterexamples (models), is inherent to the tableaux method which enumerates partial models of a finite set of formulae (in general the set of premises and the negation of the conclusion). Nevertheless they have been used mostly for proving (see for example [Fitting, 1990, Smullyan, 1968, Caferra and Zabel, 1992]). The main problem to recover models (counter-examples) from open branches is obviously that these models can be infinite. But what about finite models ?

There are numerous works on methods for finite model building; in most approaches enumeration is always subjacent (but it is also subjacent (in some way) in deduction, via Herbrand's theorem), but there are also methods for model building essentially based on inference rules (see Chapter 2). The models are represented by finite sets of ground literals, i.e. they use *atomic representation* of models (see Chapter 5 and Chapter 1). We call these methods *deductive methods* .

1, 2, 3, 4 and 5 below are examples of deductive methods. In other methods enumeration is the only machinery. We call them *enumerative methods* . 6, 7, 8 below are examples of enumerative methods. Of course, as already mentioned (see 1) combinations of both are possible (e.g. 2 and 3 below).

A lot of mathematical problems (in particular the famous SAT) as well as non trivial puzzles amount to the building of finite models for given sets of formulae. Mathematical problems and puzzles have therefore naturally furnished the examples and benchmarks, a lot of them successfully treated by systems based on enumerative methods. It is worth noticing that an approach combining enumeration with deduction has been used in order to build a model for a "big" satisfiable formula having no finite models (see [Peltier, 1998]).

A natural, obvious question is "how to compare the different approaches ?". In fact there are no simple criteria to compare model builders (as there is no simple criteria to compare theorem provers). The standard way the different authors adopt is to run examples (possibly with a probabilistic distribution) that have been customarily transformed into benchmarks (though there is no standard set of benchmarks as e.g. for theorems provers). This is of course a way of comparison, but it can hardly be considered as equivalent to, let us say, giving the algorithmic complexity of a method. Besides, studying comparative efficiency would require exhaustive tests taking into account the *last* heuristics incorporated to each solver. This is obviously out of the scope of this book.

In order to give a reasonable overview of the existing approaches, we have selected the methods we consider the most representative ones, if emphasis is put on *ideas* rather than on technicalities. To give an exhaustive list is impossible (due, among others, to the release of different improved versions of the same method) and would not increase the general understanding of the domain. By no means the authors think that the approaches not appearing in the list below are not interesting or even useless. We just mention the following approaches and systems.

1 A pioneering work by Boolos: tableaux completeness for finite satisfiability

2 Hyper-linking and model building

3 SATCHMO

4 Hypertableaux

5 The transformation method

6 FINDER

7 SEM

8 FMC

3.1 A pioneering work by Boolos: tableaux completeness for finite satisfiability

The pioneering work by Boolos [Boolos, 1984], never referenced in later works on finite model building deserves to be mentioned here.

As already pointed out in chapter 1, the tableau method, in its classical version, yields one of the most natural and elegant (though not powerful) ways of building models. We recall briefly its principles, rules and fundamental properties (see [Fitting, 1990, Smullyan, 1968, Jeffrey, 1967]). The method tries to build models of a (set of) formula(e) labelling the root of the deduction tree. In this aim it uses the rules specified in the tables below (where F and G denote well formed formulae).

Rules are classified in 4 categories: α-rules (conjunctive), β-rules (disjunctive), γ or UI-rules (UI for Universal Instantiation)-rules δ or EI-rules (EI for Existential Instantiation).

In the α-rules the formula in the node labelled α is replaced by the formulae labelled α_1 and α_2 corresponding to the *only* son of the node labelled α. In the β-rules the formula labelled β is replaced by formulae labelled β_1 and β_2 each corresponding to one of the *two* sons of the node labelled β. Once a formula F is replaced it is marked 'no more usable', except when applying γ-rules which can be used indefinitely (the γ-rules can potentially generate infinite branches). The rule corresponding to '$\Leftrightarrow$' can be omitted, but we give it because it is used in example 6.2.

α	α_1	α_2	β	β_1	β_2
$\neg\neg F$	F				
$F \wedge G$	F	G	$\neg(F \wedge G)$	$\neg F$	$\neg G$
$\neg(F \vee G)$	$\neg F$	$\neg G$	$F \vee G$	F	G
$\neg(F \Rightarrow G)$	F	$\neg G$	$F \Rightarrow G$	$\neg F$	G
			$F \Leftrightarrow G$	$F \wedge G$	$\neg F \wedge \neg G$
			$\neg(F \Leftrightarrow G)$	$F \wedge \neg G$	$\neg F \wedge G$

γ	$\gamma(t)$	δ	$\delta(t)$
$\forall x F$	$F\{x \leftarrow t\}$	$\exists x F$	$F\{x \leftarrow t\}$
$\neg\exists x F$	$\neg F\{x \leftarrow t\}$	$\neg\forall x F$	$\neg F\{x \leftarrow t\}$
	$t:$ *a closed term*		$t:$ *a* **new** *constant symbol*

The tableaux method is sound and complete for validity.

The starting point of Boolos' work is a conjecture by J. Burgess that can be stated as the following question: "Is it possible to modify the tableaux method in such a way that the method is **also** sound and complete for finite satisfiability ?" A positive answer to the question is furnished by a very simple and natural modification of the standard method. Sentences (i.e. closed formulae) are assumed to contain no function symbols but may contain equality.

Let us consider the same example as Boolos did in [Boolos, 1984]. The problem is to determine whether the formula $\forall x \exists y R(x, y)$ has finite models. Indeed this formula is true, for example, in every interpretation (D, Φ), where D contains only one element and $\Phi(R)$ is the identity. However the standard method yields the following infinite derivation:

$$\begin{array}{c} \forall x \exists y R(x,y) \\ \downarrow \\ \exists y R(a,y) \\ \downarrow \\ R(a,b) \\ \downarrow \\ \exists y R(b,y) \\ \downarrow \\ R(b,c) \\ \downarrow \\ \exists y R(c,y) \\ \downarrow \\ R(c,d) \\ \downarrow \\ \vdots \end{array}$$

The method builds a (partial) *infinite* model and "does not see" the aforementioned finite one. The explanation is that the δ-rule introduces a *new* constant with respect to that introduced by the γ-rule. The introduction of this new constant is justified by the fact that we are looking for *validity*[5]; assuming that the constant introduced by the δ-rule has the same name as that introduced by the γ-rule amounts to considering a *particular* case.

But looking for satisfiability is precisely to consider particular cases (models). This remark is the key of the extended method proposed by Boolos: Let Λ be an open branch , λ be a node on Λ; assume that λ is labelled by a formula $\exists x F(x)$. Instead of introducing *only* a new constant for instantiating x in $F(x)$ we consider the possibility of instantiating x *also* by constants already introduced in the ancestor nodes of λ in Λ. This rule yields the following derivations for the proposed example:

[5]Without this constraint, the δ-rule is *not sound*: the *premise* follows from the conclusion, and not the converse as soundness requires.

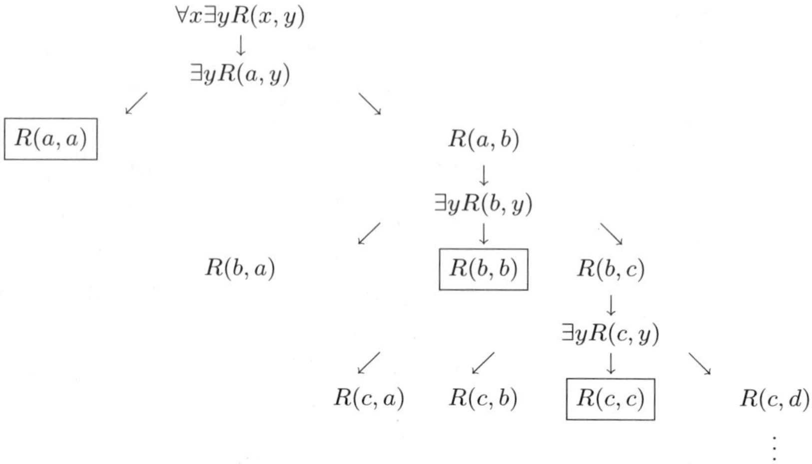

A branch is *finished* with respect to the method if no more rules applies to it. It is *open* if it contains no contradiction. As the new rule is a finitely branching rule, the new method is complete for unsatisfiability (validity). In addition, if there is a finite, finished and open branch, then the formula in the root is satisfiable.

It is intuitively clear that the method (partially) builds, as in the standard version, an infinite model on its rightmost branch. The finite model found by the method[6] means, of course, that $R(a, a)$ ($R(b, b)$, $R(c, c)$, ...) is evaluated to **true** in the interpretation. In a domain of cardinality 1 reflexivity amounts to identity (substitutivity obviously holds)[7].

For a given formula Boolos' method captures models of *all cardinalities*, but *does not* capture all the models (the method was not designed in this aim), as could be computed in a combinatory way (see next example).

For example, the model represented by

$$R(a, a),\ \underline{R(a, b)},\ R(b, a),\ R(b, b),\ R(a, c),\ \underline{R(b, c)},$$
$$\underline{R(c, a)},\ \underline{R(c, b)},\ R(c, c)$$

is not captured, but that represented by the underlined ground instances *is* (and this is enough to guarantee the existence of models of cardinality

[6] It is the only one emphasized in [Boolos, 1984] because it corresponds to the intuitive remark preceding the example in the paper.

[7] It is well known that equality can be axiomatized either as an equivalence relation plus substitutivity (for predicate and function symbols) or as a reflexive relation plus substitutivity (for predicate and function symbols).

3). Otherwise stated, the method implicitly applies the principle that to prove existence it is enough to exhibit *one* objet having a given property.

It is possible to implement (a stronger version of) Boolos' idea by imposing conditions on the branches, in the same spirit as in RAMCET (see 4). We consider the following example.

EXAMPLE 6.2 Let us consider the formula $\mathcal{A} \Rightarrow \forall x P(x)$, where $\mathcal{A}$ is an instance of Leibniz's law asserting that x, y, z are 3 *different* objects having the property denoted by P.

$$\begin{aligned} \mathcal{A}: \quad & [\exists x \exists y \exists z(\ (P(x) \wedge P(y) \wedge P(z)) \\ & \wedge(Q(x) \Leftrightarrow \neg Q(y)) \\ & \wedge(R(x) \Leftrightarrow \neg R(z)) \\ & \wedge(S(y) \Leftrightarrow \neg S(z))\ ] \end{aligned}$$

This implication is 3-valid (a formula is *n-valid*, $n \in N$, iff it is valid in all universes D, s.t. $card(D) \leq n$ and can be falsified in universes s.t. $card(D) > n$).

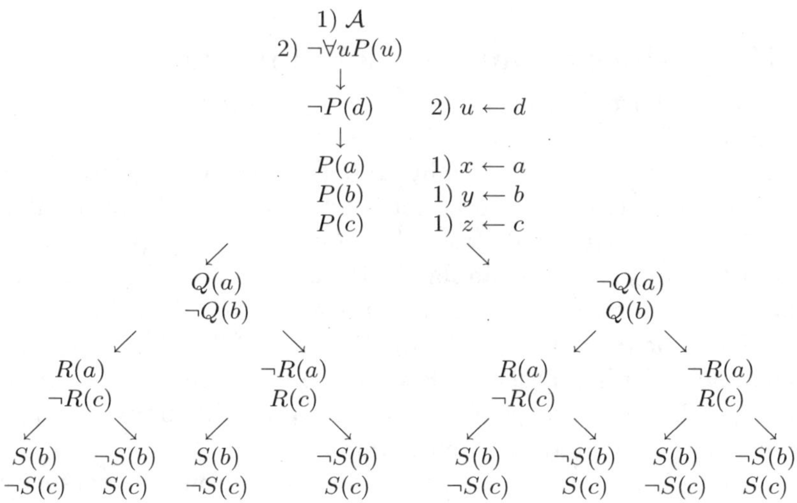

We consider universes of cardinality 1, 2, 3:

$\mathbf{card(D) = 1}$:

$a = b = c = d$ contradiction in branch 1. $\mathbf{card(D) = 2}$:

$a = b = c$ contradiction in branches 1.1 and 1.2
$a = b = d$ contradiction in branch 1.
$a = c = d$ contradiction in branch 1.
$b = c = d$ contradiction in branch 1.

$\mathbf{card(D) = 3}$:

$a = b$ contradiction in branches 1.1 and 1.2
$a = c$ contradiction in branches 1.1.1; 1.1.2; 1.2.1 and 1.2.2
$a = d$ contradiction in branch 1.
$b = c$ contradiction in branches 1.1.1.1; 1.1.1.2; 1.1.2.1; 1.1.2.2; 1.2.1.1; 1.2.1.2; 1.2.2.1 and 1.2.2.2
$b = d$ contradiction in branch 1.
$c = d$ contradiction in branch 1.

In universes D with $\mathbf{card(D)} \geq \mathbf{4}$ no branch can be closed.
With FMC_{ATINF} (on a PC Pentium 200 MHz, 64 Mb) we obtain:

Card.	Model ?	Nb of tests	Time (in s.)
1	no	1	0.00
2	no	7	0.00
3	no	50	0.00
4	yes	91	0.01

♯

3.2 Hyper-linking and model building

The work of Plaisted, Chu and Lee on hyper-linking (see [Lee and Plaisted, 1992, Chu and Plaisted, 1994a, Lee and Plaisted, 1994, Chu and Plaisted, 1994b, Plaisted and Zhu, 2000]) is an approach to theorem proving having interesting features for model building, in particular it combines saturation with direct model building[8] (see Chapter 3). Some key ideas of the approach are similar to Prawitz's method (see Chapter 2). Plaisted and co-authors mention [Prawitz et al., 1960] but neither [Prawitz, 1960] nor [Prawitz, 1969]. In [Prawitz, 1969] the presentation that Prawitz gives of his method allows to see more clearly the similarities with hyper-linking. Hyper-linking has also several common features with the method presented in [Davis, 1963].

The approach is an interesting one, particularly from the perspective of Artificial Intelligence. Given a problem described as a set of rules or *constraints* (numerous problems in AI and in computer science in general can be seen as problems of constraint solving) a *theorem prover* searches for solutions by looking for models. The use of a theorem prover is the main contribution of this work and makes this approach different from that of constraint solving in AI, Constraint Programming, Operational Research,.... The intuition behind is to consider the premises as the con-

[8]The adjective "direct" is used here to characterize methods that build models step by step instead of extracting them from a saturated set of clauses.

straints and the conclusion as a solution satisfying the constraints[9]. This intuition is analogous to the notion of logical consequence (see Chapter 2, [Etchemendy, 1988]): if an implication is a theorem, all the models of the formulae in the premises (and possibly more) are *specified* by the formula in the conclusion. As theorem provers handle consequence relations this intuition seems quite natural.

No formal definition of the meaning of "constraint" is given in Plaisted et al.'s papers, but from the context the reader can conclude that it means "formulae in clausal form" (a preserving structure translation – see [Baaz et al., 2001] – is recommended). The theorem prover implementing the approach uses the following concepts:

1 hyper-linking

2 unit simplification

3 propositional unsatisfiability test.

The meaning of each step is the following:
A *link* in a set of clauses is defined as usual, i.e. as a pair of unifiable (disregarding signs) complementary literals. The concept of link is one of the most natural and one of the first that appeared in the field of automated deduction and helped to its progress (see for example [Prawitz, 1960, Davis, 1963, Prawitz, 1969]). It has been used as the key concept in the matings (Andrews) and the connection (Bibel) methods.

A *hyper-link* of a clause C in a set of clauses S is a set of links containing a literal in C and such that there is a (simultaneous) m.g.u. (see Chapter 2) γ for all the links. More formally:

DEFINITION 6.3 Let S be a (finite) set of clauses $S = \{C_1, \ldots, C_m\}$. A *link* is a couple of literals $(L(\overline{s}), \neg L(\overline{t}))$ with $L(\overline{s}) \in C_i, \neg L(\overline{t}) \in C_j, 1 \leq i, j \leq m$, such that the equation $\overline{s} \doteq \overline{t}$ has a solution.

Let $C = \{L_1, \ldots, L_n\}$ be a clause in S. A *hyper-linking* of C is a set of links $\{(L_1, \neg L'_1), \ldots, (L_n, \neg L'_n)\}$ such that the set of solutions of links are compatible. The solution γ is called the *substitution* of the hyper-linking and $C\gamma$ is called the *instance* of the hyper-link. C is "naturally" called the *nucleus* and the literals matched with those of C, and the clauses containing them are called the *electrons*[10].

A *round* of hyperlinking is defined by adding all hyper-links for each clause, $S \leftarrow S \cup C_i\gamma_{j_i}$ (each γ_{j_i} denotes a substitution of an hyper-linking of the clause C_i). ♯

[9]It should be remarked that formulae handled by this method are closed ones and in the standard definition constraints are open formulae.

[10]Of course this terminology has been chosen by analogy with hyperresolution.

Hyper-linking is similar to hyperresolution (see 2) and therefore to semantic resolution (see [Slagle, 1967]) but there are several clear differences:

Hyper-linking	Hyperresolution
Inferences: more than 1 cl.	Inferences: more than 1 cl.
Clauses are not combined	Clauses are combined
No (interp. or order) needed	Interp. and order needed
All lit. in clause involved	Only lit. of same sign in nucleus
$\mid$instance $\mid \leq \mid$ nucleus $\mid$	$\mid$ Hyp-res $\mid$ can be $\geq \mid$ nucleus $\mid$
No duplication by comb.	Duplication by comb. possible
clause copying	no clause copying

$\mid$ Hyp-res $\mid$ means here 'number of literals in the hyperresolvent'. Similarly for $\mid$ instance $\mid$ and $\mid$ nucleus $\mid$.

A main difference to Prolog is that information expressed in negative form (the so-called "negative information") can be asserted and processed correctly. Besides, the applicability of the method is not restricted to set of Horn clauses.

- Unit clauses are used in two ways in order to simplify the set of clauses S.

  (i) unit literal deletion: Let S be a set of clauses (possibly after some rounds of hyperlinking). If there is a unit clause L in S with $\neg L$ appearing in other clauses of S then delete $\neg L$ in all clauses of S and delete all clauses containing L. This rule corresponds to the Davis and Putnam- and also to Prawitz's matrix reduction rule [Prawitz, 1969]. Obviously, this rule allows to detect unsatisfiability when there is a unit clause $\neg L$ in S.

  (ii) unit subsumption: If L is a unit clause subsuming $C \cup M$ then delete $C \cup M$.

- There is a *propositional unsatisfiability test*:
  The Davis-Putnam procedure, with a particular form of backtracking is used to test whether $Gr(S)$ is unsatisfiable, where $Gr(S)$ is obtained by replacing all variables of all clauses by the same constant symbol.

The hyperlinking prover can be specified as in figure 6.1.

When applying the Davis and Putnam's splitting rule, the method chooses the smallest clause and the literal appearing most frequently in

```
function hyper-linking prover
input: a set of clauses S_init
output: (if any) either 'S_init unsat' or 'S_init sat'
begin
% copy(S): gives a renamed copy of a clause in S (chosen in a fair way)
% Gr(S) = {Cφ | C ∈ S, φ a substitution, rg(φ) = {a constant symbol}}
% unitsimpl(S): applies unit simplification rules to S
% D&P(S): applies Davis and Putnam's method to the set of prop. clauses S
% □ ∈ S: S contains two unifiable unit clauses
S ← S_init
res ← ⊥
        while (res ≠ 'S unsat') and (res ≠ 'S sat')
        do
           S' ← S ∪ C_i γ_ji % add a round of hyperlinking
           S'' ← unitsimpl (S')
           if □ ∈ S'' then res ← 'S unsat'
                   else
                       if Gr(S'') contains neither pos. nor neg. clauses
                       then res ← 'S sat'
                       else
                           if D&P (Gr(S'')) = 'unsat'
                           then res ← 'S unsat'
                           else S ← copy(S) ∪ Gr (S'')
        od
           return res
end
```

Figure 6.1. hyper-linking prover

S (for the sake of comparison see the *linked conjunct* concept in [Davis, 1963] and the heuristic suggested there). This heuristic is similar to those used in some SAT solvers (see for example [Cocco et al., 2002, Dubois and Mondasson, 2001]).

It is not difficult to see that hyper-linking is refutationally complete (roughly, it can be proved by a direct application of Herbrand's theorem, Prawitz idea of unification and by completeness of the (propositional) Davis and Putnam's method). Different strategies like forward chaining, backward chaining and bidirectional search, can be simulated.

In order to obtain models for a set of satisfiable ground clauses the propositional decision procedure is modified (see fig. 6.2).

When the prover is used as a model builder neither (non-constant) functional terms nor equality are allowed. Some translation techniques valid on finite domains are used to avoid their use. Stated differently, the hyperlinking prover used as a model builder applies only to simple decidable classes. For example, in a domain $D = \{a_1, \ldots, a_n\}$ the formula $\exists x P(x)$ can be translated into $P(a_1)\vee, \ldots, \vee P(a_n)$. The formulae of the kind $\forall x \exists y P(x, y)$ give (over the same domain D) in a first step

$$\forall x(P(x, a_1) \vee P(x, a_2) \vee \ldots \vee P(x, a_n))$$

and in a second step

$$(P(a_1, a_1) \vee P(a_1, a_2) \vee \ldots \vee P(a_1, a_n)) \wedge \ldots \wedge$$
$$(P(a_n, a_1) \vee P(a_n, a_2) \vee \ldots \vee P(a_n, a_n)).$$

The formulae containing equality considered are those expressing the "one-to-one property":

$$\forall x \forall y \forall z \forall w.(P(x, z) \wedge P(y, w) \wedge x \neq y \Rightarrow z \neq w).$$

In finite domains they can be translated into:

$$\forall x(\neg P(a_1, x) \vee \neg P(a_2, x) \vee \ldots \vee \neg P(a_n, x))$$

i.e. into clauses **not containing** equality.

As the clauses in the input set do not contain Skolem functions of arity greater than 0 (i.e. we have clause forms representing the Bernays-Schönfinkel class), the Herbrand universe is finite and the number of hyper-linking is bounded. If the integer n is the minimal upper bound of the number of rounds, then in at most n rounds either unsatisfiability has been detected or the set of grounded instances is saturated. Once a model **represented by a conjunction (set) of ground atoms** is found the *negation* of this conjunction (i.e. a clause) is added and the prover runs again. Another solution (model), if it exists, will be found. The prover will enumerate models. When a clause corresponding to the negation of a model is added and the prover gives 'S_{init} unsatisfiable' as an answer it means that there are no more solutions (models) (the intuitive meaning is trivial: asserting that a set of clauses S is satisfiable with model $\mathcal{M}$ **is not** contradictory with the assertion that there is no (other) model $\mathcal{M}_1$ of S. If there are models of S other than $\mathcal{M}_1$, there is no contradiction). This feature is *very important* in AI, where finding as much models as possible of a set of (propositional) clauses can be of central importance (see for example [Birnbaum and Lozinskii, 1999, Bayardo and Pehoushek, 2000]).

Most of the examples given in the papers by Plaisted et al. are non-Horn problems.

The approach presented in [Chu and Plaisted, 1994b] proposes to build models using *semantically guided* hyper-linking. The difference with hyper-linking is that the models found by the propositional test

```
function hyper-linking prover/model builder
input: a set of clauses S_init, containing no other functional symbols than constants
output: either 'S_init unsat' or models of S_init represented as sets of ground atoms
begin
% copy(S): gives a copy of a clause in S (chosen in a fair way)
% Gr(S) = {Cφ | C ∈ S, φ a substitution, rg(φ) = {a constant symbol}}
% unitsimpl(S): applies unit simplification rules to S
% D&P(S): applies Davis and Putnam's method to the set of prop. clauses S
% □ ∈ S: S contains two unifiable unit clauses
S ← S_init
mbp ← false % mbp: model building phase
res ← ⊥
M ← ∅ % M: set of ground literals representing a model
% (The atoms not in the set can be evaluated arbitrarily)
        while (res ≠ 'S_init unsat')
        do
           S' ← S ∪ C_iΓ_{j_i} % add a round of hyperlinking
           if □ ∈ Gr(S') then
                          if mbp then res ← set of built models
                                 else res ← 'S_init unsat'
                          else
                              if there is no new instance w.r.t. last round
                              then
                              begin
                              mbp ← true
                                  while Gr(S') contains neither pos.
                                                nor neg. clauses
                                  do
                                     unitsimpl (Gr(S')) (say, with unit clause L)
                                     M ← M ∪ {L}
                                     D&P(Gr(S')) splitting on (say) L, ¬L;
                                     M ← M ∪ {L} xor M ← M ∪ {¬L};
                                     if Gr(S') = ∅ then S ← ¬M ∪ S
                                     % Pure literals are incorporated to the model
                                     % and the corresponding clauses deleted
                                  od
                              end
                              else
                                  if D&P (Gr(S')) = 'unsat'
                                  then res ← 'S_init unsat'
                                  else S ← copy(S) ∪ Gr (S')
        od
           return res
end
```

Figure 6.2. hyper-linking prover/model builder

are used to guide the generation of more ground instances of the input clauses. The authors call this strategy **incremental model finding**.

Roughly speaking, from a partial model of the input set of clauses (usually unit ground clauses in the input set) a round of hyper-linking is generated (i.e. ground instances in the previous partial model guides ground instances of the hyper-linking). A new ground clause is added and so on. Semantic trees (called "case trees" by the authors) are used for propositional (un)satisfiability test.

Remark: The hyper-linking method builds models only on the Bernays-Schoenfinkel class (it is therefore quite weak). RAMC and MB (the latter one after a transformation of Bernays-Schoenfinkel into PVD) also build models on this class. The relative efficiency of hyper-linking, RAMC and MB with respect to this class has not been tested so far.

3.3 SATCHMO

SATCHMO (**SAT**isfiability **CH**ecking by **MO**del generation) [Manthey and Bry, 1988, Bry and Yahya, 2000] was one of the first implementations orientated to model construction. Refutational completeness can be preserved if it is used in a level-saturation manner.

The motivating ideas are to fully exploit the features of the logic programming Prolog by overcoming some of its limitations ("It is a collection of fairly short and simple Prolog programs to be applied to different classes of problems" [Manthey and Bry, 1988], page 415).

SATCHMO is an interpreter working bottom-up. This is a key feature because it is intended to treat the so called *range restricted clauses*, i.e. clauses (the H_i, $1 \leq i \leq m$ form the *head* of the clause, the B_j, $1 \leq j \leq n$ its *body*):

$$C\colon\ H_1, \ldots, H_m \leftarrow B_1, \ldots, B_n$$

with $Var(C) \subseteq Var(B_1) \cup \ldots \cup Var(B_n)$. When reasoning forward, from ground facts, disjunctions of *ground* literals will be split (see 3.25).

The occurs check in the unification algorithm (see section 2) is not needed with this strategy, since at least one of the literals to be unified is always ground. This kind of splitting has been used in resolution at least since 1972 by Chang's splitting rule (see [Chang, 1972] and also [Henschen and Wos, 1974]) (SATCHMO's authors mention the tableaux splitting rule, see [Smullyan, 1968]). The Prawitz's matrix reduction rule [Prawitz, 1960, Prawitz, 1969] is another example.

As a consequence of these rules, and unlike standard Prolog, SATCHMO can handle negative clauses and clauses containing more than one positive literal, i.e. non-Horn clauses. The implementational clever ideas used in SATCHMO allow finite model generation with Prolog. As the Herbrand universe can be infinite, infinite models can *potentially* be generated but *cannot* be finitely represented in this formalism (i.e. via sets of ground literals). Model generation with SATCHMO can be justified on the basis of hyperresolution. Range restriction is not a severe limitation from the theoretical point of view because there is a satisfiability preserving transformation from arbitrary set of clauses into range restricted set of clauses. The idea of this transformation is simple (it has been also used for instance in [Winker, 1982] to specify the domain of the model):

- a new predicate $Dom(x)$ meaning "x is in the domain of the problem", is added to the initial set of clauses S
- it is also asserted that all constants in S are in the domain (if there is none in S, it is asserted that the constant "a" is in the domain). Formally we include the unit clauses $D(a)$.
- for each function symbol f in S a new Horn clause is added with the intended meaning that if each argument of f: $x_1, \ldots, x_n$ is in the domain, then $f(x_1, \ldots, x_n)$ is too. The corresponding clause is

$$D(f(x_1, \ldots, x_n)) \leftarrow D(x_1), \ldots, D(x_n).$$

- Finally, for each non range restricted clause with the variable x in its head but not in his body, $Dom(x)$ is added to its body.

This is a way to specify $H(S)$ (i.e. the Herbrand universe of S) using Horn clauses.

The consequences of these changes are that in a bottom-up (or forward) strategy all generated facts will be ground. Stated differently, clauses are exhaustively instantiated on the Herbrand universe generated with the help of the *Dom* predicate. This is an "enumerative feature" of SATCHMO. Of course this transformation decreases the efficiency of the method. The following example of model generation with SATCHMO is taken from [Manthey and Bry, 1988] page 419 (as well known, in the sequent formalism $A, B \leftarrow$ means $A \vee B$ and $\leftarrow A, B$ means $\neg A \vee \neg B$).

Let us consider the set of clause S below, where a, b denote constants, f a Skolem function, and x a variable.

1 $P(a), Q(b) \leftarrow;$

2 $S(f(x)) \leftarrow Q(x);$

3 $S(x) \leftarrow R(x);$

4 $Q(x), R(x) \leftarrow P(x)$;

5 $\leftarrow Q(x), S(y)$;

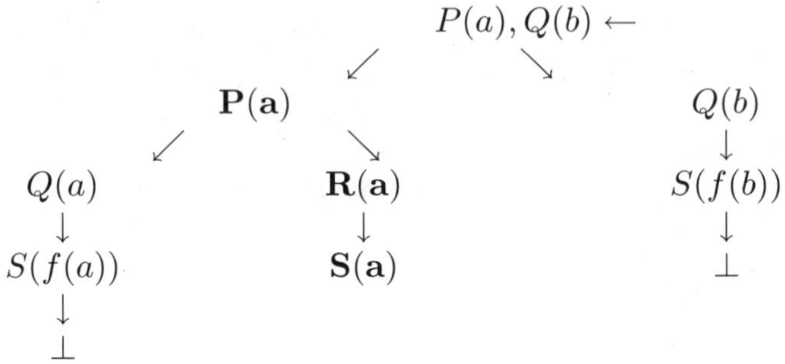

The rule applied at the root is the splitting rule on 1 (giving $P(a)$ **or** $Q(b)$), then on the leftmost branch resolution among $P(a)$ and 4 is applied, then splitting again, giving $Q(a)$ **or** $R(a)$. Resolution among $Q(a)$ and 2 gives $S(f(a)))$. $Q(a)$ and $S(f(a)))$ contradict 5 (by hyperresolution). In the middle branch resolution of $R(a)$ with 3 gives $S(a)$. At this point, no additional facts can be produced, and the open branch *represents the model* $\{P(a), R(a), S(a)\}$.

Several authors have proposed systems improving SATCHMO. We mention here three representative works:

- MGTP: (**M**odel **G**eneration **T**heorem **P**rover) [Fujita and Hasegawa, 1991], a SATCHMO implementation in a parallel logic language programming.
- SATCHMORE: (**SATCHMO** with **RE**levance) [Loveland et al., 1995], in which only relevant clauses are used in forward chaining (in order to control it).
- I-SATCHMO: (**I**ntelligent **SATCHMO**) [He, 2001], proposes an intelligent backtracking in order to reduce the model generation search space of SATCHMO.

3.4 Hyper Tableaux

In [Baumgartner et al., 1996] a variant of clausal tableaux called *hyper tableaux* is introduced. Given a set of clauses $S = \{C_1, C_2, \ldots, C_n\}$ a clausal tableau is a tableau containing only *or* nodes (i.e. nodes corresponding to applications of a β-rule) labelled by instances $C_i\sigma$ for $C_i \in S$ $(1 \leq i \leq n)$ (with exception of the root).

Hyper tableaux can be seen as an extension (and formalization) of SATCHMO-like approaches. It combines analytic tableaux with positive hyperresolution (see 2). Of course, as ordinary tableaux, hyper tableaux can be used for model building, with some advantages. The authors also introduce a subsumption technique, more powerful than that used in standard tableaux, whose application yields a decision procedure for Datalog[11]. The main difference to subsumption is that variables in hyper tableaux can be replaced by *any* term, while ordinary tableaux variables (called *rigid*) can be replaced by only *one* term.

Hyper tableaux offers some interesting features, as for example the possibility of controlling hyperresolution by deletion of irrelevant clauses. Unlike SATCHMO, hyper tableaux is not limited to range restricted clause (of course without use of domain predicates). Extension rules can be applied on single branches isolated from the others (and the variables there). Hyper tableaux is a confluent calculus.

The concept of *pure clause*[12] (necessary for the extension rule) is defined as:

DEFINITION 6.4 A clause

$$C\colon\ A_1, \ldots, A_m \leftarrow B_1, \ldots, B_n \ (m, n \geq 0)$$

is called *pure* iff

$$V(A_i) \bigcap V(A_j) = \emptyset \ (1 \leq i, j \leq m).$$

A substitution π is called a *purification substitution* for C iff $C\pi$ is pure. Obviously such a purifying substitution always exists (it is enough to substitute common variables by ground terms). ♯

Remark: Pure clauses allow for splitting, because if the clause $A \vee B$ is pure then we have

$$\forall(A \vee B) \Leftrightarrow (\forall A \vee \forall B).$$

♯

DEFINITION 6.5 Given a branch (i.e. a sequence of literals) $\mathcal{B}$ in a hyper tableau, an *hyper-extension* step of a clause C as above can be applied under the following two conditions:

- there is a most general substitution σ such that $\forall(B_1\wedge, \ldots, \wedge B_n)\sigma$ is a logical consequence of the least Herbrand model of $\mathcal{B}$.

[11] i.e. set of Horn clauses not containing function symbols other than constants.

[12] The terminology is not so well chosen: it may lead to a misunderstanding induced by the meaning of 'pure literal'. Maybe *separable clause* would have been better.

- π is a purifying substitution for $C\sigma$. Then attach to the leaf of $\mathcal{B}$ the following child nodes:

$$A_1\sigma\pi, \ldots, A_m\sigma\pi, \neg B_1\sigma\pi, \ldots, \neg B_n\sigma\pi$$

  marking the leaves of the first m branches leaves *open* and that of the last n *closed*.

$\sharp$

After this definition the similarities to SATCHMO are apparent.

An example of an application of hyper-extension given in [Baumgartner et al., 1996] is the following:

Assume that a branch contains only $R(f(x_1))$ and a clause

$$C\colon\ P(x), Q(x, y) \leftarrow R(x).$$

then $\sigma = \{x \leftarrow f(x_1)\}$ and

$$C\sigma = \ P(f(x_1)), Q(f(x_1), y) \leftarrow R(f(x_1)).$$

The head of $C\sigma$ is impure. But it can be made pure by applying the substitution $\pi = \{x_1 \leftarrow a\}$ (a being a constant). After application of the hyper extension step the hyper tableau has the two open branches given below:

$$\begin{array}{ll} R(f(x)) & R(f(x)) \\ P(f(a)) & Q(f(a), y) \end{array}$$

The intended model candidates for C are the least Herbrand models of each of these branches.

Experimental comparisons with with OTTER and SATCHMO shows that hyper tableaux performs well for satisfiable propositional and first-order non-Horn problems.

3.5 The Transformation Method

In the methods 1,2,3 and 4 the process of model building is not separable from the process of refutation. This is a common feature with RAMC, where refutation and construction of (Herbrand models) is carried out simultaneously. The model building methods MB and MBEQ in Chapter 3 are of a different type: Herbrand models are constructed as a *postprocessing* step of resolution decision procedures. Like for RAMC and RAMCET the output of MB and MBEQ consists of ARMs (not of CARMs as for the former ones) which represent Herbrand models. In [Fermüller and Leitsch, 1996] Fermüller and Leitsch showed that this need not be the end point of a model building procedure. Indeed for some syntactic classes of ARMs the representations can be *transformed* into finite models without any form of enumeration of search. As deductive procedures in clause logic typically produce Herbrand models such a postprocessing leads to a *purely deductive* construction of finite models. In the subsection 1.4 we have demonstrated how linear atomic representations of models are transformed into finite realizations (i.e. under preservation of the truth values for predicates). The advantage of the transformation method is twofold:

- The computational expense is largely independent of the domain size. Indeed the number of domain elements depends largely on the syntactic form of the ARMS. As the interpretation of predicate- and function symbols are *derived* from the ARM via congruence closure, there is no search involved. Thus finite models with *large domains* can be obtained by this method.

- The method is robust w.r.t. signature. While the number of possible finite interpretations increases dramatically with the signature, this is not the case for the transformation method. Thus finite models for problems with a high number of function- and predicate symbols can be built this way.

Clearly the disadvantage of the transformation method is also twofold:

- We do not always have an ARM.

- The ARM need not be linear.

But in the same way as automated theorem proving should be considered *also* as a discipline of proof transformation this is the case for model building as well. The transformation of the Herbrand models into finite models in Chapter 3 is a step into this direction.

Finally we would like to mention Tammet's method of finite model building on the monadic and on the Ackermann class [Tammet, 1991]. On these decidable classes of first-order logic (or more precisely on the corresponding clause classes) hyperresolution does not terminate and we obtain no ARMs. On the other hand specific ordering refinements terminate on these classes (for details see [Fermüller et al., 1993]). Tammet uses the finite fixed points $R_x^*(\mathcal{C})$ as a raw material for model building. The characteristic feature of this method, like of this mentioned above, is that no backtracking is required in the construction process. Though no representation of a Herbrand model appears in this procedure it is a transformation method in the sense that it makes use of a fixed point $R_x^*(\mathcal{C})$ not containing $\Box$, i.e. of a proof of the *satisfiability* of $\mathcal{C}$.

3.6 FINDER

In his system SCOTT (**S**emantically **C**onstrained **OTT**ER) J. Slaney incorporates his model generator FINDER (**FIN**ite **D**omain **E**numerato**R**) to the very well known and widely used theorem prover OTTER (see [Slaney, 1993]).

The idea underlying FINDER is extremely simple and corresponds to recursive enumerability of satisfiable formulae. In fact, if a theory has finite models, a natural way of finding them is to consider domains of increasing cardinality, all possible functions and relations that can be defined on them and *check* whether these functions and relations satisfy the axioms of the theory. The key point is of course to discover efficient or at least feasible enumerations for domains of "reasonable" size (practically it appears that 10 to 15 is a reasonable cardinality). FINDER [Slaney, 1992] performs a backtracking search, avoiding to backtrack twice for the same reason. As Slaney remarks [Slaney, 1993], FINDER works in a way different to logic programming. Thus this approach essentially differs from SATCHMO.

FINDER is not refutationally complete and the models it finds are not Herbrand models: it specifies functions on *arbitrary* domains and their values are given *explicitly*, e.g. $f(a,b) = c$. This allows to incorporate expert knowledge about the problem at hand in a simple way , e.g. how the searched model should look like or how to avoid searching too many isomorphic subspaces.

In summary, FINDER implements cleanly very simple ideas, resulting in a powerful system.

3.7 SEM

In [Zhang and Zhang, 1995] model building is regarded as a special case of the Constraint Satisfaction Problem (CSP). The treatment of the CSP is different from that of Plaisted et al. in 3.2.

SEM (a **S**ystem for **E**numerating finite **M**odels) is based upon two main techniques: an efficient implementation of constraint propagation and a heuristic which prevents considering many isomorphic (potential) partial models[13]. SEM, that combines features of FALCON and SATO, two older systems by the same authors, is presently considered as one of the most powerful systems for finite model building. SEM's inputs are set of clauses. As in FINDER, the models built by SEM are not Herbrand models.

[13]A model M_1 is *isomorphic* to a model M_2 if M_1 can be obtained by permuting element names in M_1.

The predicate- and function symbols are sorted (where sorts are finite). If

$$f \ : \ s_1 \times s_2, \ldots, \times s_n \rightarrow s; \quad s, s_i \in S$$

where S is a set of sorts, the terms $f(e_1, \ldots, e_n)$ for $e_i \in s_i$ $(1 \leq i \leq n)$ are called *cells* (a cell corresponds to a row of the multiplication table of f). By assigning elements in $D_1 \times D_2 \times, \ldots, \times D_{n+1}$ (obviously, elements in D_{n+1} correspond to the values assigned to the function for each cell) of the appropriate sort to cells, a partial *interpretation* of the corresponding function symbol is obtained.

In particular the model building process consists in enumerating the assignments to cells, partially instantiating clauses, and evaluating and propagating constraints. J. Zhang and H. Zhang study heuristics in order to improve the efficiency of SEM. Obviously, the crucial problem, how to choose the *next* cell, is considered. Particular attention is payed to the *reduction of the cardinality* of the D_i.

When two values e and e' are symmetric in the constraints (input clauses), given a potential model containing e and e', another potential model can *surely* be obtained by exchanging e and e' and *it is not necessary* to check it. In fact *Isomorphism rejection is crucial to the efficiency of finite model generation* ([Zhang and Zhang, 1995], page 300).

The authors borrow a heuristic from FALCON called LNH (**L**east **N**umber **H**euristic). Basically LNH eliminates from the candidates to be tested elements playing symmetric roles in the set of clauses[14].

The experiments on non trivial problems (some even on formerly open problems) show that clever implementation of enumerative methods can be very useful as an assistant mathematician (logician,...).

3.8 FMC

The method FMC (**F**inite **M**odel **C**onstruction) due to N. Peltier is an enumerative method using techniques similar to those used by FINDER and SEM. It is more general, much more formalized and can be adapted to logics other than first-order logic.

The method applies to first-order *formulae* and *no transformation is needed.* This is an important point, because though exponential growth in transformations to clausal form can be avoided by a structure preserving transformation (see [Baaz et al., 2001]), the latter one introduces new predicate symbols increasing the number of cells (see 3.7) to be checked.

[14]No formal study of symmetries is done, in particular concerning complexity. The authors appeal to the reader's intuition, see also 3.8.

Moreover FMC uses information from previous failures in order to prune the search espace (i.e. avoiding checking all the possible interpretations) and symmetries between elements of the domain.

The method works on sorted formulae and a total order $<$ is defined on cells. The set of partial interpretations on a set of cells can be ordered by lexicographically extending $<$.

The notion of *jump function* is introduced. The aim of a jump function is to avoid considering all the interpretations which might become a model. The jump function exploits information on previous failures in testing satisfiability. As an example of how jump functions are defined we present an example that N. Peltier called "refutation jump function" (refutation means here "the tested interpretation is not a model").

Let us consider a formula

$$F\colon \forall x.x \neq f(x), \quad \text{a domain } \{1,2\},$$
$$\text{and the interpretation } \mathcal{I} = \{f(1) = 1, f(2) = 1\}.$$

Once $f(1) = 1$ fails to satisfy F, it is *not necessary* to check any other interpretation containing $f(1) = 1$ because, clearly, all of them will falsify F.

A more sophisticated jump function is the *covering refutation* jump function. The idea of this new notion is to store cells responsible of previous failures. It is shown that this set has interesting properties allowing to simplify other refutations. It is impossible to impart here all the details and theoretical justifications of this work. The interested reader may consult [Peltier, 1997b].

The experimental results with FMC are comparable to those of FINDER and SEM.

The use of symmetries to simplify and speed up reasoning has been recognized already a long time ago, in particular by mathematicians (see [Polya, 1973] and [Rosen, 1995]). Its importance in the automation of reasoning, similarly to reasoning by analogy (that can also be seen as symmetry, see [Rosen, 1995]), using semantics, counterexamples, ... was highlighted since the very beginning of automated deduction (see for example [Gelernter, 1959]). A symmetry rule as an extension of resolution was introduced in [Krishnamurthy, 1985][15].

Only a few systems (either for model building or for automated theorem proving) incorporate symmetry (see 3.6), and one of the strong features of FMC is its extensive use of symmetry.

[15] As for reasoning by analogy, the problem of symmetry strongly depends on the *description* of the problem.

There is no doubt that the search for finite models can profit from an analysis of the search space in terms of *isomorphism classes*, since isomorphic interpretations are always equivalent w.r.t. formulae truth values. But the structure of the isomorphism class of an interpretation is rather complex, depending first on the signature (see [Boy de la Tour, 2000a]), and next on its inner symmetries. Few symmetries means many isomorphic interpretations, whence the idea to restrict the search space to only one interpretation per isomorphism class (see [Boy de la Tour, 2000b]). This novel approach relying on techniques from combinatorics, and more specifically from computational group theory, could bring a deeper understanding of finite model building procedures, their theoretical limits and potentials.

4. Finite models in different fields

Problems whose solution require techniques for finite model building (possibly in slightly different forms) exist in many areas, among the most important ones we have chosen:

- Mathematics and Logic
- Operations Research
- Constraint Programming (Constraint Solving)
- Artificial Intelligence

Albeit the problems and their solutions are highly related to each other, research and applications in some of these areas have not interacted as would have been desirable. Such interactions can be very fruitful as shown, e.g. by that of constraint programming and operations research (see for example [Marriott and Stuckey, 1998]).

4.1 Mathematics and Logic

The investigation of finite algebraic structures seems to be the mathematical area in which automated model building has a "natural" application and where several of its more striking results have been obtained (see 4.5.1).

It is the author's opinion that the investigation of *spectra* is, independent of its intrinsic mathematical importance, a challenge from the point of view of automated model building, and deserves some attention. But, to the best of our knowledge, there have been no attempts in this direction.

The concept of spectra has been introduced, according to [Fagin, 1975a], by H. Scholz in 1952. It is defined in [Ebbinghaus and Flum, 1999] (see also [Grätzer, 1979]) in the following way:

DEFINITION 6.6 Given a sentence F let the *spectrum* $Spec(F)$ be the set

$$Spec(F) = \{m \geq 1 \mid \text{ there is an } \mathcal{M} \text{ s.t. } \mathcal{M} \models F \text{ and } |\mathcal{M}| = m\}.$$

where $|\mathcal{M}|$ denotes the domain size of an interpretation $\mathcal{M}$. ♯

The theorem below expresses an important property of spectra (see [Fagin, 1976, Ebbinghaus and Flum, 1999]):

THEOREM 4.1 *For any first-order sentence F at least one of the sets $Spec(F)$ or $Spec(\neg F)$ is cofinite, i.e. there exists an n_0 s.t. $\{n \mid n \geq n_0\} \subseteq Spec(F)$ or $\{n \mid n \geq n_0\} \subseteq Spec(\neg F)$.*

Several problems on spectra of equational classes (varieties) are listed in [Grätzer, 1979] (see also [Fagin, 1975a, Fagin, 1975b, Mckenzie, 1975]). Some efforts have been made to investigate varieties V with $Spec(V) = \{1\}$, i.e. with only trivial finite models.

The only models $\mathcal{M}\colon (D, \Phi)$ of the following identity on a groupoid:

$$((y^2 \cdot y) \cdot x)((y^2 \cdot (y^2 \cdot y)) \cdot z) = x$$

are those fulfilling either $|D| = 1$ or $|D| = \infty$ (for $\Phi(\cdot)\colon D \times D \to D$). Proofs can be found in [Stein, 1963, Austin, 1965]; see also [Grätzer, 1979]).

This kind of problems suggests the use of symbolic methods able to handle finite and infinite models within the same framework (obviously, enumerative methods can only capture the trivial model of this groupoid).

We guess that the attempt to compute (fragments of) spectra can clearly point out some of the bottlenecks in finite model building, thus offering natural and arbitrarily difficult benchmarks.

It is not very difficult to give schemas of set of formulae with the spectra

$$2n,\ 3n,\ 5n,\ 7n,\ 11n,\ \ldots (n \in \mathbb{N} \setminus \{0\}).$$

The idea is to specify sets D using a binary relation $R \subseteq A \times B$ s.t.

$$\begin{aligned} D &= A \cup B \text{ and} \\ |D| &= |A \cup B| = i \cdot n \text{ where } n \in \mathbb{N} \setminus \{0\} \text{ and } i \text{ is prime.} \end{aligned}$$

To this aim we postulate the following properties:

- Each element in A is in relationship with exactly $i-1$ elements in B.
- If x in A is in relationship with $y_1, y_2, \ldots y_{i-1}$ in B and y in A (for $y \neq x$) is in relationship with $z_1, z_2, \ldots z_{i-1}$ in B then

$$\{y_1, y_2, \ldots y_{i-1}\} \cap \{z_1, z_2, \ldots z_{i-1}\} = \emptyset.$$

- $A \cap B = \emptyset$.
- All elements in B are in relationship with elements of A.
- If a bound m on the size of models is desired (i.e. we want finite spectra), a statement asserting that the cardinality of the domains is at most m must be added (see statement F_6 below).

For example, the spectrum of the set of the formulae F_1 to F_5 below is $3 \times n$:

F_1. $\forall x \exists y \exists z. P(x,y) \wedge P(x,z) \wedge y \neq z$.
% each element in A is in relationship with at least 2 elements in B.

F_2. $\forall x \forall y \forall u \forall z. P(x,u) \wedge P(x,y) \wedge P(x,z) \Rightarrow y = u \vee y = z \vee u = z$.
% each element is in relationship with at most 2 elements.

F_3. $\forall x \forall y \forall u \forall v \forall w \forall z. P(x,u) \wedge P(x,v) \wedge u \neq v \wedge x \neq y \wedge P(y,w) \wedge P(y,z) \wedge w \neq z \Rightarrow u \neq w \wedge u \neq z \wedge v \neq w \wedge v \neq z$.
% the set of elements in B corresponding to different elements in A are pairwise disjoint.

F_4. $\forall x \forall v \forall u \forall y. P(x,u) \wedge P(v,y) \Rightarrow u \neq v$.
% elements in A and in B are disjoint.

F_5. $\forall x \exists y. P(y,x)$
% there are no other elements in B than those in relationship with elements in A.

If the formula F_6 below is added, then $Spec\{F_1, \ldots, F_6\} = \{3, 6, 9\}$.

F_6. $\forall x \forall y \forall z \forall u_1 \forall u_2 \forall v_1 \forall v_2 \forall w_1 \forall w_2 \forall s \forall t. P(x,u_1) \wedge P(x,u_2) \wedge P(y,v_1) \wedge P(y,v_2) \wedge P(z,w_1) \wedge P(z,w_2) \wedge x \neq y \wedge x \neq z \wedge y \neq z \wedge P(s,t) \Rightarrow s = x \vee s = y \vee s = z$.
% There are at most 9 elements in D.

The results that can be obtained with SEM and FMC_{ATINF} for the set of formulae F_1 to F_5 are the following:

FMC_{ATINF} (on a PC Pentium 200 MHz, 64 Mb):

Card.	Model ?	Nb of tests	Time (in s.)
1	no	1	0
2	no	1	0
3	yes	37	0
4	no	11300	1.22
5	no	905500	58.22
6	yes	3302	0.66
7	?	?	out of memory
8	?	?	out of memory
9	?	?	out of memory

and with *SEM* (on a PC Athlon 1GHz, 256 MB) :

Card.	Model ?	Nb of tests	Time (in s.)
1	?	-	Instant. impossible
2	no	18	0.01
3	yes	11	0.01
4	no	284790	1.03
5	no	11254074	123.46
6	yes	65	0.98
7	?	?	out of memory
8	?	?	out of memory
9	?	?	out of memory

From these figures at least three facts deserve to be pointed out:

- The rather low bound in the computed fragment of the spectra.
- The huge number of tests for the negative case (as for $n = 5$, $n = 7$, $n = 8$ and $n = 9$.)
- The importance of having powerful formalism to express infinite (or very large) sets of models following a particular pattern.

Concerning the first point things are, of course, worse for set of formulae with more scattered spectra like $7 \times n$, $11 \times n, \ldots$ for which even models with $n = 1$ can be missed.

The second point is not negligible (and can easily become crippling). It confirms that to know *cardinalities* of models (i.e. fragments of a spectrum) can be of great *practical* help in the effective construction of finite models.

Sometimes, as in the (very simple) example below the use of constraints allows to compute cardinalities for which a problem has a solution.

The third point provides additional evidence of the importance of model representation formalisms, already evoked in detail in chapter 5.

EXAMPLE 6.7 In order to refute the claim that the conclusion F_8 follows from the premises F_1 to F_7 below, we use semantic tableaux, i.e. we try to build models of $\{G_1, \dots G_7\} \cup \{\neg G_8\}$.

Of course resolution would work as well on this example. But the point here is to illustrate that, in investigating cardinalities of models, we should develop techniques for computing (or bounding) cardinalities of domains in which one can expect to solve a set of constraints extractable from deductions by tableaux, resolution,...; in these computations we should avoid solving the constraints effectively.

If cardinalities can be computed exactly, the problem of computing (fragments of) spectra is solved. Otherwise (i.e. cardinalities are bounded) important information can be given to finite model builders that can be used in the standard way.

The formulae are

$$\begin{aligned}
&G_1\colon\ \exists x(\neg P(x) \wedge \neg Q(x) \wedge R(x)),\\
&G_2\colon\ \exists x(\neg P(x) \wedge Q(x) \wedge \neg R(x)),\\
&G_3\colon\ \exists x(\neg P(x) \wedge Q(x) \wedge R(x)),\\
&G_4\colon\ \exists x(\neg Q(x) \wedge \neg R(x)),\\
&G_5\colon\ \exists x(P(x) \wedge \neg Q(x)),\\
&G_6\colon\ \exists x(P(x) \wedge \neg R(x)),\\
&G_7\colon\ \exists x(P(x) \wedge Q(x)),\\
&G_8\colon\ \forall x(P(x) \vee Q(x) \vee R(x)).
\end{aligned}$$

and the tableau is

$$\begin{array}{c} \neg P(a) \\ \neg Q(a) \\ R(a) \\ \neg P(b) \\ Q(b) \\ \neg R(b) \\ \neg P(c) \\ Q(c) \\ R(c) \\ \neg Q(d) \\ \neg R(d) \\ P(e) \\ \neg Q(e) \\ P(f) \\ \neg R(f) \\ P(g) \\ Q(g) \\ \downarrow \\ \neg P(x) \\ \neg Q(x) \\ \neg R(x) \end{array}$$

Below we define the constraints allowing to build a counterexample (we implicitly assume that all constants are different).

The constraints induced by $P(x)$: $\underline{x \neq e} \wedge \underline{x \neq f} \wedge \underline{x \neq g}$.

The constraints induced by $Q(x)$: $\underline{x \neq b} \wedge \underline{x \neq c} \wedge x \neq g$.

The constraints induced by $R(x)$: $\underline{x \neq a} \wedge x \neq c$.

In order to be able to solve (simultaneously) this set of constraints it is necessary to consider universes D with cardinalities $\mid D \mid \geq 7$ (this results immediately from the set of underlined disequations). Stated differently, the reasoning is correct only if domains with cardinalities less or equal to 6 are considered. $\sharp$

4.2 Operational Research

The celebrated SAT problem (i.e. the problem of looking for a model of a given propositional formula) and optimization problems in mathematical programming, or more precisely, discrete or integer programming[16], furnish concepts and techniques that can be used (and has been used) in finite model building. Both of them are related to *enumeration*[17], the basic technique of all but one approaches to finite model building described in this chapter.

Of course we will not deal with SAT or mathematical programming, they are well established research areas extensively explored elsewhere. We will talk instead of a method in which the links between finite model building (and decision procedures) in first-order logic and mathematical programming can be seen clearly. Jeroslow seems to have been the first to suggest such kind of methods (see [Jeroslow, 1988]). More recently J. Hooker [Hooker, 1993a] has proposed a generalization of Jeroslow's approach, and (according to [Hooker, 1993a]) a dual approach for Datalog formulae to that of Gallo and Rago (in 1990). These approaches apply to the following decidable fragments of first-order logic (without function symbols): $\forall*, \exists*, \exists * \forall*, \forall*$Horn (the latter corresponds to a subset of Datalog -where constants are allowed). In [Jeroslow, 1988] an example gives hints as how to adapt the method to $\forall * \exists*$ formulae.

We present the appproach borrowing from [Jeroslow, 1988, Hooker, 1993a] what seems to be the most relevant features of it.

Jeroslow considers formulae in prenex normal form, Hooker requires clausal form which simplifies somewhat the treatment[18].

Given a formula F in one of the admitted fragments the method considers *partial instantiation* of variables (i.e. instantiates only a subset of $V(F)$) and assign truth values to the partially instantiated atomic formulae in order to evaluate F to **true**. As atomic formulae are partially instantiated, the assigned truth values can be inconsistent. Otherwise stated, on some points of the universe an evaluation to **true and** to **false** of the same predicate has been proposed.

In this case the valuation is *blocked* (Jeroslow). This blockage can be detected by using an unification algorithm[19].

An *unblocking* algorithm is proposed. The idea is that the less instantiated of the two partially instantiated atom causing the blocked

[16]Given a function f on the variables $x_1, \ldots, x_n$ and a set of constraints to be satisfied by these variables, find the integer values of $x_1, \ldots, x_n$ that minimizes (or maximizes) f.

[17]Avoiding as much as possible, "brute force" methods.

[18]A linear transformation to clausal form is proposed in [Hooker, 1993a] . No mention is made to the well known results in automated deduction on these transformations.

[19]In [Hooker, 1993a] a linear algorithm to compute the m.g.u. of two literals that is a particular case of Martelli-Montanari's (or Huet's) unification algorithm (see Chapter 2) is proposed. No mention is made to these or other unification algorithms. The terminology is inaccurate w.r.t. the usual one: for example the literal $P(a, b)$ is called a unifier of $P(x, y)$ and $P(x, b)$ (instead of their *common instance*); $P(x, b)$ is called a most general unifier of $P(x, y)$ and $P(x, b)$ (instead of their *most general common instance*).

valuation inheritates the truth value from the more instantiated one after application of the mgu.

The approach has similarities and differences with the usual approaches in automated deduction.

In [Jeroslow, 1988] page 191 they are clearly pointed out:

> Quite clearly, this approach to removing blockage via, in essence, examining possible unifications, is reminiscent of many other techniques for theorem proving based on resolution. However, it differs in:
>
> 1 Being guided by truth valuations, and being aimed directly at removing obstacles to the satisfaction of the formula.
>
> 2 Having tests (...) which allows termination long before all unifications are performed.
>
> 3 Not being restricted to clausal form.
>
> 4 Utilizing the $\forall *$ fragment in an optimal way, as regards complexity.

The method considers iteratively a growing conjunction of partially instantiated matrices, until either unsatisfiability is detected or the formula is proven satisfiable by building a model of it. In order to test satisfiability of the growing conjunction of matrices it is proposed in [Hooker, 1993a] to use *incremental satisfiability algorithms* as in [Hooker, 1993b]. In [Jeroslow, 1988] time complexity results of the partial instantiation procedure are proven. No experimental results are given neither in [Jeroslow, 1988] nor in [Hooker, 1993a].

The following very simple example from [Jeroslow, 1988] illustrates the main feature of the method.

EXAMPLE 6.8 The aim is to decide the (un)satisfiability of the formula:

F: $\forall x \forall y \forall z (P(a, x) \Rightarrow P(y, z))$

(Where a denotes a constant, x, y, z denote variables).

The method starts by assigning truth values to partially instantiated atoms in order to evaluate F to **true**

$$M_1 : \underbrace{P(a,x)}_{\textbf{false}} \Rightarrow \underbrace{P(y,z)}_{\textbf{true}}$$

But this valuation is *blocked*, because they assign contradictory truth values to the same atom in one point of the universe, i.e. the atom under the unifier

$$\{x \leftarrow a,\ y \leftarrow a,\ z \leftarrow a\}.$$

In order to remove the blockage the m.g.u of $P(a, x)$ and $P(y, z)$ is applied and truth valuation is inherited from the more instantiated atom, to yield:

$$M_2 : \underbrace{P(a,x)}_{\textbf{false}} \Rightarrow \underbrace{P(a,x)}_{\textbf{false}}$$

$M_1 \wedge M_2$ is satisfiable, the method builds a model (D, Φ) with $D = \{a\}$ and $\Phi(P(a,a)) = \mathbf{false}$. ♯

4.3 Constraint Programming (Constraint Solving)

The ubiquity of constraint satisfaction problems and the naturalness constraints allow in specifying problems, easily explain the present importance of constraints in programming languages. The three most important types of constraints in constraint programming are arithmetic constraints, tree constraints and finite domain constraints (see for example [Marriott and Stuckey, 1998]). We are interested here in *finite domain constraints*[20].

Given a set of variables $\mathcal{V}$, a finite domain D and k-ary relations $\mathcal{R}_i$ on D^k (denoted by the predicate symbol R_i), are there assigments of values of D to variables in $\mathcal{V}$, i.e. a substitution $\sigma : \mathcal{V} \to D$ such that the formula

$$(*) \qquad \bigwedge_{1 \leq i \leq n} R_i(x_{i_1}, \ldots, x_{i_n})$$

is satisfied ? Looking for this assignment is called *solving constraints.*

Otherwise stated, given a finite set D (possibly specified as a union of finite sets in which different variables must take their values), are there models of the formula (*) with domain D (i.e. *finite models*) ?

In solving constraints a very useful notion is that of *projection*, that is to say a *simplification* with respect to distinguished variables (often called *variables of interest*, for obvious reasons). Of course any variable in the constraint can be considered as a variable of interest. The aim is to *eliminate* variables. It is required that all the solutions of the initial constraint be solutions of the projected one, and that each solution of the projected constraint be a *partial* solution (i.e. when restricted to the new set of variables both solution coincides) of the original one. More formally:

DEFINITION 6.9 The *projection* of a constraint C onto a variable set *vars*, where $vars \subseteq V(C)$ is a constraint C_P with $V(C_P) = vars$, such that:

If σ is a solution of C the σ is a solution of C_P.

If $\sigma_R = \sigma \mid_{vars}$ and σ_R is a solution of C_P then σ_R is a partial solution of C (i.e. there exists an extension of σ_R that is a solution of C). ♯

[20] Solving finite domain constraints can be reduced in the obvious way to the SAT-problem.

This notion and the related algorithm is one of those allowing to identify relationships between methods in finite model building.

We have chosen the rule given in [Marriott and Stuckey, 1998], pages 56-57 to solve Boolean constraints by projection.

It relies on the observation that for eliminating a variable x from a Boolean constraint C (thus obtaining the projection of C on $V(C)-\{x\}$) the following equivalence can be used:

$$C \Leftrightarrow (x \wedge C_x) \vee (\neg x \wedge C_{\neg x})$$

where C_x is obtained by replacing all occurrences of x by 1, and $C_{\neg x}$ is obtained by replacing all occurrences of $\neg x$ by 0.

This rule corresponds exactly to the Davis and Putnam's splitting rule[21].

In the following example, taken from [Marriott and Stuckey, 1998], S' and S" are the two matrices generated by the splitting rule, C_Y and $C_{\neg Y}$ are implicit operations in Davis and Putnam's method.

$S : (\neg X \vee Y) \wedge (\neg X \vee Z) \wedge (X \vee \neg Y \vee \neg Z)$
$C_Y : (\neg X \vee 1) \wedge (\neg X \vee Z) \wedge (X \vee \neg 1 \vee \neg Z)$
simplifies to:
$S' : (\neg X \vee Z) \wedge (X \vee \neg Z)$
$C_{\neg Y} : (\neg X \vee 0) \wedge (\neg X \vee Z) \wedge (X \vee \neg 0 \vee \neg Z)$
simplified to:
$S'' : \neg X \wedge (\neg X \vee Z)$
that, by subsumption, simplifies to $\neg X$.

4.4 Artificial Intelligence

Naturally models play an important role in Artificial Intelligence. Handling models of propositional formulae is important in the study of abduction, common sense reasoning, default reasoning, inference in Bayesian belief networks,... A lot of works exits (see for example [Khardon and Roth, 1994, Roth, 1996, Birnbaum and Lozinskii, 1999]). The difficult point in these problems is to be able to master complexity (most of them have been shown untractable). Some positive results exits nevertheless (e.g. [Zhang, 1996]). For example those using counting of models ([Bayardo and Pehoushek, 2000, Birnbaum and Lozinskii, 1999]) for approximate reasoning. In [Khardon and Roth, 1994] the authors address common sense reasoning. They characterize propositional theories (Knowlede Bases, or KB for short) by sets of (some) models of them,

[21] Davis and Putnam's method is not mentioned in [Marriott and Stuckey, 1998].

the so called *model-based representation*. The aim is to reason (i.e. to implement a consequence relation) in reasonable time on these KB.

The underlying idea in [Khardon and Roth, 1994] is to use a set of models of the KB (called *test sets*) such that if they are not countermodels of a given query, it can be concluded that the query is true in KB. The authors define such sets allowing to answer on the basis of these particular KB representation *for a restricted set of queries*. An order on $\{0,1\}^n$ is used to compare assignments of truth values and *basis* (with the usual meaning) for class of interpretations are defined. The authors also shown how to perform abduction (in propositional logic) using a model-based representation.

Kowalski and others pointed out the duality between abduction and model building. Given a logic program $\mathcal{P}$ and a fact to prove, an explanation, i.e. the formulae inferred by abduction, corresponds exactly to a model of the "only-if" part of $\mathcal{P}$ (see [Console et al., 1991, Denecker and de Schreye, 1994]).

Therefore the capability of building models leads naturally to a method for abductive reasoning. Such a method (an extension of RAMC) is proposed in [Caferra and Peltier, 1998]. It is more powerful than other methods for abduction in logic programming.

4.5 Practical applications of model building: some results

Model builders (and theorem provers used like that) have been applied with nice results.

We mention only some of the many results obtained with these systems in order to give a flavour of the kind of problems one can expect to handle with these programs. Results have been obtained basically in the following fields:

1 Mathematics and Logics

2 Automated Deduction

3 Program verification

4.5.1 Mathematics and Logics

Wos' group at Argonne was pionnier in looking at theorem provers *also* as model generators. Winker [Winker, 1982] (see also [Wos, 1988]) solved *open* questions in ternary algebras (i.e. algebras with a ternary operator) using interactively[22] a theorem prover.

In a given 5-axioms axiomatization the problem was to know if any of 2 axioms is independent of the remaining 4. By building small (cardinality 4 or less) models of 4 axioms, that are countermodels of the fifth one, Winker answered positively to the open question.

In [Winker et al., 1981] (see also [Wos, 1993]) it is answered positively (in an interactive way) by building finite models to an open question on the existence of finite semigroups with automorphism but no involution.

J. Slaney has used his program FINDER in order to prove results about semigroups.

J. Zhang and H. Zhang have used SEM to prove results about semigroups (see [Zhang and Zhang, 1995, Zhang, 1994]).

W. McCune has answered to a question on orthomodular and ortholattices (problems arising from quantum logic) using his program MACE (**M**odels **A**nd **C**ounter-**E**xamples) (see [McCune, 1998, McCune and Padmanabham, 1996]).

In [Rose and Wilkinson, 2001] Rose and Wilkinson have used MACE and SEM to study problems in lattice theory.

In [Fèvre and Peltier, 1997] Fèvre and Peltier present an automated proof of the incompleteness of a constructive geometry of von Plato and of the independence of its axioms by using FMC_{ATINF} and $RAMC_{\mathrm{ATINF}}$.

In [Peltier, 1998] N. Peltier uses the same systems in order to build (in *almost* automatic way) an *infinite* model of the non finitely satisfiable formula introduced by Goldfarb in order to prove the undecidability of the Gödel class with identity (thus invalidating a Gödel's conjecture).

4.5.2 Automated Deduction

Research on model building in the field of automated deduction have naturally been applied to research *on* automated deduction.

Once a system is able to build models an obvious question is: "How to use these models in order to *guide* proof search ?". The most natural answer seems to be the use of semantic resolution (see [Slagle, 1967]). Slaney uses FINDER capabilities in order to build, when possible, models required by semantic resolution to select clauses to be potentially used. SCOTT can be considered as a good answer to Slagle's wish [Slagle, 1967]:

[22] The user made major decisions, like what kind of model to look for (cyclic, symmetric,...) and what equalities looked interesting or useful. The program provided the calculating power to generate and explore consequences of the user decisions ([Winker, 1982], page 283).

> A disadvantage of a semantic strategy is that the program must at present be given a model along with the theorem to be proved.
>
> One might hope that someday the program could derive its own models.

J. Slaney, who has obtained some striking practical results with FINDER and SCOTT (difficult problems in Łukasiewicz's propositional calculus for pure implication, extending intuitionist pure implication, axiomatizations of many-valued logics) lucidly wrote in his nice paper [Slaney, 1993]:

> In theorem proving there are no magic bullets. Not one technique gives easy solutions to all problems.

This way of automatizing semantic resolution should be compared with the one by Caferra and Peltier ([Caferra and Peltier, 1995]). In order to construct the models that will guide the search constraints are used to impose conditions (coded in the rules of the method) to force a c-clause (see 4) to be (or not to be) evaluated to false in the partial (finite or infinite) model being (automatically) built.

The method, implemented as an extension of OTTER allowed to shorten proofs as well as to reduce the search space in some examples (for details see [Caferra and Peltier, 1995]).

But, what does it mean to *fully automatize* semantic resolution ? For finite interpretations a full automatization is not a real problem. Very likely Slagle also emphasized the need to do semantic resolution over non trivial infinite models. In this sense the results in this book represent a step into the right direction. Clearly in a very general sense there can be no full automatization of semantic resolution.

4.5.3 Program verification

The examples in this part are not of the same kind of those in parts 1 and 2 above. Model building there concerns problems whose solution would have been very likely too difficult or impossible to compute by hand. The examples below, though coming from real applications (teaching programming, verification of simple programs) by no means correspond to "big" applications. Nevertheless they are embryonic applications of model building to simple cases of *hard* practical problems when programs of bigger sizes, or using more complex data structures, are considered. Besides, they show evidence that there is room for symbolic model building methods in practical applications.

In [Caferra and Peltier, 1997a] the authors present a way of merging Presburger arithmetic with equational problems in order to obtain a method allowing to verify and help to correct logic programs w.r.t. first-order logic specifications via *counterexamples* building, i.e. models of the program that are not models of its specification[23]. As the basis of the method is model building it can be also used to correct logic programs also w.r.t. *informal* specifications (such as those used in teaching programming). The ability to construct models could be used in testing (of course informally) whether formal specifications fit the specifier's *intention* as well as consistency of specifications. Frequently partial models suffice to detect errors. As in the previous section we refer the interested reader to the publications describing these works. In unpublished works (see also [Peltier, 1997b]) techniques of (finite and infinite) model building have been applied in order to detect errors (and to help the programmer or the student to correct them) in some simple imperative programs. We give below some details of unpublished running (interactively or automatically) examples.

EXAMPLE 6.10 (MISSING ANSWERS)

The following program is *intended* to detect all the cycles of length 2 in a graph with n nodes. In fact, for each node N it detects 1 cycle containing N and the others (if any) are missing. We show how $RAMC_{ATINF}$ could be applied in program verification for imperative programs.

```
begin
  p := 1;
  while p ≤ n
  do
    q := 1;
    cyc := false;
    while (q ≤ n and not cyc)
    do
      if (arc[p, q] and arc[q, p]) then
        cyc := true;
        cycle[p,q] := true;
      (*)
      q := q + 1;
    od
    p := p + 1;
  od
end
```

For the sake of clarity the formulae $I(p)$ and $I'(p, q)$ are introduced.

[23] A logic program is *correct* if the objects it deduces (computes) fit the specification (see, for ex. [Hogger, 1984]).

$cycle[p,q]$ is correct iff $(arc[p,q] \wedge arc[q,p]) \Leftrightarrow cycle[p,q]$.
$I(p)$ means that $cycle[p',q]$ is correct if $p' < p$, and **false** if $p' > p$.
$I'(p,q)$ means that $cycle(p,q')$ is correct if $q' < q$ **false** otherwise.

$$
\begin{array}{lllll}
I(p) & \forall q', p' \in [1,\ldots,n]^2 & & & \\
 & \quad (1 \leq p' < p) & \Rightarrow & (cycle(p',q') \Leftrightarrow (arc(p',q') \wedge arc(q',p'))) \\
 & \quad (p < p' \leq n) & \Rightarrow & (\neg cycle(p',q')) \\
 & & & \\
I'(p,q) & \forall q' \in [1,\ldots,n] & & \\
 & \quad (1 \leq q' < q) & \Rightarrow & (cycle(p,q') \Leftrightarrow (arc(p,q') \wedge arc(q',p))) \\
 & \quad (q \leq q' \leq n) & \Rightarrow & (\neg cycle(p,q'))
\end{array}
$$

After generation of pre and post-condition we get:

```
begin
  {I(1) ∧ I'(1,1)}
  p := 1;
  {I(p) ∧ I'(p,1)}
  while p ≤ n
  do
    {I(p) ∧ I'(p,1) ∧ p ≤ n}
    {I(p) ∧ I'(p,1)}
    q := 1;
    cyc := false;
    while (q ≤ n and not cyc)
    do
      {I(p) ∧ I'(p,q) ∧ q ≤ n ∧ ¬cyc}
      if (arc[p,q] and arc[q,p]) then
        cyc := true;
       cycle[p,q] := true;
      (*)
      {I(p) ∧ I'(p,q+1)}
      q := q + 1;
      {I(p) ∧ I'(p,q)}
    {I(p) ∧ I'(p,q) ∧ ¬(q ≤ n ∧ ¬cyc)}
    {I(p+1) ∧ I'(p+1,1)}
    p := p + 1;
    od
    {I(p) ∧ I'(p,1)}
  od
  {I(p) ∧ I'(p,1) ∧ p > n}
end
```

From the pre and post-conditions of the program, the following formulae are obtained:

$$I(p) \wedge I'(p,q) \wedge \neg(q \leq n \wedge \neg cyc) \wedge \neg(I(p+1) \wedge I'(p+1,1))$$

And after transformation (by hand) into clausal form:

1	$[\![cyc : (q \leq n)]\!]$
2	$[\![-cycle(x,y) \vee arc(x,y) : (1 \leq x < p) \wedge (1 \leq y \leq n)]\!]$
3	$[\![-cycle(x,y) \vee arc(y,x) : (1 \leq x < p) \wedge (1 \leq y \leq n)]\!]$
4	$[\![-arc(x,y) \vee -arc(y,x) \vee cycle(x,y) : (1 \leq x < p) \wedge (1 \leq y \leq n)]\!]$
5	$[\![-cycle(x,y) : (p < x \leq n) \wedge (1 \leq y \leq n)]\!]$
6	$[\![-cycle(p,x) \vee arc(x,y) : (1 \leq x < q)]\!]$
7	$[\![-cycle(p,x) \vee arc(y,x) : (1 \leq x < q)]\!]$
8	$[\![cycle(p,x) \vee -arc(y,x) \vee -arc(x,y) : (1 \leq x < q)]\!]$
9	$[\![-cycle(p,x) : (q \leq x \leq n)]\!]$
10	$[\![-cycle(i,j) \vee -arc(i,j) \vee -arc(j,i) \vee cycle(p+1,j) : (i \leq p+1)]\!]$
11	$[\![cycle(i,j) \vee arc(i,j) \vee cycle(p+1,j) : \top]\!]$
12	$[\![cycle(i,j) \vee arc(j,i) \vee cycle(p+1,j) : \top]\!]$
13	$[\![cycle(i,j) \vee cycle(p+1,j) : (i \geq p+1)]\!]$
14	$[\![\square : 1 > i \vee i > n]\!]$
15	$[\![\square : 1 > j \vee j > n]\!]$
16	$[\![-cyc \vee cycle(p,q-1) : \top]\!]$
17	$[\![cyc \vee -cycle(p,q-1) : \top]\!]$

Remark:
Here i, j, n, p, q are integer *constants* (uninterpreted) and x, y are *integer variables* (see [Caferra and Peltier, 1997b] for more details on the handling of integer constant in constraint solving).

RAMC gives[24] the clauses below: [25]

[24]This result *could* be an automated output of the RAMC system. We give here a hand simulation because Presburger arithmetic has not been implemented.
[25]Clauses 14 and 15 are used to simplify constants (as for ex. in clause 19)

19	(res,5,13)	$[\![cycle(i,j) : (i \geq p+1)]\!]$
20	(dis-res,5,13)	$[\![cycle(i,j) \vee cycle(p+1,j) : \bot]\!]$
21	(res,5,19)	$[\![\Box : (i \geq p+1)]\!]$
22	(dis-res,5,19)	$[\![cycle(i,j) : \bot]\!]$
23	(res,5,10)	$[\![\neg cycle(i,j) \vee \neg arc(i,j) \vee -arc(j,i) : (1 \leq p+1)]\!]$
24	(dis-res,5,10)	$[\![\neg cycle(i,j) \vee \neg arc(i,j) \vee -arc(j,i) \vee -cycle(p+1,j) : \bot]\!]$
25	(dis-res,4,23)	$[\![\neg cycle(i,j) \vee \neg arc(i,j) \vee -arc(j,i) : (i \leq p+1) \wedge (i \geq p)]\!]$
26	(dis-res,4,23)	$[\![\neg cycle(i,j) \vee \neg arc(i,j) \vee -arc(j,i) : (i < p)]\!]$
27	(res,4,10)	$[\![\neg arc(i,j) \vee \neg arc(j,i) : (i < p)]\!]$
28	(res,5,11)	$[\![cycle(i,j) \vee arc(i,j) : \top]\!]$
29	(dis-res,5,11)	$[\![cycle(i,j) \vee arc(i,j) \vee cycle(p+1,j) : \bot]\!]$
30	(res,2,28)	$[\![arc(i,j) : (i < p)]\!]$
31	(res,5,12)	$[\![cycle(i,j) \vee arc(j,i) : \top]\!]$
32	(dis-res,5,12)	$[\![cycle(i,j) \vee arc(j,i) \vee cycle(p+1,j) : \bot]\!]$
33	(res,3,31)	$[\![arc(j,i) : (i < p)]\!]$
34	(res,30,27)	$[\![\neg arc(j,i) : (i < p)]\!]$
35	(res,33,24)	$[\![\Box : (i < p)]\!]$
36	(simp,21,35)	$[\![\Box : i \neq p]\!]$
37	(res,8,25)	$[\![\neg arc(p,j) \vee \neg arc(j,p) : (1 \leq j < q)]\!]$
38	(res,6,38)	$[\![arc(p,j) : (1 \leq j < q)]\!]$
39	(res,7,31)	$[\![arc(j,p) : (1 \leq j < q)]\!]$
40	(res,37,38)	$[\![\neg arc(j,p) : (1 \leq j < q)]\!]$
41	(res,39,40)	$[\![\Box : (1 \leq j < q)]\!]$
42	(simp,15,41)	$[\![\Box : j < q]\!]$
43	(res,9,28)	$[\![arc(p,j) : \top]\!]$
44	(dis-res,9,28)	$[\![cycle(i,j) \vee arc(i,j) : \bot]\!]$
45	(res,9,31)	$[\![arc(j,p) : \top]\!]$
46	(dis-res,9,28)	$[\![cycle(i,j) \vee arc(j,i) : \bot]\!]$
47	(simp,35,15,1)	$[\![cyc : \top]\!]$
48	(res,47,16)	$[\![cycle(p,q-1) : \top]\!]$
49	(res,48,6)	$[\![arc(p,q-1) : \top]\!]$
50	(res,48,7)	$[\![arc(q-1,p) : \top]\!]$

The following model is obtained

41	(resolution,39,40)	$[\![\Box : (1 \leq j < q)]\!]$
45	(resolution,9,31)	$[\![arc(j,p) : \top]\!]$
47	(simplify,35,15,1)	$[\![cyc : \top]\!]$
48	(resolution,47,16)	$[\![cycle(p,q-1) : \top]\!]$
49	(resolution,48,6)	$[\![arc(p,q-1) : \top]\!]$
50	(resolution,48,7)	$[\![arc(q-1,p) : \top]\!]$
36	(simplify,21,35)	$[\![\Box : i \neq p]\!]$

This model indicates *why* the program is not correct, thus helping to correct the initial program. Here $cyc :=$ **false** must replace (*). In fact the boolean *cyc* becomes useless because it will be always **false**, independently of the input graph.

It should be pointed out that pre and post-conditions are introduced by the user and the choice of formulae on which RAMC was applied (i.e. the strategy) was not automatic. ♯

EXAMPLE 6.11 [26]

In [Couchot et al., 2003] the authors deal with the problem of proving invariants in a decidable fragment of set theory. The characteristic function of sets are encoded by Booleans arrays indexed on the elements of the set.

Proof obligations are translated to fist-order logic with equality and augmented with a particular theory: the theory of array with extensionality.

The example chosen by the authors is that of a specification of a simple scheduler.

What is relevant to model building in this work is that when the proof of the formula corresponding to the invariant preservation fails, a subformula can be extracted whose model (if built) can "explain" *why* the (negation of) the formula is not valid. The authors tried to build a model for such a formula with the well known *MACE* and *SEM* but failed. We used FMC_{ATINF} that allowed to build the model given below.

```
%% Clauses

list(sos).

% Boolean theory

tt != ff.

% Array theory

(all A,X,E. (read(write(A, X, E), X) = E)).
(all X,J,A,E. X = J | read(write(A, X, E), J) = read(A, J)).

% Set theory

(all X. read(arrayEmpty, X) != tt) .
arrayK_25 = write(arrayEmpty, pw, tt) .

(all X. read(arrayInter_17, X) != tt | read(arrayActive, X) = tt).
(all X. read(arrayInter_17, X) != tt | read(arrayWaiting, X) = tt).
(all X. read(arrayActive, X) != tt | read(arrayWaiting, X) != tt |
```

[26] We thank Silvio Ranise (Silvio.Ranise@loria.fr) who communicated this problem to us.

```
read(arrayInter_17, X) = tt).

(all X. read(arrayInter_14, X) != tt | read(arrayReady, X) = tt).
(all X. read(arrayInter_14, X) != tt | read(arrayWaiting, X) = tt).
(all X. read(arrayReady, X) != tt | read(arrayWaiting, X) != tt |
read(arrayInter_14, X) = tt).

(all X. read(arrayInter_11, X) != tt | read(arrayReady, X) = tt).
(all X. read(arrayInter_11, X) != tt | read(arrayActive, X) = tt).
(all X. read(arrayReady, X) != tt | read(arrayActive, X) != tt |
read(arrayInter_11, X) = tt).

(all X. read(arrayInter_35, X) != tt | read(arrayActive_p, X) = tt).
(all X. read(arrayInter_35, X) != tt | read(arrayWaiting_p, X) = tt).
(all X. read(arrayActive_p, X) != tt | read(arrayWaiting_p, X) != tt |
read(arrayInter_35, X) = tt).

(all X. read(arrayEmpty, X) = tt | read(arrayEmpty, X) = ff).

(all X. read(arrayActive, X) = read(arrayEmpty, X)) .
(all X. read(arrayWaiting_p, X) = read(arrayWaiting, X)) .
(all X. read(arrayReady_p, X) = read(arrayReady, X)) .
(all X. read(arrayActive_p, X) = read(arrayK_25, X)) .
(all X. read(arrayWaiting, pw) = tt).
(all X. read(arrayInter_17, X) = read(arrayEmpty, X)) .
(all X. read(arrayInter_14, X) = read(arrayEmpty, X)) .
(all X. read(arrayInter_11, X) = read(arrayEmpty, X)) .
(all X. read(arrayInter_35, sk_index) != read(arrayEmpty, sk_index)).

% end_of_list.
```

Model built by FMC:

```
0 write(0,0,0)=0
1 read(0,0)=0
2 write(0,0,1)=1
3 write(0,1,0)=0
4 write(0,1,1)=2
5 write(1,0,0)=0
6 write(1,0,1)=1
7 write(1,1,0)=1
8 write(1,1,1)=3
9 read(0,1)=0
10 read(1,0)=1
11 read(1,1)=0
12 write(2,0,0)=2
```

```
13 write(2,0,1)=3
14 write(2,1,0)=0
15 write(2,1,1)=2
16 read(2,0)=0
17 read(2,1)=1
18 write(3,0,0)=2
19 write(3,0,1)=3
20 write(3,1,0)=1
21 write(3,1,1)=3
22 read(3,0)=1
23 read(3,1)=1
24 arrayEmpty=0
25 sk_index=0
26 arrayInter_35=0
27 arrayInter_11=0
28 arrayInter_14=0
29 arrayInter_17=0
30 tt=1
31 arrayWaiting=0
32 pw=0
33 arrayK_25=1
34 arrayActive_p=1
35 arrayReady=0
36 arrayReady_p=0
37 arrayWaiting_p=0
38 arrayActive=0
39 ff=0

Run Time:      0.02

Tested: 301
```

♯

Chapter 7

CONCLUSION

In the same sense as proofs are more that just the fact of provability, models provide more information than that of satisfiability. Therefore the discipline of automated model building can be considered as a "logical" continuation of the enterprise to investigate theorem provers as decision procedures. So the main theme of this book is the analysis and the increase of semantic information generated by computational inference systems.

We demonstrated how well established inference systems on clause logic, like hyperresolution and ordered paramodulation, can be adapted for the construction of model representations on satisfiable sets of clauses. We also showed how and under which conditions Herbrand models can be transformed into finite models. As there are cases where the model representations cannot be transformed into finite models (as finite models do not exist at all) methods of *symbolic* model building clearly go beyond model generation via finite domain search. In a second step we presented inference systems, like RAMC, which are specifically designed for refutation *and* model building on clause logic. In these systems we used the language of equational constraints in order to partition the set of ground instances of a clause and to delete redundant information. Both systems produce (on satisfiable sets and in case of success) finite sets of atoms (or of constrained atoms), which represent Herbrand models of the original set of clauses. It is this point where the current limitations of symbolic model building methods become most apparent: the expressivity of the model representations produced by current systems is quite weak; even the construction of some very simple Herbrand models requires formalisms beyond clause logic (without and with constraints). This problem is addressed in a specific chapter on model representation and evaluation where we presented different extensions of clause logic, which are good candidates for new and more expressive model building systems. It seems that methods like inductive generalization and schematizations (using the expressivity of natural numbers) will play a major role in future model building and in automated deduction as a whole.

Finally, the authors hope that this work contributes to the understanding of logical information in automated deduction and supports the claim that automated theorem provers *can* be and *should* be *more* than fast and clever proof generators. Indeed the model building methods shed more light on the very nature of inference as a transformation of formulae or set of formulae into some kind of "normal form"; by this we mean a form which allows the extraction of a particular information (like a model of a satisfiable formula) in a simple way. In this sense model building is an important and natural inference activity, like inductive and nonmonotonic inference.

Still much more ambitious, our aim is to deepen the *understanding* of inference in general:

> *... Thus proof and evidence are the same. And what is it that makes a judgement evident to you? Before you have understood or grasped the judgement, it is not evident to you, and, when you have grasped it, it is obvious or evident to you. Thus is simply your act of understanding or grasping it which confers evidence on the judgement, that is, which makes it evident to you. This is one way of seeing that the proof of a judgement is nothing but the act of knowing, or, perhaps better, the act of understanding or grasping, and that what you grasp, namely, the object of knowledge, is the same as what you prove, namely, the assertion or judgement. [Martin-Löf, 1987].*

Notation Index

Subject Index

References

[Amaniss et al., 1993] Amaniss, A., Hermann, M., and Lugiez, D. (1993). Etude comparative des méthodes de schématisation de séquences infinies de termes du premier ordre. Research Report 93-R-114, Centre de Recherche en Informatique de Nancy.

[Appel and Haken, 1978] Appel, K. and Haken, W. (1978). The Four-color Problem. In Steen, L. A., editor, *Mathematics Today, Twelve Informal Essays*, pages 153–180. Springer-Verlag.

[Austin, 1965] Austin, A. K. (1965). A note on models of identities. *Proceedings of the American Mathematical Society*, 16:522–523.

[Avron, 1991] Avron, A. (1991). Simple consequence relations. *Information and Computation*, 92:105–139.

[Baader and Snyder, 2001] Baader, F. and Snyder, W. (2001). Unification theory. In Robinson, A. and Voronkov, A., editors, *Handbook of Automated Reasoning*, volume I, chapter 8, pages 445–532. Elsevier Science.

[Baaz, 1996] Baaz, M. (1996). Personal communication.

[Baaz et al., 2001] Baaz, M., Egly, U., and Leitsch, A. (2001). Normal form transformations. In Robinson, A. and Voronkov, A., editors, *Handbook of Automated Reasoning*, volume I, chapter 5, pages 273–333. Elsevier Science.

[Baaz and Leitsch, 1992] Baaz, M. and Leitsch, A. (1992). Complexity of resolution proofs and function introduction. *Annals of Pure and Applied Logic*, 20:181–215.

[Bachmair and Ganzinger, 1994] Bachmair, L. and Ganzinger, H. (1994). Rewrite-based equational theorem proving with selection and simplification. *Journal of Logic and Computation*, 3(4):217–247.

[Ballantyne and Bledsoe, 1982] Ballantyne, A. M. and Bledsoe, W. W. (1982). On generating and using examples in proof discovery. In Hayes, J.E., Michie, D., and Pao, Y-H., editors, *Machine Intelligence 10*, pages 3–39. Ellis Horwood.

[Barendregt, 1984] Barendregt, H. P. (1984). *The Lambda calculus. Its Syntax and Semantics*, volume 103 of *Studies in Logic and The Foundations of Mathematics.* North Holland.

[Barwise and Etchemendy, 1994] Barwise, J. and Etchemendy, J. (1994). *Hyperproof*, volume 42 of *CSLI Lecture Notes.* CSLI Publications.

[Barwise and Feferman, 1985] Barwise, J. and Feferman, S., editors (1985). *Model-theoretic logics.* Perspectives in Mathematical Logic. Springer-Verlag.

[Baumgartner et al., 1996] Baumgartner, P., Furbach, U., and Niemelä, I. (1996). Hyper-tableaux. In *Logics in AI, JELIA'96.* Springer.

[Bayardo and Pehoushek, 2000] Bayardo, R.J.Jr. and Pehoushek, J.D. (2000). Counting models using connected components. In *Proc. of AAAI 2000.*

[Beeson, 1988] Beeson, M. (1988). Computerizing mathematics: Logic and computation. In Herken, Rolf, editor, *The Universal Turing Machine, A Half-Century Survey*, pages 191–225. Oxford University Press.

[Beeson, 1985] Beeson, M. J. (1985). *Foundations of constructive mathematics*, volume 6 of *Modern surveys in Mathematics.* Springer-Verlag.

[Bell, 1986] Bell, E. T. (1986). *Men of Mathematics. The lives and achievements of the great mathematicians from Zeno to Poincaré.* Simon and Schuster.

[Birnbaum and Lozinskii, 1999] Birnbaum, E. and Lozinskii, E. L. (1999). The good old Davis-Putnam procedure helps counting models. *Journal of Artificial Intelligence Research*, 10:457–477.

[Bledsoe, 1984] Bledsoe, W. (1984). Some automatic proofs in analysis. In Bledsoe, W. W. and Loveland, Donald W., editors, *Automated Theorem Proving: After 25 Years*, pages 89–118. American Mathematical Society.

[Bledsoe, 1977] Bledsoe, W. W. (1977). Non-resolution theorem proving. *Journal of Artificial Intelligence*, 9:1–35.

[Bledsoe, 1986] Bledsoe, W. W. (1986). Some thoughts on proof discovery. In *Proc. of the 1986 Symposium on Logic Programming*, pages 2–10. The Computer Society of the IEEE.

[Bledsoe and Loveland, 1984] Bledsoe, W.W. and Loveland, D.W. (1984). *Automated Theorem Proving after 25 years*, volume 29 of *Contemporary Mathematics.* American Mathematical Society, Providence, RI, USA.

[Bogaert and Tison, 1992] Bogaert, B. and Tison, S. (1992). Equality and disequality constraints on direct subterms in tree automata. In *Symposium on Theoretical Aspects of Computer Science*, pages 436–449. LNCS 577.

[Boolos, 1984] Boolos, G. (1984). Trees and Finite Satisfiability: Proof of a conjecture of Burgess. *Notre Dame Journal of Formal Logic*, 25(3):193–197.

[Boolos and Jeffrey, 1974] Boolos, G.S. and Jeffrey, R.C. (1974). *Computability and logic.* Cambridge University Press.

[Börger et al., 1997] Börger, E., Grädel, E., and Gurevich, Y. (1997). *The Classical Decision Problem.* Perspectives in Mathematical Logic. Springer.

[Boy de la Tour, 2000a] Boy de la Tour, Thierry (2000a). On the complexity of finite sorted algebras. In Caferra, Ricardo and Salzer, Gernot, editors, *Automated Deduction in Classical and Non-Classical Logics*, Lecture Notes in Artificial Intelligence 1761, pages 95–108. Springer Verlag.

[Boy de la Tour, 2000b] Boy de la Tour, Thierry (2000b). Some techniques of isomorphism-free search. In *Artificial Intelligence and Symbolic Computation, International Conference AISC'2000*, pages 240–252. Springer Verlag. LNAI 1930.

[Boyer and Moore, 1979] Boyer, R. S. and Moore, J. S. (1979). *A computational logic.* Academic Press.

[Bry and Yahya, 1996] Bry, François and Yahya, Adnan (1996). Minimal model generation with positive unit hyper-resolution tableaux. In *Proceeding of Tableaux'96*, LNAI 1071, pages 143–159. Springer.

[Bry and Yahya, 2000] Bry, François and Yahya, Adnan (2000). Positive unit hyper-resolution. *Journal of Automated Reasoning*, 25(1):35–82.

[Buchberger, 1985] Buchberger, B. (1985). Gröbner-bases: An algorithmic method in polynomial ideal theory. In Bose, N.K., editor, *Multidimensional Systems Theory*, pages 184–232. Reidel Publ. Comp.

[Bullock and Schneider, 1973] Bullock, A. and Schneider, H. (1973). On Generating the Finitely Satisfiable Formulas. *Notre Dame Journal of Formal Logic*, 14(3):373–376.

[Caferra and Herment, 1995] Caferra, R. and Herment, M. (1995). A generic framework for combining inference tools and editing proofs and formulae. *Journal of Symbolic Computation*, 19:217–243.

[Caferra and Peltier, 1995] Caferra, R. and Peltier, N. (1995). Model Building and Interactive Theory Discovery. In *Proceeding of Tableaux'95*, LNAI 918, pages 154–168. Springer.

[Caferra and Peltier, 1997a] Caferra, R. and Peltier, N. (1997a). Combining inference and disinference rules with enumeration for model building. Workshop on model-based reasoning. IJCAI'97.

[Caferra and Peltier, 1997b] Caferra, R. and Peltier, N. (1997b). A new technique for verifying and correcting logic programs. *Journal of Automated Reasoning*, 19(3):277–318.

[Caferra and Peltier, 1998] Caferra, R. and Peltier, N. (1998). *Disinference rules, model building and abduction*, chapter 20, pages 331–353. "Logic at work. Essays dedicated to the memory of Helena Rasiowa" (Part 5:Logic in Computer Science, Chap. 20). Physica-Verlag.

[Caferra and Peltier, 2000a] Caferra, R. and Peltier, N. (2000a). Combining enumeration and deductive techniques in order to increase the class of constructible infinite models. *Journal of Symbolic Computation*, 29:177–211.

[Caferra and Peltier, 2000b] Caferra, R. and Peltier, N. (2000b). The Connection Method, Constraints and Model Building. In Holldöbler, Steffen, editor, *Intellectics and Computational Logic, papers in honor of Wolfgang Bibel*, pages 67–84. Kluwer Academic Publishers. Applied Logic Series, Volume 19.

[Caferra and Zabel, 1992] Caferra, R. and Zabel, N. (1992). A method for simultaneous search for refutations and models by equational constraint solving. *Journal of Symbolic Computation*, 13:613–641.

[Caferra and Zabel, 1993] Caferra, R. and Zabel, N. (1993). Building models by using tableaux extended by equational problems. *Journal of Logic and Computation*, 3:3–25.

[Caicedo, 1978] Caicedo, Xavier (1978). A Formal System for the Non-Theorems of the Propositional Calculus. *Notre Dame Journal of Formal Logic*, 19(1):147–151.

[Catach, 1991] Catach, L. (1991). TABLEAUX: a general theorem prover for modal logic. *Journal of Automated Reasoning*, 7(4):489–510.

[Cerri et al., 1990] Cerri, S., Gottlob, G., and Tanca, L. (1990). *Logic programming and databases.* Springer.

[Chang, 1972] Chang, C. L. (1972). Theorem proving with variable-constrained resolution. *Information Sciences*, 4:217–231.

[Chang and Keisler, 1990] Chang, C.C. and Keisler, H. J. (1990). *Model Theory*, volume 73 of *Studies in Logic and the Foundations of Mathematics.* North-Holland. Third edition.

[Chang and Lee, 1973] Chang, Chin-Liang and Lee, Richard Char-Tung (1973). *Symbolic Logic and Mechanical Theorem Proving.* Computer Science and Applied Mathematics. Academic Press.

[Chen and Hsiang, 1991] Chen, H. and Hsiang, J. (1991). Logic programming with recurrence domains. In *Automata, Languages and Programming (ICALP'91)*, pages 20–34. Springer, LNCS 510.

[Chen et al., 1990] Chen, H., Hsiang, J., and Kong, H.C. (1990). On finite representations of infinite sequences of terms. In *Conditional and Typed Rewriting Systems, 2nd International Workshop*, pages 100–114. Springer, LNCS 516.

[Chu and Plaisted, 1994a] Chu, H. and Plaisted, D. A. (1994a). Model finding in semantically guided instance-based theorem-proving. *Fundamenta Informaticae*, 21:221–235.

[Chu and Plaisted, 1994b] Chu, H. and Plaisted, D. A. (1994b). Semantically guided first-order theorem-proving using hyper-linking. In *Proc. of CADE-12*, pages 192–206. Springer. LNAI 814.

[Cocco et al., 2002] Cocco, S., Dubois, O. Mandler, J., and Mondasson, R. (2002). Au seuil de la complexité combinatoire. *Pour la Science*, 295:52–60.

[Comon, 1990] Comon, H. (1990). Equational formulas in order-sorted algebras. In *ICALP'90*, pages 674–688. Springer, LNCS 443.

[Comon, 1991] Comon, H. (1991). Disunification: A survey. In Lassez, J.-L. and Plotkin, G., editors, *Computational Logic: Essays in Honor of Alan Robinson*, pages 322–359. MIT Press, Cambridge, MA.

[Comon, 1992] Comon, H. (1992). On unification of terms with integer exponents. Technical report, LRI, Orsay, France.

[Comon and Lescanne, 1989] Comon, H. and Lescanne, P. (1989). Equational problems and disunification. *Journal of Symbolic Computation*, 7:371–475.

[Comon, 1988] Comon, Hubert (1988). *Unification et Disunification. Théorie et Applications.* PhD thesis, INPG, Grenoble.

[Comon, 1993] Comon, Hubert (1993). Complete axiomatizations of some quotient term algebras. *Theoretical Computer Science*, 118(2):167–191.

[Comon, 1995] Comon, Hubert (1995). On unification of terms with integer exponents. *Mathematical System Theory*, 28:67–88.

[Comon and Delor, 1994] Comon, Hubert and Delor, C. (1994). Equational formulae with membership constraints. *Information and Computation*, 112(2):167–216.

[Comon et al., 1994] Comon, Hubert, Haberstrau, Marianne, and Jouannaud, Jean-Pierre (1994). Syntacticness, cycle-syntacticness and shallow theories. *Information and Computation*, 111(1):154–191.

[Console et al., 1991] Console, L., Dupre, D. Theseider, and Torasso, P. (1991). On the Relationship between Abduction and Deduction. *Journal of Logic and Computation*, 1(5):661–690.

[Constable, 1986] Constable, R. L. (1986). *Implementing Mathematics with the Nuprl proof development system.* Prentice-Hall.

[Constable, 1993] Constable, Robert L. (1993). Formal theories and software systems: Fundamental connections between computer science and logic. Department of Computer Science Cornell University.

[Coquand and Huet, 1988] Coquand, T. and Huet, G. (1988). The calculus of construction. *Information and Computation*, 76:95–120.

[Couchot et al., 2003] Couchot, J-F., Dadeau, F., Déharbe, D., Giorgetti, A., and Ranise, S. (2003). Proving and debugging set-based specifications. In *Proc. of the 6th Workshop on Formal Methods.*

[Davis, 1963] Davis, M. (1963). Eliminating the irrelevant from mechanical proofs. In *Proc. Symposia of Applied Mathematics, Vol. 15*, pages 15–30. American Mathematical Society.

[Davis, 1983] Davis, M. (1983). A computer program for Presburger arithmetic. In Siekmann, J. and Wrighston, G., editors, *Automation of Reasoning 1*, pages 41–48. Springer. First published in 1957.

[Davis, 2001] Davis, M. (2001). The early history of automated deduction. In Robinson, A. and Voronkov, A., editors, *Handbook of Automated Reasoning*, chapter 1, pages 3–18. North-Holland.

[Davis and Hersh, 1981] Davis, Philip J. and Hersh, Reuben (1981). *The mathematical experience.* Birkhäuser.

[De Bruijn, 1980] De Bruijn, N. G. (1980). A survey of the project AUTOMATH. In Seldin, J.P. and Hindley, J.R., editors, *To H.B. Curry: Essays on Combinatory Logic, Lambda Calculus and Formalism*, pages 579–606. Academic Press.

[De Bruijn, 1991] De Bruijn, N. G. (1991). A plea for weaker frameworks. In Huet, G. and Plotkin, G., editors, *Logical Frameworks*, pages 40–67. Cambridge University Press.

[Denecker and de Schreye, 1994] Denecker, M. and de Schreye, D. (1994). On the duality of abduction and model generation in a framework for model generation with equality. *Theoretical Computer Science*, 122:225–262.

[Dreben and Goldfarb, 1979] Dreben, B. and Goldfarb, W. D. (1979). *The Decision Problem, Solvable Classes of Quantificational Formulas.* Addison-Wesley.

[Dubois and Mondasson, 2001] Dubois, O. and Mondasson, R. (2001). Phase transition in combinatorial problems. *Theoretical Computer Science*, 265(1,2).

[Ebbinghaus and Flum, 1999] Ebbinghaus, H-D. and Flum, J. (1999). *Finite Model Theory.* Perspectives in Mathematical Logic. Springer. Second Revised and Enlarged Edition.

[Eder, 1990] Eder, E (1990). *Relative Complexities of First-Order Logic Calculi.* Vieweg.

[Enderton, 1972] Enderton, H. (1972). *A Mathematical Introduction to Logic.* Academic Press.

[Epstein and Levy, 1995] Epstein, D. and Levy, S. (1995). Experimentation and Proof in Mathematics. *Notices of the AMS*, 42(6):670–674.

[Etchemendy, 1988] Etchemendy, J. (1988). Tarski on truth and logical consequence. *The Journal of Symbolic Logic*, 53(1):51–79.

[Fagin, 1975a] Fagin, R. (1975a). Monadic generalized spectra. *Zeitschr. f. math. Logik und Grundlagen d. Math.*, 21:89–96.

[Fagin, 1975b] Fagin, R. (1975b). A spectrum hierarchy. *Zeitschr. f. math. Logik und Grundlagen d. Math.*, 21:123–134.

[Fagin, 1976] Fagin, R. (1976). Probabilities on Finite Models. *Journal of Symbolic Logic*, 41(1):50–58.

[Fagin, 1993] Fagin, R. (1993). Finite model theory, a personal perspective. *Theoretical Computer Science*, 116:3–31.

[Fagin et al., 1995] Fagin, R., Halpern, J. Y., Moses, Y., and Vardi, M. Y. (1995). *Reasoning About Knowledge.* The MIT Press.

[Fagin et al., 1992] Fagin, R., Halpern, J. Y., and Vardi, M. Y. (1992). What is an inference rule ? *The Journal of Symbolic Logic*, 57(3):1018–1045.

[Fermüller and Leitsch, 1996] Fermüller, C. and Leitsch, A. (1996). Hyperresolution and automated model building. *Journal of Logic and Computation*, 6(2):173–203.

[Fermüller and Leitsch, 1998] Fermüller, C. and Leitsch, A. (1998). Decision procedures and model building in equational clause logic. *Journal of the IGPL*, 6(1):17–41.

[Fermüller et al., 1993] Fermüller, C., Leitsch, A., Tammet, T., and Zamov, N. (1993). *Resolution Methods for the Decision Problem.* LNAI 679. Springer.

[Fermüller et al., 2001] Fermüller, C. G., Leitsch, A., Hustadt, U., and Tammet, T. (2001). Resolution decision procedures. In Robinson, A. and Voronkov, A., editors, *Handbook of Automated Reasoning*, chapter 25, pages 1791–1849. North-Holland.

[Fermüller and Leitsch, 1992] Fermüller, Christian G. and Leitsch, Alexander (1992). Model building by resolution. In *Computer Science Logic, CSL'92*, pages 134–148. Springer, LNCS 702.

[Fèvre and Peltier, 1997] Fèvre, S. and Peltier, N. (1997). Two problems in geometry solved by using automated model builders. AAR Newsletter 38.

[Fitting, 1990] Fitting, M. (1990). *First-Order Logic and Automated Theorem Proving.* Texts and Monographs in Computer Science. Springer-Verlag.

[Fujita and Hasegawa, 1991] Fujita, M. and Hasegawa, R. (1991). A model generation theorem prover in KL1 using a ramified stack algorithm. In *Proceedings of 8th International Conference Symp. Logic Programming*, pages 1070–1080.

[Gelernter, 1959] Gelernter, H. (1959). A note on syntactic symmetry and the manipulation of formal systems by machine. *Information and Control*, 2:80–89.

[Gelernter, 1983] Gelernter, H. (1983). Realization of a Geometry-Theorem Proving Machine. In Siekmann, J. and Wrightson, G., editors, *Automation of Reasoning*, chapter Vol. 1, pages 99–122. Springer-Verlag. First published in 1959.

[Gilmore, 1960] Gilmore, P.C. (1960). A Proof Method for Quantification Theory: its Justification and Realization. *The IBM Journal*, 4(1):28–35.

[Girard, 2003] Girard, J.-Y. (2003). From foundations to Ludics. *The Bulletin of Symbolic Logic*, 9(2):131–168.

[Goguen, 2005] Goguen, J. (2005). *Theorem Proving and Algebra.* MIT-Press. To appear.

[Goguen et al., 2000] Goguen, J., Lin, K., Rosu, G., Mori, A., and Warinschi, B. (2000). An overview of the Tatami project. In Futatsugi, Kokichi, editor, *Cafe: An Industrial-Strength Algebraic Formal Method*, pages 61–78. Elsevier.

[Goodman, 1984] Goodman, N. D. (1984). The knowing mathematician. *Synthèse*, 60:21–38.

[Goodman, 1990] Goodman, N. D. (1990). Mathematics as natural science. *The Journal of Symbolic Logic*, 55(1):182–193.

[Gordon et al., 1979] Gordon, M., Milner, R., and Wadsworth, C. (1979). *Edinburgh LCF: a mechanized logic of computation.* Springer-Verlag. LNCS 78.

[Gottlob and Pichler, 2001] Gottlob, G. and Pichler, R. (2001). Working with ARMs: Complexity results on atomic representations of Herbrand models. *Information and Computation*, 165:183–207.

[Grätzer, 1979] Grätzer, G. (1979). *Universal Algebra.* Springer-Verlag. Second Edition.

[Gurevich, 1984] Gurevich, Y (1984). Toward logic tailored for computational complexity. In *Computation and Proof Theory, M.M. Richter et al. eds.*, pages 175–216. Springer-Verlag. LNM 1104.

[Gurevich, 1988] Gurevich, Y. (1988). Logic and the challenge of computer science. In Börger, E., editor, *Current trends in theoretical computer science*, pages 1–57. Computer Science Press.

[Hailperin, 1961] Hailperin, T. (1961). A Complete Set of Axioms for Logical Formulas Invalid in Some Finite Domain. *Zeitschr. f. math. Logik und Grundlagen d. Math.*, 7:84–96.

[Hájek and Havránek, 1978] Hájek, P. and Havránek, T. (1978). *Mechanizing hypothesis formation. Mathematical foundations for a general theory.* Springer-Verlag.

[Hauchecorne, 1988] Hauchecorne, Bertrand (1988). *Les Contre-exemples en Mathématiques.* Ellipses.

[He, 2001] He, Li (2001). I-SATCHMO: An improvement of SATCHMO. *Journal of Automated Reasoning*, 27(3):313–322.

[Henschen and Wos, 1974] Henschen, L. and Wos, L. (1974). Unit Refutations and Horn Sets. *Journal of the Association for Computing Machinery*, 21(4):590–605.

[Herken, 1988] Herken, Rolf (1988). *The Universal Turing Machine, A Half-Century Survey.* Oxford University Press.

[Hermann, 1994] Hermann, M. (1994). Divergence des systèmes de réécriture et schématisation des ensembles infinis de termes. Habilitation, Université de Nancy I, and CRIN-CNRS Inria-Lorraine, Nancy, France.

[Hermann and Galbavý, 1997] Hermann, M. and Galbavý, R. (1997). Unification of Infinite Sets of Terms schematized by Primal Grammars. *Theoretical Computer Science*, 176(1–2):111–158.

[Hermann, 1992] Hermann, Micki (1992). On the relation between primitive recursion, schematization, and divergence. In *Proceeding 3rd Conference on Algebraic and Logic Programming*, pages 115–127. Springer, LNCS 632.

[Hermerén, 1974] Hermerén, G. (1974). Models. In Stenlund, S., editor, *Logical theory and semantic analysis, essays dedicated to Stig Kanger on his fiftieth birthday*, pages 175–191. D. Reidel.

[Hodges, 1983] Hodges, W. (1983). Elementary Predicate Logic. In Gabbay, D. and Guenthner, F., editors, *Handbook of Philosophical Logic*, chapter I.1, pages 2–131. D. Reidel. Vol. 1, Elements of Classical Logic.

[Hodges, 1993] Hodges, W. (1993). *Model theory*, volume 42 of *Encyclopedia of Mathematics and its Applications.* Cambridge University Press.

[Hogger, 1984] Hogger, C.J. (1984). *Introduction to Logic Programming*, volume 21 of *A.P.I.C. Studies in Data Processing.* Academic Press.

[Hooker, 1993a] Hooker, J.N. (1993a). New methods for computing inferences in first order logic. *Annals of Operations Research*, 43:479–492.

[Hooker, 1993b] Hooker, J.N. (1993b). Solving the incremental satisfiability problem. *The Journal of Logic Programming*, 15(1 & 2):177–186.

[Horgan, 1993] Horgan, J. (1993). L'ordinateur en mathématiques. *Pour la Science*, 194:84–93.

[Hsiang and Rusinowitch, 1991] Hsiang, J. and Rusinowitch, M. (1991). Proving refutational completeness of theorem proving strategies: The transfinite semantic tree method. *Journal of the ACM*, 38(3):559–587.

[Jeffrey, 1967] Jeffrey, R. C. (1967). *Formal logic: its scope and limits.* McGraw-Hill.

[Jeroslow, 1988] Jeroslow, R. (1988). Computation-oriented reductions of predicate to propositional logic. *Decision Support Systems*, 4:183–197.

[Jouannaud and Kirchner, 1991] Jouannaud, J.P. and Kirchner, C. (1991). Solving equations in abstract algebras: a rule based survey of unification. In Lassez, J.-L. and Plotkin, G., editors, *Essays in Honor of Alan Robinson*, pages 91–99. The MIT-Press.

[Joyner, 1976] Joyner, W.H. (1976). Resolution strategies as decision procedures. *Journal of the ACM*, 23:398–417.

[Khardon and Roth, 1994] Khardon, R. and Roth, D. (1994). Reasoning with models. In *AAAI-94*, pages 1148–1153. The MIT Press.

[Kleene, 1952] Kleene, S.C. (1952). *Introduction to Metamathematics.* North-Holland.

[Knuth and Bendix, 1970] Knuth, D. and Bendix, P. (1970). Simple word problems in universal algebra. In Leech, John, editor, *Computational Problems in Abstract Algebra*, pages 263–297. Pergamon Press.

[Krishnamurthy, 1985] Krishnamurthy, B. (1985). Short proofs for tricky formulas. *Acta Informatica*, 22:253–275.

[Lam, 1990] Lam, C. W. H. (1990). How reliable is a computer-based proof? *The Mathematical Intelligencer*, 12(1):8–12.

[Lam, 1991] Lam, C. W. H. (1991). The search for a finite projective plane of order 10. *The American Mathematical Monthly*, 98(4):305–318.

[Lassez et al., 1986] Lassez, J.L., Maher, M.J., and Marriott, K. (1986). Unification Revisited. In *Foundations of Logic and Functional Programming*, pages 67–113. Springer. LNCS 306.

[Lee and Plaisted, 1992] Lee, S. and Plaisted, D. A. (1992). Eliminating duplication with the hyper-linking strategy. *Journal of Automated Reasoning*, 9:25–42.

[Lee and Plaisted, 1994] Lee, S. and Plaisted, D. A. (1994). Problem solving by searching for models with a theorem prover. *Journal of Artificial Intelligence*, 69:205–233.

[Leitsch, 1997] Leitsch, A. (1997). *The resolution calculus.* Springer. Texts in Theoretical Computer Science.

[Leitsch, 1993] Leitsch, Alexander (1993). Deciding clause classes by semantic clash resolution. *Fundamenta Informaticae*, 18:163–182.

[Lloyd, 1990] Lloyd, G.E.R. (1990). *Origines et Développement de la Science Grecque*, volume 365 of *Champs*. Flammarion.

[Loveland et al., 1995] Loveland, D., Reed, D., and Wilson, D. S. (1995). SATCHMORE: SATCHMO with RElevancy. *Journal of Automated Reasoning*, 14(2):325–351.

[Loveland, 1978] Loveland, Donald W. (1978). *Automated Theorem Proving: A Logical Basis*, volume 6 of *Fundamental Studies in Computer Science*. North Holland.

[Loveland, 1984] Loveland, Donald W. (1984). Automated theorem proving: a quarter century review. In Bledsoe, Woody W. and Loveland, Donald W., editors, *Automated Theorem Proving: After 25 Years*, pages 1–45. American Mathematical Society.

[Łukasiewicz, 1972] Łukasiewicz, Jan (1972). *La Syllogistique d'Aristote*. Philosophies pour l'âge de la science. Armand Colin.

[Mal'cev, 1971] Mal'cev, A.I. (1971). Axiomatizable classes of locally free algebra of various type. In *The Metamathematics of Algebraic Systems: Collected Papers 1936–1967*, pages 262–281. Benjamin Franklin Wells editor, North Holland. Chapter 23.

[Manthey and Bry, 1988] Manthey, R. and Bry, F. (1988). SATCHMO: A theorem prover implemented in Prolog. In *Proc. of CADE-9*, pages 415–434. Springer, LNCS 310.

[Marriott and Stuckey, 1998] Marriott, K. and Stuckey, J. (1998). *Programming with Constraints, an Introduction.* The MIT Press.

[Martelli and Montanari, 1982] Martelli, A. and Montanari, U. (1982). An efficient unification algorithm. *ACM Transactions on Programming Languages and Systems*, 4(2):258–282.

[Martin-Löf, 1987] Martin-Löf, Per (1987). Truth of a proposition, evidence of a judgement, validity of a proof. *Synthèse*, 73:407–420.

[Maslov, 1968] Maslov, S.Y. (1968). The inverse method for establishing deducibility for logical calculi. *Proc. Steklov Inst. Math.*, 98:25–96.

[Matzinger, 1997] Matzinger, R. (1997). Computational representations of Herbrand models using grammars. In *Computer Science Logic, CSL'96*, pages 334–348. Springer. LNCS 1258.

[Matzinger, 2000] Matzinger, R. (2000). *Computational Representations of Models in First-Order Logic*. PhD thesis, TU-Vienna.

[McCarthy, 1962] McCarthy, John (1962). Computer programs for checking mathematical proofs. In *Recursive Function Theory*, pages 219–228, Providence, Rhode Island. Proc. of Symposia in Pure Mathematics, Volume 5, American Mathematical Society.

[McCune, 1998] McCune, W. (1998). Automatic proofs and counterexamples for some ortholattice identities. *Information Processing Letter*, 65(6):285–291.

[McCune and Padmanabham, 1996] McCune, W. and Padmanabham, R. (1996). *Automated Deduction in Equational Logic and Cubic Curves*, volume 1095 of *LNAI*. Springer.

[Mckenzie, 1975] Mckenzie, R. (1975). On spectra, and the negative solution of the decision problem for identities having a finite nontrivial model. *The Journal of Symbolic Logic*, 40(2):186–196.

[Meseguer, 1989] Meseguer, J. (1989). General logics. In *Proc. of Logic Colloquium'87*, pages 275–330. Nort-Holland.

[Milner, 1985] Milner, R. (1985). The use of machines to assist in rigorous proof. In Hoare, C.A.R. and J.C. Shepherdson, Editors, editors, *Mathematical Logic and Programming Languages*, pages 77–88. Prentice-Hall.

[Mongy, 1981] Mongy, J. (1981). *Transformation de noyaux reconnaissables d'arbres. Forêts RATEG*. PhD thesis, Laboratoire d'Informatique Fondamentale de Lille, Villeneuve d'Ascq, France.

[Newell et al., 1983] Newell, A., Shaw, J., and Simon, H. (1983). Empirical explorations with the logic theory machine. In Siekmann, J. and Wrightson, G., editors, *Automation of Reasoning. Classical papers on computational logic*, volume 1, pages 49–73. Springer-Verlag. First published in 1957.

[Ohlbach et al., 2001] Ohlbach, H. J., Nonnegart, A., De Rijke, M., and Gabbay, D. M. (2001). Encoding two-valued nonclassical logics in classical logic. In Robinson, A. and Voronkov, A., editors, *Handbook of Automated Reasoning*, chapter 21, pages 1403–1486. North-Holland.

[Paulson, 1989] Paulson, L. C. (1989). The foundation of a generic theorem prover. *Journal of Automated Reasoning*, 5(3):363–398.

[Peltier, 1997a] Peltier, N. (1997a). Increasing the capabilities of model building by constraint solving with terms with integer exponents. *Journal of Symbolic Computation*, 24:59–101.

[Peltier, 1997b] Peltier, N. (1997b). *Nouvelles Techniques pour la Construction de Modèles finis ou infinis en Déduction Automatique*. PhD thesis, Institut National Polytechnique de Grenoble. ftp://ftp.imag.fr/pub/Mediatheque.IMAG/theses/97-Peltier.Nicolas/.

[Peltier, 1997c] Peltier, N. (1997c). Simplifying formulae in tableaux. Pruning the search space and building models. In *Proceeding of Tableaux'97*, pages 313–327. Springer LNAI 1227.

[Peltier, 1997d] Peltier, N. (1997d). Tree automata and automated model building. *Fundamenta Informaticae*, 30(1):59–81.

[Peltier, 1998] Peltier, N. (1998). An equational constraints solver. In *Proceedings of CADE-15*, pages 119–123. Springer.

[Peltier, 1999] Peltier, N. (1999). Pruning the search space and extracting more models in tableaux. *Logic Journal of the IGPL*, 7(2):217–251.

[Peltier, 2001a] Peltier, N. (2001a). A General Method for Using Terms Schematizations in Automated Deduction. In *Proceedings of the International Joint Conference on Automated Reasoning (IJCAR'01)*, pages 578–593. Springer LNCS 2083.

[Peltier, 2001b] Peltier, N. (2001b). On the decidability of the PVD class with equality. *Logic Journal of the IGPL*, 9(4):601–624.

[Pichler, 1998] Pichler, R. (1998). Algorithms on atomic representations of Herbrand models. In *Logic in AI, JELIA'98*, pages 199–215. Springer. LNAI 1489.

[Pichler, 1999] Pichler, R. (1999). Solving equational problems efficiently. In Ganzinger, Harald, editor, *Automated Deduction – CADE-16, 16th International Conference on Automated Deduction*, LNAI 1632, pages 97–111, Trento, Italy. Springer-Verlag.

[Pichler, 2000] Pichler, R. (2000). *H-Subsumption in Automated Deduction.* PhD thesis, TU-Vienna.

[Plaisted and Zhu, 2000] Plaisted, D. A. and Zhu, Y. (2000). Ordered semantic hyperlinking. *Journal of Automated Reasoning*, 25(3):167–217.

[Plotkin, 1971] Plotkin, D. (1971). A Further Note on Inductive Generalization. In Meltzer, B. and Michie, D., editors, *Machine Intelligence 6*, pages 101–124. Edinburgh University Press.

[Polya, 1973] Polya, G. (1973). *How to Solve It, a New Aspect of Mathematical Method.* Princeton University Press. Second Edition.

[Prawitz, 1960] Prawitz, D. (1960). An improved proof procedure. *Theoria*, 26:102–139.

[Prawitz, 1969] Prawitz, D. (1969). Advances and problems in mechanical proof procedure. In Michie, D. and Meltzer, R., editors, *Machine Intelligence*, volume 4, pages 59–71. Edinburg U. Press.

[Prawitz et al., 1960] Prawitz, D., Prawitz, H., and Voghera, N. (1960). A mechanical proof procedure and its realization in an electronic computer. *Journal of the Association for Computing Machinery*, 7(1-2):102–128.

[Prawitz, 1980] Prawitz, Dag (1980). Proof theory and the processing of proofs. Summary draft of talk at the Fifth Conference on Automated Deduction.

[Rasiowa and Sikorski, 1970] Rasiowa, H. and Sikorski, R. (1970). *The mathematics of metamathematics*, volume 41 of *monografie Matematyczne*. PWN Polish Scientific Publishers.

[Reeves, 1987] Reeves, S. (1987). Semantic tableaux as framework for automated theorem-proving. In *Proc. of AISB Conference*, pages 125–139.

[Richman, 1990] Richman, F. (1990). Intuitionism as Generalization. *Philosophia Mathematica*, 5(1/2):124–128.

[Robinson and Voronkov, 2001] Robinson, A. and Voronkov, A., editors (2001). *Handbook of Automated Reasoning*. North-Holland.

[Robinson and Wos, 1969] Robinson, G. and Wos, L. (1969). Paramodulation and theorem-proving in first-order theories with equality. In Michie, D. and Meltzer, R., editors, *Machine Intelligence*, volume 4, pages 135–150. Edinburg U. Press.

[Robinson, 1965a] Robinson, J. A. (1965a). Automatic deduction with hyperresolution. *Intern. Journal of Computer Math.*, 1:227–234.

[Robinson, 1965b] Robinson, J. A. (1965b). A machine-oriented logic based on the resolution principle. *J. Assoc. Comput. Mach.*, 12:23–41.

[Robinson, 2000] Robinson, J. A. (2000). Proof = Guarantee + Explanation. In Holldöbler, Steffen, editor, *Intellectics and Computational Logic, papers in honor of Wolfgang Bibel*, pages 277–294. Kluwer Academic Publishers. Applied Logic Series, Volume 19.

[Rose and Wilkinson, 2001] Rose, M. and Wilkinson, K. (2001). Application of model search to lattice theory. *AAR Newsletter*, 52:3–7.

[Rosen, 1995] Rosen, J. (1995). *Symmetry in Science. An Introduction to the General Theory*. Springer-Verlag.

[Roth, 1996] Roth, D. (1996). On the hardness of approximate reasoning. *Journal of Artificial Intelligence*, 82(1-2):273–302.

[Rudlof, 2000] Rudlof, T. (2000). SHR tableaux - A Framework for Automated Model Generation. *Journal of Logic and Computation*, 10(6):107–155.

[Salzer, 1992] Salzer, G. (1992). The unification of infinite sets of terms and its applications. In *Logic Programming and Automated Reasoning (LPAR'92)*, pages 409–429. Springer, LNAI 624.

[Salzer, 1994] Salzer, G. (1994). Primal grammar and unification modulo a binary clause. In *Proc. of CADE-12*, pages 72–86. Springer. LNAI 814.

[Siekmann and Wrightson, 1983a] Siekmann, J. and Wrightson, G. (1983a). *Automation of Reasoning. Classical Papers on Computational Logic 1957-1966*, volume 1. Springer-Verlag.

[Siekmann and Wrightson, 1983b] Siekmann, J. and Wrightson, G. (1983b). *Automation of Reasoning. Classical Papers on Computational Logic 1967-1970*, volume 2. Springer-Verlag.

[Slagle, 1967] Slagle, J. R. (1967). Automatic theorem proving with renamable and semantic resolution. *Journal of the ACM*, 14(4):687–697.

[Slaney, 1992] Slaney, J. (1992). Finder (FINite Domain EnumeratoR): Notes and guides. Technical report, Australian National University Automated Reasoning Project, Canberra.

[Slaney, 1993] Slaney, J. (1993). SCOTT: a model-guided theorem prover. In *Proceedings IJCAI-93*, volume 1, pages 109–114. Morgan Kaufmann.

[Smoryński, 1991] Smoryński, C. (1991). *Logical number theory I. An Introduction.* Springer-Verlag.

[Smullyan, 1968] Smullyan, R. M. (1968). *First-Order Logic.* Springer.

[Stein, 1963] Stein, S. K. (1963). Finite models of identities. *Proceedings of the American Mathematical Society*, 14:216–222.

[Stenlund, 1971] Stenlund, S. (1971). *Combinators, λ-terms and proof theory.* Reidel.

[Stolzenburg, 1999] Stolzenburg, Frieder (1999). Loop-detection in hyper-tableaux by powerful model generation. *Journal of Universal Computer Science*, 5(3):135–155.

[Suppes, 1966] Suppes, Patrick (1966). A Comparison of the Meaning and Uses of Models in Mathematics and Empirical Sciences. *Synthèse*, 12:287–301.

[Tammet, 1991] Tammet, T. (1991). Using resolution for deciding solvable classes and building finite models. In *Proc. of Baltic Computer Science*, pages 33–64. Springer, LNCS 502.

[Taton, 1995] Taton, R. (1995). *La science moderne de 1450 à 1800*, volume 187 of *Quadrige.* Presses Universitaires de France.

[The Coq Development Team, 2001] The Coq Development Team (2001). *The Coq Proof Assistant Reference Manual V7.1.* INRIA-Rocquencourt, CNRS-ENS Lyon (France). `http://coq.inria.fr/doc/main.html`.

[Trahtenbrot, 1963] Trahtenbrot, B. A. (1963). Impossibility of an Algorithm for the Decision Problem in Finite Classes. *American Mathematical Society Translations*, 23(Series 2):1–5. First published (Russian): Doklady Akadémii Nauk SSSR, vol. 70 (1950), pp. 569–572.

[Tymoczko, 1979] Tymoczko, T. (1979). The four-color problem and its philosophical significance. *The Journal of Philosophy*, 76(2):57–83.

[Vorobyov, 1996] Vorobyov (1996). An improving lower bound for the elementary theories of trees. In *Proc. of CADE-13, LNAI 1104*, pages 275–287. Springer.

[Weidenbach, 2001] Weidenbach, C. (2001). Combining superposition, sorts and splitting. In Robinson, A. and Voronkov, A., editors, *Handbook of Automated Reasoning*, volume II, chapter 27, pages 1965–2013. Elsevier Science.

[Winker, 1982] Winker, S. (1982). Generation and verification of finite models and counter-examples using an automated theorem prover answering two open questions. *Journal of the ACM*, 29(2):273–284.

[Winker et al., 1981] Winker, S.W., Wos, L., and Lusk, E.L. (1981). Semigroups, antiautomorphisms, and involutions: a computer solution to an open problem, I. *Mathematics of Computation*, 37(156):533–545.

[Wos, 1988] Wos, L. (1988). *Automated Reasoning, 33 Basic Research Problems.* Prentice Hall.

[Wos, 1993] Wos, L. (1993). Automated reasoning answers open questions. *Notices of the American Mathematical Society*, 40(1):15–26.

[Wos and et al., 1985] Wos, L. and et al. (1985). An overview of automated reasoning and related fields. *Journal of Automated Reasoning*, 1(1):5–48.

[Zhang, 1994] Zhang, J. (1994). Problems on the generation of finite models. In *Proc. of CADE-12*, pages 753–757. Springer. LNAI 814.

[Zhang and Zhang, 1995] Zhang, J. and Zhang, H. (1995). SEM: a system for enumerating models. In *Proc. IJCAI-95*, volume 1, pages 298–303. Morgan Kaufmann.

[Zhang, 1996] Zhang, W. (1996). Number of models and satisfiability of sets of clauses. *Theoretical Computer Science*, 155(1):277–288.

APPLIED LOGIC SERIES

1. D. Walton: *Fallacies Arising from Ambiguity.* 1996 ISBN 0-7923-4100-7
2. H. Wansing (ed.): *Proof Theory of Modal Logic.* 1996 ISBN 0-7923-4120-1
3. F. Baader and K.U. Schulz (eds.): *Frontiers of Combining Systems.* First International Workshop, Munich, March 1996. 1996 ISBN 0-7923-4271-2
4. M. Marx and Y. Venema: *Multi-Dimensional Modal Logic.* 1996 ISBN 0-7923-4345-X
5. S. Akama (ed.): *Logic, Language and Computation.* 1997 ISBN 0-7923-4376-X
6. J. Goubault-Larrecq and I. Mackie: *Proof Theory and Automated Deduction.* 1997 ISBN 0-7923-4593-2
7. M. de Rijke (ed.): *Advances in Intensional Logic.* 1997 ISBN 0-7923-4711-0
8. W. Bibel and P.H. Schmitt (eds.): *Automated Deduction - A Basis for Applications.* Volume I. Foundations - Calculi and Methods. 1998 ISBN 0-7923-5129-0
9. W. Bibel and P.H. Schmitt (eds.): *Automated Deduction - A Basis for Applications.* Volume II. Systems and Implementation Techniques. 1998 ISBN 0-7923-5130-4
10. W. Bibel and P.H. Schmitt (eds.): *Automated Deduction - A Basis for Applications.* Volume III. Applications. 1998 ISBN 0-7923-5131-2
    (Set vols. I-III: ISBN 0-7923-5132-0)
11. S.O. Hansson: *A Textbook of Belief Dynamics.* Theory Change and Database Updating. 1999 Hb: ISBN 0-7923-5324-2; Pb: ISBN 0-7923-5327-7
    Solutions to exercises. 1999. Pb: ISBN 0-7923-5328-5
    Set: (Hb): ISBN 0-7923-5326-9; (Pb): ISBN 0-7923-5329-3
12. R. Pareschi and B. Fronhöfer (eds.): *Dynamic Worlds from the Frame Problem to Knowledge Management.* 1999 ISBN 0-7923-5535-0
13. D.M. Gabbay and H. Wansing (eds.): *What is Negation?* 1999 ISBN 0-7923-5569-5
14. M. Wooldridge and A. Rao (eds.): *Foundations of Rational Agency.* 1999 ISBN 0-7923-5601-2
15. D. Dubois, H. Prade and E.P. Klement (eds.): *Fuzzy Sets, Logics and Reasoning about Knowledge.* 1999 ISBN 0-7923-5911-1
16. H. Barringer, M. Fisher, D. Gabbay and G. Gough (eds.): *Advances in Temporal Logic.* 2000 ISBN 0-7923-6149-0
17. D. Basin, M.D. Agostino, D.M. Gabbay, S. Matthews and L. Viganò (eds.): *Labelled Deduction.* 2000 ISBN 0-7923-6237-3
18. P.A. Flach and A.C. Kakas (eds.): *Abduction and Induction.* Essays on their Relation and Integration. 2000 ISBN 0-7923-6250-0
19. S. Hölldobler (ed.): *Intellectics and Computational Logic.* Papers in Honor of Wolfgang Bibel. 2000 ISBN 0-7923-6261-6

APPLIED LOGIC SERIES

20. P. Bonzon, M. Cavalcanti and Rolf Nossum (eds.): *Formal Aspects of Context.* 2000 ISBN 0-7923-6350-7
21. D.M. Gabbay and N. Olivetti: *Goal-Directed Proof Theory.* 2000 ISBN 0-7923-6473-2
22. M.-A. Williams and H. Rott (eds.): *Frontiers in Belief Revision.* 2001 ISBN 0-7923-7021-X
23. E. Morscher and A. Hieke (eds.): *New Essays in Free Logic.* In Honour of Karel Lambert. 2001 ISBN 1-4020-0216-5
24. D. Corfield and J. Williamson (eds.): *Foundations of Bayesianism.* 2001 ISBN 1-4020-0223-8
25. L. Magnani, N.J. Nersessian and C. Pizzi (eds.): *Logical and Computational Aspects of Model-Based Reasoning.* 2002 Hb: ISBN 1-4020-0712-4; Pb: ISBN 1-4020-0791-4
26. D.J. Pym: *The Semantics and Proof Theory of the Logic of Bunched Implications.* 2002 ISBN 1-4020-0745-0
27. P.B. Andrews: *An Introduction to Mathematical Logic and Type Theory: To Truth Through Proof.* Second edition. 2002 ISBN 1-4020-0763-9
28. F.D. Kamareddine: *Thirty Five Years of Automating Mathematics.* 2003 ISBN 1-4020-1656-5
29. F. Kamareddine, T. Laan and R. Nederpelt: *A Modern Perspective on Type Theory.* From its Origins until Today. 2004 ISBN 1-4020-2334-0
30. H. Leitgeb: *Inference on the Low Level.* An Investigation into Deduction, Nonmonotonic Reasoning, and the Philosophy of Cognition. 2004 ISBN 1-4020-2492-4
31. R. Caferra, A. Leitsch and P. Nicholas: *Automated Model Building.* 2004 ISBN 1-4020-2652-8

KLUWER ACADEMIC PUBLISHERS – DORDRECHT / BOSTON / LONDON